Computational Linear Algebra

Courses on linear algebra and numerical analysis need each other. Often NA courses have some linear algebra topics, and LA courses mention some topics from numerical analysis/scientific computing. This text merges these two areas into one introductory undergraduate course. It assumes students have had multivariable calculus. A second goal of this text is to demonstrate the intimate relationship of linear algebra to applications/computations.

A rigorous presentation has been maintained. A third reason for writing this text is to present, in the first half of the course, the very important topic on singular value decomposition, SVD. This is done by first restricting consideration to real matrices and vector spaces. The general inner product vector spaces are considered starting in the middle of the text.

The text has a number of applications. These are to motivate the student to study the linear algebra topics. Also, the text has a number of computations. MATLAB® is used, but one could modify these codes to other programming languages. These are either to simplify some linear algebra computation, or to model a particular application.

Textbooks in Mathematics

Series editors:
Al Boggess, Kenneth H. Rosen

Geometry and Its Applications, Third Edition
Walter J. Meyer

Transition to Advanced Mathematics
Danilo R. Diedrichs and Stephen Lovett

Modeling Change and Uncertainty
Machine Learning and Other Techniques
William P. Fox and Robert E. Burks

Abstract Algebra
A First Course, Second Edition
Stephen Lovett

Multiplicative Differential Calculus
Svetlin Georgiev, Khaled Zennir

Applied Differential Equations
The Primary Course
Vladimir A. Dobrushkin

Introduction to Computational Mathematics: An Outline
William C. Bauldry

Mathematical Modeling the Life Sciences
Numerical Recipes in Python and MATLAB™
N. G. Cogan

Classical Analysis
An Approach through Problems
Hongwei Chen

Classical Vector Algebra
Vladimir Lepetic

Introduction to Number Theory
Mark Hunacek

Probability and Statistics for Engineering and the Sciences with Modeling using R
William P. Fox and Rodney X. Sturdivant

Computational Optimization: Success in Practice
Vladislav Bukshtynov

Computational Linear Algebra: with Applications and MATLAB® Computations
Robert E. White

https://www.routledge.com/Textbooks-in-Mathematics/book-series/CANDHTEXBOOMTH

Computational Linear Algebra

with Applications and MATLAB®
Computations

Robert E. White

CRC Press
Taylor & Francis Group
Boca Raton London New York

CRC Press is an imprint of the
Taylor & Francis Group, an **informa** business

A CHAPMAN & HALL BOOK

First edition published 2023
by CRC Press
6000 Broken Sound Parkway NW, Suite 300, Boca Raton, FL 33487-2742

and by CRC Press
4 Park Square, Milton Park, Abingdon, Oxon, OX14 4RN

CRC Press is an imprint of Taylor & Francis Group, LLC

ISBN: 978-1-032-30246-1 (hbk)
ISBN: 978-1-032-30245-4 (pbk)
ISBN: 978-1-003-30412-8 (ebk)

DOI: 10.1201/9781003304128

Typeset in CMR10
by KnowledgeWorks Global Ltd.

Contents

List of Figures   xi

Preface   xiii

Introduction   xv

Author Biography   xvii

1 Solution of $Ax = d$ 1
  1.1 Matrix Models   1
    1.1.1 Column vectors and $\mathbb{R}^n$ 1
    1.1.2 Matrices   5
    1.1.3 Application to visualization of minimum cost   7
    1.1.4 Application to two-bar truss   8
    1.1.5 Application to two-loop circuit   9
    1.1.6 Exercises   11
  1.2 Matrix Products   12
    1.2.1 Matrix-vector products   13
    1.2.2 Matrix-matrix products   16
    1.2.3 Application to heat conduction   18
    1.2.4 Matrix computations using MATLAB®   21
    1.2.5 Exercises   21
  1.3 Special Cases of $Ax = \mathbf{d}$ 22
    1.3.1 Five possible classes of "solutions"   22
    1.3.2 Triangular matrices   24
    1.3.3 Application to heat in wire with current   29
    1.3.4 Matrix computations using MATLAB®   29
    1.3.5 Exercises   31
  1.4 Row Operations and Gauss Elimination   31
    1.4.1 Introductory illustration   32
    1.4.2 Three types of row operations   33
    1.4.3 Gauss elimination for solving $Ax = \mathbf{d}$ 34
    1.4.4 Application to six-bar truss   37
    1.4.5 Gauss elimination using MATLAB®   39
    1.4.6 Exercises   40
  1.5 Inverse Matrices   41
    1.5.1 Examples of inverse matrices   41

      1.5.2   Gauss–Jordan method to find inverse matrices 44

      1.5.3   Properties of inverse matrices 49

      1.5.4   Inverse matrices and MATLAB® 53

      1.5.5   Exercises 54

  1.6 Determinants and Cramer's Rule 57

      1.6.1   Determinants for 2×2 and 3×3 matrices 57

      1.6.2   Determinant of an $n \times n$ matrix 59

      1.6.3   Cramer's rule and inverses 62

      1.6.4   Determinants using MATLAB® 66

      1.6.5   Exercises 67

2 Matrix Factorizations   **69**

  2.1 The Schur Complement 69

      2.1.1   Heat diffusion in fin with two directions 73

      2.1.2   Exercises 74

  2.2 $PA = LU$ and A Nonsingular 75

      2.2.1   Exercises 78

  2.3 $A = LU, A^{-1} \geq 0$ and M-Matrix 78

      2.3.1   Exercises 82

  2.4 $A = GG^T$ and A SPD 82

      2.4.1   SPD and minimization 85

      2.4.2   Exercises 86

3 Least Squares and Normal Equations   **89**

  3.1 Normal Equations 89

      3.1.1   Exercises 92

  3.2 MATLAB® Code price_expdata.m 93

      3.2.1   Exercises 95

  3.3 Basis of Subspace 96

      3.3.1   Exercises 99

  3.4 Projection to Subspace 100

      3.4.1   Exercises 102

4 $Ax = d$ with $m < n$ **105**

  4.1 Examples in $\mathbb{R}^3$ 105

  4.2 Row Echelon Form 106

      4.2.1   Solutions in $\mathbb{R}^4$ 107

      4.2.2   General solution of $Ax = d$ 110

      4.2.3   Exercises 112

  4.3 Relationship of $R(A), N(A^T)$ and $R(A^T), N(A)$ 113

      4.3.1   Construction of bases 116

      4.3.2   Exercises 117

  4.4 Null Space Method for Equilibrium Equations 118

      4.4.1   Block Gauss elimination method 119

      4.4.2   Null space method for equilibrium equations 121

	4.4.3	Application to three-loop circuit	123
	4.4.4	Application to six-bar truss	124
	4.4.5	Application to fluid flow	126
	4.4.6	Exercises	128

5 Orthogonal Subspaces and Bases | **129**
5.1	Orthogonal Subspace	129
	5.1.1	Exercises	131
5.2	Fundamental Theorem: $\mathbb{R}^n = N(A) \oplus R(A^T)$	131
	5.2.1	Exercises	133
5.3	$A = QR$ Factorization	133
	5.3.1	MATLAB® code qr_col.m	136
	5.3.2	Exercises	138
5.4	Orthonormal Basis	138
	5.4.1	Exercises	140
5.5	Four Methods for QR Factors	141
	5.5.1	Classical Gram–Schmidt	141
	5.5.2	Givens transform	142
	5.5.3	Householder transform	143
	5.5.4	Exercises	146

6 Eigenvectors and Orthonormal Basis | **147**
6.1	Eigenvectors of Symmetric Matrix	147
	6.1.1	Exercises	150
6.2	Approximation of Eigenvalues	150
	6.2.1	Gerschgorin circles	151
	6.2.2	Power iterations	151
	6.2.3	QR iteration	152
	6.2.4	Exercises	153
6.3	Spectral Theorem Factors $AQ = QD$	153
	6.3.1	Exercises	155
6.4	Applications	155
	6.4.1	Nonsingular $Ax = d$	155
	6.4.2	Singular value decomposition	156
	6.4.3	Exercises	159

7 Singular Value Decomposition | **161**
7.1	"Small" SVD	161
	7.1.1	Exercises	164
7.2	"Full" SVD	164
	7.2.1	MATLAB® code svd_ex.m	166
	7.2.2	Exercises	169
7.3	"Truncated" SVD	169
	7.3.1	Exercises	172

8 Three Applications of SVD **173**
  8.1 Image Compression 173
      8.1.1 MATLAB® code svdimage.m 175
  8.2 Search Engines 177
      8.2.1 MATLAB® codes sengine.m, senginesparse.m 179
  8.3 Noise Filter 184
      8.3.1 MATLAB® code Image1dsvd.m 185

9 Pseudoinverse of A **189**
  9.1 $\Sigma^\dagger$ and $A^\dagger = V\Sigma^\dagger U^T$ 189
      9.1.1 Exercises 192
  9.2 $A^\dagger$ and Least Squares 193
      9.2.1 Exercises 195
  9.3 Ill-Conditioned Least Squares 196
      9.3.1 Exercises 199
  9.4 Application to Hazard Identification 199
      9.4.1 MATLAB® code hazidsvd1.m 201

10 General Inner Product Vector Spaces **209**
  10.1 Vector Spaces 209
      10.1.1 Exercises 214
  10.2 Inner Products and Orthogonal Vectors 214
      10.2.1 General inner products 215
      10.2.2 Orthonormal vectors 217
      10.2.3 Norms on vector spaces 219
      10.2.4 Exercises 221
  10.3 Schur Decomposition 221
      10.3.1 Norms and spectral radius 224
      10.3.2 Normal matrices 225
      10.3.3 Cayley–Hamilton theorem 227
      10.3.4 Exercises 229
  10.4 Self-Adjoint Differential Operators 229
      10.4.1 Linear operators 230
      10.4.2 Sturm-Liouville problem 233
      10.4.3 Exercises 235
  10.5 Self-Adjoint Positive Definite BVP 235
      10.5.1 Exercises 243

11 Iterative Methods **245**
  11.1 Inverse Matrix Approximations 245
      11.1.1 Exercises 247
  11.2 Regular Splittings for M-Matrices 247
      11.2.1 Exercises 252
  11.3 P-Regular Splittings for SPD Matrices 253
      11.3.1 SOR for diffusion in 3D 255

      11.3.2 MATLAB® implementation of SOR 256

      11.3.3 Exercises 258

  11.4 Conjugate Gradient for SPD Matrices 259

      11.4.1 MATLAB® implementations of CG 262

      11.4.2 Exercises 266

  11.5 Generalized Minimum Residual 267

      11.5.1 MATLAB® implementations of GMRES 271

      11.5.2 Exercises 275

12 Nonlinear Problems and Least Squares   **277**

  12.1 Picard Approximation 277

      12.1.1 MATLAB® code piccool.m 281

  12.2 Newton Method 283

      12.2.1 MATLAB® code newtcool.m 286

  12.3 Levenberg-Marquardt Method 288

      12.3.1 MATLAB® code levmarqprice.m 291

  12.4 SIRD Epidemic Models 293

      12.4.1 MATLAB® code sird_parid.m 295

  12.5 The Cumulated Infection Version of SIRD 300

      12.5.1 US COVID-19: An aggregated model 302

Bibliography   **305**

Index   **307**

List of Figures

1.1.1	Box with fixed volume.	7
1.1.2	Cost of a box.	8
1.1.3	Two-bar truss.	9
1.1.4	Two-loop circuit.	10
1.2.1	Thin cooling fin.	18
1.4.1	Six-bar truss.	38
1.5.1	Five-bar truss.	56
2.1.1	Three-loop circuit.	72
3.2.1	Parameter identification using least squares	93
4.4.1	Bar e with four forces.	125
4.4.2	Navier–Stokes grid.	127
5.2.1	All least squares solutions.	133
8.1.1	USA matrix via mesh().	174
8.1.2	USA jpg picture.	174
8.1.3	Negative of USA jpg picture.	175
8.1.4	Image compression using SVD.	175
8.2.1	All singular values.	182
8.2.2	Search using truncated SVD.	182
8.3.1	Vectors from SVD of K.	186
8.3.2	Noise filter with truncation k = 40.	187
9.4.1	Hazard Id with osites = [10 30 60 90].	204
9.4.2	Hazard Id with osites = [10 30 60 70].	206
9.4.3	Hazard Id with osites = [10 30].	207
11.3.1	Steady state temperature.	258
11.5.1	GMRES output.	272
12.1.1	Picard iterations.	282
12.3.1	Levenberg-Marquardt prediction of price.	291
12.4.1	Variable contact parameter.	295
12.4.2	Parameter identification.	296

12.5.1 US COVID-19 data.   301
12.5.2 Covid projection with 21-day data.   303

Preface

Courses on linear algebra and numerical analysis need each other. Often NA courses have some linear algebra topics, and LA courses mention some topics from numerical analysis/scientific computing. One reason for writing this text is to merge these two areas into one introductory undergraduate course for students with multivariable calculus. A second goal is to demonstrate the intimate relationship of linear algebra to applications/computations. A third objective is to present, in the first half of the course, the very important topic of singular value decomposition, SVD.

Of course some topics have not been fully covered. Here the rigorous presentation has been maintained. This is done by first restricting consideration to real matrices and vector spaces in $\mathbb{R}^n$. The general inner product vector spaces are considered starting in the middle of the text, Chapter 10.

The text has a number of applications. These are to motivate the student to study the linear algebra topics. The purpose of an application is not to necessarily teach all of the application, but it is to make a case for the importance and utility of linear algebra.

Also, the text has a number of computations. Here MATLAB® [18] has been used, but one could modify these codes to other programming languages. These are either to simplify some linear algebra computation, or to model a particular application. One could invest a lot of time examining these codes, but they should be viewed as supplements and not a replacement to linear algebra topics.

Matrices with inverses are covered in the first two chapters. Least squares and under-determined systems are in Chapters 3 and 4. Orthogonality is presented in Chapters 5 and 6. Chapters 7–9 are related to the SVD. Chapter 10 covers general inner product spaces and includes topics related to boundary value problems. Basic iterative methods are introduced in Chapter 11: matrix splitting, conjugate gradient (CG) and generalized minimum residual method (GMRES). Finally, not all problems are linear, and in the last chapter introductions are given to the basic nonlinear methods: Picard, Newton and Levenberg-Marquardt.

Some topics in this text have been taken from my lecture notes and [26] [28]. These select topics are not comprehensive and neither are the remaining contents of this text. In this text I cited a number of authors and mention several here. A classic text is volume one by R. Courant and D. Hilbert, [6], and the preface and Chapters 5 and 6 are very interesting. The book by G. Golub and C. Van Loan, [9], gives a comprehensive study of matrix analysis as it relates to numerical linear algebra. The popular book by G. Strang, [23], has a number of applications of linear algebra. The very comprehensive book by C. Meyer, [17], is a valuable resource for further study in linear algebra and applications.

Robert E White, July 28, 2022

Introduction

This text is a continuation of vector, matrices and algebraic linear systems with small dimension. Because of the technical difficulties of by-hand calculations, "small" means $n = 2, 3$ or 4. One goal is to extend this to higher dimensions and the use of computing tools. For example, fluid flow models in 3D space can have algebraic systems with very large numbers of unknowns. Consider finding air pressures inside of small cubes with 100 edges in each direction; this gives $n = 10^6$ unknowns! Linear algebra and computation can be combined to solve such problems.

The first four chapters contain extensions of some topics you have studied. Chapter 1 extends vector and matrix operations to vectors with n real numbers and to $n \times n$ matrices with n^2 components. Algebraic systems with n unknowns and n equations will be solved by Gauss elimination (row operations), inverse matrices and determinants. Four very special matrices with inverse are introduced in the second chapter. The third and fourth chapters are a study of over-determined (least squares) and under-determined (multiple solutions) algebraic systems.

Another view of the course is the study of seven factorizations of a matrix: $P^T LU$, LU, $G^T G$, QR, QDQ^T, $U\Sigma V^T$ and UTU^*. The first three will be done in Chapter 2, the next two will be presented in Chapters 5 and 6. The singular value decomposition (SVD) of an $m \times n$ matrix is $U\Sigma V^T$, and it will be studied in Chapters 7–9. The SVD has been known for well over a century, but with the use of computers it has proven useful in a number of areas such as image compression and search engines. The last factorization, UTU^*, is for matrices with complex numbers as components.

The notion of vectors is extended to functions or objects that can be added and multiplied by scalers. This is done in Chapter 10 and will be applied to boundary value problems where matrices are now replaced by differential operators. Here there are parallels between symmetric matrices and self-adjoint boundary value problems.

For very large algebraic systems, direct methods such as Gauss elimination require large storage and large computer operations. An alternative is to use iterative methods which are introduced in Chapter 11. The last chapter deals with another problem: not all problems are linear. For example, finding the square root means solve $x^2 - d = 0$. Fortunately, these can be approximated by a sequence of linear solves as in Newton's method to find the square root.

This one-semester course is just the beginning for additional studies. Mathematics courses on numerical analysis, numerical solutions of differential equations, numerical linear algebra and functional analysis might be interesting. Computer science has a number of courses that will help with using particular computer architectures, artificial intelligence and algorithms. Statistical modeling and computing have always been important. Finally, there are many application courses in the science/engineering programs.

The computer codes and updates for this book can be found at the website:

white.math.ncsu.edu/filenames

The MathWorks, Inc.
3 Apple Hill Drive
Natick, MA 01760-2098, USA
Tel: 508-647-7000
Fax: 508-647-7001
E-mail: info@mathworks.com
www.mathworks.com

Author Biography

Robert E. White (on the web at white.math.ncsu.edu) is Professor Emeritus of Mathematics, North Carolina State University. During his tenure from 1973 to 2013, his research-teaching areas were in partial differential equations, numerical analysis and scientific computation. Since retiring from teaching in 2013, he has continued research and completed two textbooks. He is also the author of *Computational Mathematics: Models, Methods, Analysis with MATLAB® and MPI*, Second Edition and *Elements of Matrix Modeling and Computing with MATLAB®*, both published by CRC Press.

1

Solution of $Ax = d$

This chapter contains the basic methods for solving matrix equations of the form $Ax = d$ where A is an matrix and $\mathbf{x}$ and $\mathbf{d}$ are column vectors. The possibilities of a unique solution, no solution, multiple solutions and least square solutions are discussed. In this chapter most of the algebraic systems are assumed to be square so that the number rows (equations) and columns (variables) are equal. Row operations and elementary matrices are used to do by-hand computations of the Gauss elimination, inverse matrix and determinant methods. These methods are also implemented in MATLAB®. Applications to steady state circuits, heat conduction and support trusses are given.

1.1 Matrix Models

The extension from vectors with two or three components and from 2×2 or 3×3 matrices to higher dimensions will be motivated by graphics and a variety of applications. In this section the simplest models of cost, circuits and trusses will be presented. We will return to these models in the remainder of the text so as to increase their complexity, data dependence and accuracy.

1.1.1 Column vectors and $\mathbb{R}^n$

Vectors may either be row or column vectors and usually are an ordered list of real or complex numbers. However, they could have more general components such as a phone book where the components are a triples having a name, address and phone number. In the remainder of this chapter, we will assume the vector is a column and will use a bold lower case font to denote vectors.

Definition. *An $n \times 1$ column vector is an ordered list of n real or complex numbers. This will be described in two ways:*

$$\mathbf{a} = \begin{bmatrix} a_1 \\ \vdots \\ a_n \end{bmatrix} \text{ or } \mathbf{a} = [a_i] \text{ where } i = 1, \cdots, n.$$

DOI: 10.1201/9781003304128-1

The set of all $n \times 1$ *column vectors with real components is denoted by* $\mathbb{R}^n$, *and* $\mathbb{C}^n$ *is the set of all* $n \times 1$ *column vectors with complex components.*

Four basic operations with vectors are scalar product, addition, augmentation and transpose.

Definition. *Let* **a** *and* **b** *be* $n \times 1$ *column vectors and let* s *be a real or complex number.*

Scalar product $s\mathbf{a}$ *is another* $n \times 1$ *column vector whose components are* sa_i

$$s\mathbf{a} \equiv \begin{bmatrix} sa_1 \\ \vdots \\ sa_n \end{bmatrix} = [sa_i] \text{ where } i = 1, \cdots, n.$$

Vector addition is $\mathbf{a} + \mathbf{b}$ *is another* $n \times 1$ *column vector whose components are* $a_i + b_i$

$$\mathbf{a} + \mathbf{b} \equiv \begin{bmatrix} a_1 + b_1 \\ \vdots \\ a_n + b_n \end{bmatrix} = [a_i + b_i] \text{ where } i = 1, \cdots, n.$$

Vector augmentation of **a** *and* **b** *denoted by* [**a** **b**] *is an* $n \times 2$ *matrix with two column vectors where* **a** *is the first column of the matrix*

$$[\mathbf{a} \ \mathbf{b}] = \begin{bmatrix} a_1 & b_1 \\ \vdots & \vdots \\ a_n & b_n \end{bmatrix} = [a_i \ b_i] \text{ where } i = 1, \cdots, n.$$

One can augment m $n \times 1$ *column vectors to form an* $n \times m$ *matrix, which will be denoted by upper case fonts.*

The transpose of **a** *is a* $1 \times n$ *row vector*

$$\mathbf{a}^T = [a_1 \cdots a_n].$$

Example 1.1.1. Consider the three 4×1 column vectors

$$\mathbf{a} = \begin{bmatrix} 1 \\ 6 \\ 3 \\ 5 \end{bmatrix}, \ \mathbf{b} = \begin{bmatrix} 3 \\ 2 \\ -1 \\ 3 \end{bmatrix} \text{ and } \mathbf{c} = \begin{bmatrix} 0 \\ 1 \\ 4 \\ -2 \end{bmatrix}.$$

One can combine scalar multiplication and vector addition to form a linear combination of these vectors

$$2\mathbf{a}+3\mathbf{b}-\mathbf{c} = 2\begin{bmatrix} 1 \\ 6 \\ 3 \\ 5 \end{bmatrix} + 3\begin{bmatrix} 3 \\ 2 \\ -1 \\ 3 \end{bmatrix} - \begin{bmatrix} 0 \\ 1 \\ 4 \\ -2 \end{bmatrix}$$

$$= \begin{bmatrix} 2(1) + 3(3) - 0 \\ 2(6) + 3(2) - 1 \\ 2(3) + 3(-1) - 4 \\ 2(5) + 3(3) - (-2) \end{bmatrix} = \begin{bmatrix} 11 \\ 17 \\ -1 \\ 21 \end{bmatrix}.$$

This is in contrast to the augmentation of the three column vectors, which is a 4×3 matrix or array

$$A = \begin{bmatrix} 1 & 3 & 0 \\ 6 & 2 & 1 \\ 3 & -1 & 4 \\ 5 & 3 & -2 \end{bmatrix}.$$

The set of $n \times 1$ column vectors $\mathbb{R}^n$ is called a vector space because it has the properties listed in the next theorem. There are five properties for addition and five properties for scaler product. The proofs are a consequence of the usual rules for single numbers. For example, consider $\mathbf{a} + \mathbf{b} = \mathbf{b} + \mathbf{a}$.

$$\mathbf{a} = \begin{bmatrix} a_1 \\ a_2 \\ \vdots \\ a_n \end{bmatrix} \text{ and } \mathbf{b} = \begin{bmatrix} b_1 \\ b_2 \\ \vdots \\ b_n \end{bmatrix}$$

$$\mathbf{a} + \mathbf{b} = \begin{bmatrix} a_1 + b_1 \\ a_2 + b_2 \\ \vdots \\ a_n + b_n \end{bmatrix} = \begin{bmatrix} b_1 + a_1 \\ b_2 + a_2 \\ \vdots \\ b_n + a_n \end{bmatrix} = \mathbf{b} + \mathbf{a}.$$

Theorem 1.1.1. *(Vector Space Properties of $\mathbb{R}^n$)* *Let $\mathbf{a}$, $\mathbf{b}$ and $\mathbf{c}$ be in $\mathbb{R}^n$ and let s and t be real numbers. Let $\mathbf{a} + \mathbf{b}$ and $s\mathbf{a}$ denote vector addition and scalar product. Then the following hold:*

$\mathbf{a} + \mathbf{b} \in \mathbb{R}^n$, $\mathbf{a} + \mathbf{0} = \mathbf{a}$ *where* $\mathbf{0} = 0_{n\times1} \in \mathbb{R}^n$ *has zeros as components,*
$\mathbf{a} + (-\mathbf{a}) = \mathbf{0}$, $\mathbf{a} + \mathbf{b} = \mathbf{b} + \mathbf{a}$, $\mathbf{a} + (\mathbf{b} + \mathbf{c}) = (\mathbf{a} + \mathbf{b}) + \mathbf{c}$;
$s\mathbf{a} \in \mathbb{R}^n$, $1\mathbf{a} = \mathbf{a}$, $s(\mathbf{a}+\mathbf{b}) = s\mathbf{a}+s\mathbf{b}$, $(s+t)\mathbf{a} = s\mathbf{a}+t\mathbf{a}$ *and* $s(t\mathbf{a}) = (st)\mathbf{a}$.

The vector space properties hold for many sets with addition and scaler operations such as the continuous real valued functions, see Chapter 10. Any subset of $\mathbb{R}^n$, $S \subset \mathbb{R}^n$, which has two properties $\mathbf{a} + \mathbf{b} \in S$ and $s\mathbf{a} \in S$ when $\mathbf{a}, \mathbf{b} \in S$ and $s \in \mathbb{R}$, will inherit all the other vector space properties. Examples include the xy-plane in $\mathbb{R}^3$, and the set of points in $\mathbb{R}^3$ that satisfy the equation $ax + by + cz = 0$. Sets with these two properties are called *subspaces* of $\mathbb{R}^n$.

The definitions of dot product and norms for vectors in $\mathbb{R}^2$ and $\mathbb{R}^3$ have the following generalizations and analogous properties.

Definition. *Let $\mathbf{a}$ and $\mathbf{b}$ be given vectors in $\mathbb{R}^n$. The dot product of $\mathbf{a}$ and $\mathbf{b}$ is a real number defined by*

$$\mathbf{a} \bullet \mathbf{b} \equiv \mathbf{a}^T\mathbf{b} = a_1 b_1 + a_2 b_2 + \cdots + a_n b_n.$$

The vectors $\mathbf{a}$ and $\mathbf{b}$ called orthogonal if and only if their dot product is zero

$$\mathbf{a} \bullet \mathbf{b} = \mathbf{a}^T\mathbf{b} = 0.$$

The Euclidean norm of $\mathbf{a}$ is a real number defined by

$$\|\mathbf{a}\| \equiv (\mathbf{a} \bullet \mathbf{a})^{1/2} = (\mathbf{a}^T\mathbf{a})^{1/2} = (a_1^2 + a_2^2 + \cdots + a_n^2)^{1/2}.$$

In order to illustrate the dot product and norm, consider Example 1.1.1 where $n = 4$

$$\mathbf{a} \bullet \mathbf{b} = 1(3) + 6(2) + 3(-1) + 5(3) = 27 \text{ and}$$
$$\|\mathbf{a}\| = (1^2 + 6^2 + 3^2 + 5^2)^{1/2} = \sqrt{71}.$$

Many properties of the dot product and norm, that hold for vectors in the plane and space, also hold for vectors in $\mathbb{R}^n$. The proofs of these are similar except for the Cauchy inequality.

Theorem 1.1.2. *(Properties of Dot Product) Let $\mathbf{a}$, $\mathbf{b}$ and $\mathbf{c}$ be given vectors in $\mathbb{R}^n$ and let s be a real number. Then the following hold*

$$\mathbf{a} \bullet \mathbf{b} = \mathbf{b} \bullet \mathbf{a}, \; s(\mathbf{a} \bullet \mathbf{b}) = (s\mathbf{a}) \bullet \mathbf{b},$$
$$\mathbf{a} \bullet (\mathbf{b} + \mathbf{c}) = \mathbf{a} \bullet \mathbf{b} + \mathbf{a} \bullet \mathbf{c} \text{ and}$$
$$|\mathbf{a} \bullet \mathbf{b}| \le \|\mathbf{a}\| \, \|\mathbf{b}\| \; \textit{(Cauchy inequality)}. \tag{1.1.1}$$

Theorem 1.1.3. *(Properties of Norm) Let $\mathbf{a}$ and $\mathbf{b}$ be given vectors in $\mathbb{R}^n$ and let s be a real number. Then the following hold*

$$\|\mathbf{a}\| \ge 0; \; \mathbf{a} = \mathbf{0} = \mathbf{0}_{n \times 1} \text{ if and only if } \|\mathbf{a}\| = 0,$$
$$\|s\mathbf{a}\| = |s| \, \|\mathbf{a}\| \text{ and}$$
$$\|\mathbf{a} + \mathbf{b}\| \le \|\mathbf{a}\| + \|\mathbf{b}\| \; \textit{(triangle inequality)}. \tag{1.1.2}$$

The *Cauchy inequality* for vectors in $\mathbb{R}^2$ and $\mathbb{R}^3$ follows from the cosine identity. The proof for vectors in $\mathbb{R}^n$ makes use of the dot product properties and a clever observation. Let t be a real number and define the following function with $\mathbf{b} \ne \mathbf{0}$

$$f(t) \equiv (\mathbf{a} + t\mathbf{b}) \bullet (\mathbf{a} + t\mathbf{b}) \ge 0$$

Note, it can be written as a quadratic function of t

$$f(t) \equiv \mathbf{a} \bullet \mathbf{a} + 2t\mathbf{a} \bullet \mathbf{b} + t^2 \mathbf{b} \bullet \mathbf{b}.$$

The first and second derivatives are

$$\frac{df}{dt} = 0 + 2\mathbf{a} \bullet \mathbf{b} + 2t\mathbf{b} \bullet \mathbf{b} \text{ and}$$
$$\frac{d^2 f}{dt^2} = 0 + 2\mathbf{b} \bullet \mathbf{b}.$$

Choose t_0 so that $f(t_0)$ is a minimum, that is,

$$\frac{df}{dt} = 0 + 2\mathbf{a} \bullet \mathbf{b} + 2t_0 \mathbf{b} \bullet \mathbf{b} = 0.$$

Thus,

$$t_0 = \frac{-\mathbf{a} \bullet \mathbf{b}}{\mathbf{b} \bullet \mathbf{b}} \text{ and}$$
$$f(t_0) = \mathbf{a} \bullet \mathbf{a} + 2t_0 \mathbf{a} \bullet \mathbf{b} + t_0^2 \mathbf{b} \bullet \mathbf{b}$$
$$= \mathbf{a} \bullet \mathbf{a} - \frac{(\mathbf{a} \bullet \mathbf{b})^2}{\mathbf{b} \bullet \mathbf{b}} \geq 0.$$

The inequality is equivalent to the *Cauchy* inequality.

The proof of the triangle inequality uses the *Cauchy* inequality:

$$\|\mathbf{a} + \mathbf{b}\|^2 = (\mathbf{a} + \mathbf{b}) \bullet (\mathbf{a} + \mathbf{b})$$
$$= \mathbf{a} \bullet \mathbf{a} + 2\mathbf{a} \bullet \mathbf{b} + \mathbf{b} \bullet \mathbf{b}$$
$$\leq \|\mathbf{a}\|^2 + 2\|\mathbf{a}\|\|\mathbf{b}\| + \|\mathbf{b}\|^2$$
$$= (\|\mathbf{a}\| + \|\mathbf{b}\|)^2.$$

The dot product is a special case of the inner product for general vector spaces, see Chapter 10. The Cauchy inequality was formulated in 1821, and generalized in 1859 by Bunyakovsky and later in 1888 by Schwarz. The Cauchy inequality is often called the CBS inequality. It is a very important generalization. For example, if $\mathbf{a} \bullet \mathbf{b} = 0$ (the vectors are "perpendicular"), then $\|\mathbf{a} + \mathbf{b}\|^2 = \|\mathbf{a}\|^2 + \|\mathbf{b}\|^2$.

1.1.2 Matrices

Matrices can be used to store a variety of information, which often requires visualizing or modifications. There are four element-wise or array operations that will be very useful.

Definition. *An augmentation of n $m \times 1$ column vectors with $1 \leq j \leq n$ of the form*

$$\begin{bmatrix} a_{1j} \\ \vdots \\ a_{mj} \end{bmatrix}$$

is an $m \times n$ matrix A where the first index $1 \leq i \leq m$ is the row number, the second index $1 \leq j \leq n$ is the column number and

$$A = \begin{bmatrix} a_{11} & a_{12} & \cdots & a_{1n} \\ \vdots & \vdots & \cdots & \vdots \\ a_{m1} & a_{n2} & \cdots & a_{mn} \end{bmatrix} = [a_{ij}].$$

Let A and B be $m \times n$ matrices. Define the following four array operations, which generate additional $m \times n$ matrices:

$$f(A) \equiv [f(a_{ij})],$$
$$A.*B \equiv [a_{ij}b_{ij}],$$
$$A./B \equiv [a_{ij}/b_{ij}] \text{ and}$$
$$A.\hat{\ }k \equiv [a_{ij}\hat{\ }k].$$

Example 1.1.2. Example Consider the 4×3 matrix in Example 1.1.1 formed by augmenting the three column vectors. The component in the third row and first column is $a_{31} = 3$, and $a_{23} = 1$ is the component in the second row and third column. Array operations with exponents are not the same as matrix products

$$A.\hat{\ }2 = \begin{bmatrix} 1^2 & 3^2 & 0^2 \\ 6^2 & 2^2 & 1^2 \\ 3^2 & (-1)^2 & 4^2 \\ 5^2 & 3^2 & (-2)^2 \end{bmatrix} = \begin{bmatrix} 1 & 9 & 0 \\ 36 & 4 & 1 \\ 9 & 1 & 16 \\ 25 & 9 & 4 \end{bmatrix}.$$

Example 1.1.3. This example illustrates how array operations and matrices can be used to create a graph in MATLAB® of a surface given by $z = f(x,y) = 100 - x^2 - 4y^2$. In order to graph the surface, a collection of points in the xy-plane must be selected and then the values of $f(x,y)$ at these points must be computed and stored. Suppose three points in the x-direction are $x = 0, 4$ and 8, and four points in the y-direction are $y = 0, 2, 4$ and 6. There are a total of 12 points in the xy-plane; the 12 values of x can be stored in a 4×3 matrix X, and the 12 values of y can be stored in a 4×3 matrix Y. The corresponding 12 values of $f(x,y)$ can be computed and stored in a 4×3 matrix F. The following MATLAB® commands do this. Note, the row 4 and column 3 component of F is $100 - 8^2 - 4(6^2) = -108$. By using smaller increments in the MATLAB® command meshgrid(), one can generate much larger matrices and a more accurate depiction of the surface, which is graphed by the MATLAB® command mesh(X,Y,F).

```
>> [X Y] = meshgrid(0:4:8,0:2:6)
   X =
         0 4 8
         0 4 8
         0 4 8
         0 4 8
   Y =
         0 0 0
         2 2 2
         4 4 4
         6 6 6
>> F = 100 - X.^2 - 4*Y.^2
   F =
         100   84   36
         084   68   20
         036   20  -28
         -44  -60 -108
>> mesh(X,Y,F)
```

1.1.3   Application to visualization of minimum cost

The objective is to find the dimensions of a 3D box with fixed volume and with minimum cost. In order to approximate the solution, the cost function will be graphed, and we will inspect the graph for a minimum cost. Suppose the volume must be 1000 ft^3, the bottom cost 3 \$/ft^2, the four sides cost 1 \$/ft^2 and there is no top. Let the bottom have edges equal to x and y, and let the height of the box be z. This is illustrated by Figure 1.1.1. The cost of the bottom is $3xy$, and the cost of the sides is $1(yz + yz + xz + xz)$. Since the volume must be 1000, $1000 = xyz$ or $z = 1000/(xy)$. Because the total cost

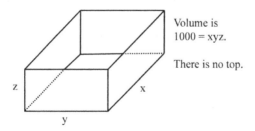

Volume is
1000 = xyz.

There is no top.

FIGURE 1.1.1
Box with fixed volume.

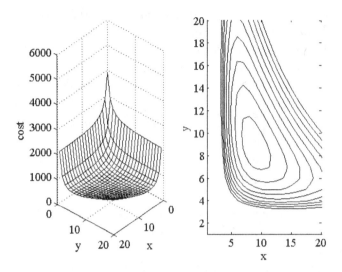

FIGURE 1.1.2
Cost of a box.

equals the sum of the cost of the bottom and the four sides,

$$\begin{aligned}
C(x, y) &= 3xy + 2(yz + xz) \\
&= 3xy + 2(y1000/(xy) + x1000/(xy)) \\
&= 3xy + 2(1000/x + 1000/y).
\end{aligned}$$

The following MATLAB® commands create the surface and contour plots of this cost function as given in Figure 1.1.2. The three matrices X, Y and C are 20×20 and note how array operations are used to compute the possible costs in the matrix C. An estimate of the minimum cost is given by $x = 9$ and $y = 9$ so that $C(9, 9) = 243 + 4000/9 \approx 687.4444$. A better estimate is given by using multivariable calculus where one sets the partial derivatives equal to zero and solve for x and y to obtain $x = y = 8.7358$ and minimum cost equal to 686.8285.

```
>> [X Y ] = meshgrid(1:1:20,1:1:20);
>> C = 3*X.*Y + 2000./Y + 2000./X;
>> subplot(1,2,1)
>> mesh(X,Y,C)
>> subplot(1,2,2)
>> contour(X,Y,C, 650:25:900)
```

1.1.4   Application to two-bar truss

Consider the two-bar truss in Figure 1.1.3. This truss is attached to a wall on the left and can be used to support a mass on the node joining the two

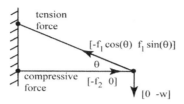

FIGURE 1.1.3
Two-bar truss.

bars. In the truss, the bar ends are always pin connected. Consequently, the forces in the bars are always along the axis. Assume there is no deformation and the bars are in a static state. The horizontal bar will have compression force $f_2 < 0$, and the diagonal bar will have a tension force $f_1 > 0$. At the joining node, there are three force vectors whose sum must be the zero vector to satisfy the equilibrium equations

$$[-f_1 \cos(\theta) \quad f_1 \sin(\theta)] + [-f_2 \quad 0] + [0 \quad -w] = [0 \quad 0] \text{ or}$$
$$[-f_1 \cos(\theta) - f_2 \quad f_1 \sin(\theta) - w] = [0 \quad 0].$$

By equating the first and second components of the vector equation, we obtain two scalar equations

$$-f_1 \cos(\theta) - f_2 = 0 \text{ and}$$
$$f_1 \sin(\theta) = w.$$

The matrix version of this is

$$\begin{bmatrix} -\cos(\theta) & -1 \\ \sin(\theta) & 0 \end{bmatrix} \begin{bmatrix} f_1 \\ f_2 \end{bmatrix} = \begin{bmatrix} 0 \\ w \end{bmatrix}.$$

The solution is easy to compute. For example, if $w = 100$ and $\theta = \pi/6$, then

$$\begin{bmatrix} f_1 \\ f_2 \end{bmatrix} = \begin{bmatrix} 200 \\ -100\sqrt{3} \end{bmatrix}.$$

If θ decreases, then the tension and compressive force magnitudes will increase. If there are more than two bars, then the number of joints will increase as well as the number of force vectors and force vector equations.

1.1.5   Application to two-loop circuit

Consider the two-loop circuit in Figure 1.1.4 with three resistors and two batteries. The current going through resistors R_1, R_2 and R_3 will be denoted by i_1, i_2 and i_3 and will have directions indicated by the arrows in the figure. Kirchhoff's current law requires the sum of the currents at any node be zero

$$i_1 - i_2 + i_3 = 0.$$

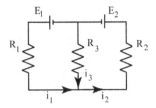

FIGURE 1.1.4
Two-loop circuit.

Ohm's law states that the voltage drop across each resistor is the resistance times the current. Kirchhoff's voltage law requires the sum of the voltage drops in each loop be equal to zero

$$E_1 - R_1 i_1 + R_3 i_3 = 0$$
$$-E_2 - R_2 i_2 - R_3 i_3 = 0.$$

The Kirchhoff current law requires the three currents to satisfy

$$-i_1 + i_2 - i_3 = 0.$$

The matrix version for the three unknown currents and the above three scalar equations is

$$\begin{bmatrix} 1 & -1 & 1 \\ R_1 & 0 & -R_3 \\ 0 & -R_2 & -R_3 \end{bmatrix} \begin{bmatrix} i_1 \\ i_2 \\ i_3 \end{bmatrix} = \begin{bmatrix} 0 \\ E_1 \\ E_2 \end{bmatrix}.$$

An alternative is to use the Kirchhoff voltage law which requires the voltage drops in the three vertical portions of the loops to be equal

$$x_1 + E_1 - R_1 i_1 = 0$$
$$x_1 - R_3 i_3 = 0$$
$$x_1 - (-i_2)R_2 + E_2 = 0.$$

This gives the system for four unknowns

$$\begin{bmatrix} R_1 & 0 & 0 & -1 \\ 0 & R_2 & 0 & 0 \\ 0 & 0 & R_3 & -1 \\ -1 & 1 & -1 & 0 \end{bmatrix} \begin{bmatrix} i_1 \\ i_2 \\ i_3 \\ x_1 \end{bmatrix} = \begin{bmatrix} E_1 \\ -E_2 \\ 0 \\ 0 \end{bmatrix}$$

The solution of this 3×3 algebraic system can be found in a number of ways. For example, if $R_1 = 1$, $R_2 = 2$, $R_3 = 3$, $E_1 = 10$ and $E_2 = 20$, then one could use Cramer's rule, see Section 1.6, to find the currents. Let

$$A = \begin{bmatrix} 1 & -1 & 1 \\ 1 & 0 & -3 \\ 0 & -2 & -3 \end{bmatrix} \text{ and } \mathbf{d} = \begin{bmatrix} 0 \\ 10 \\ 20 \end{bmatrix} \text{ and note}$$

$$\det(A) = 1(0 - (-3)(-2)) - (-1)(1(-3) - 0) + 1(1(-2) - 0) = -11.$$

The solutions can be computed using Cramer's Rule and determinants

$$i_1 = \det(\begin{bmatrix} 0 & -1 & 1 \\ 10 & 0 & -3 \\ 20 & -2 & -3 \end{bmatrix})/\det(A) = 10/(-11),$$

$$i_2 = \det(\begin{bmatrix} 1 & 0 & 1 \\ 1 & 10 & -3 \\ 0 & 20 & -3 \end{bmatrix})/\det(A) = 50/(-11) \text{ and}$$

$$i_3 = \det(\begin{bmatrix} 1 & -1 & 0 \\ 1 & 0 & 10 \\ 0 & -2 & 20 \end{bmatrix})/\det(A) = 40/(-11).$$

Most circuits have many more loops so that the resulting algebraic systems are very large. In these cases it is not practical to solve them using Cramer's rule. In the following sections viable alternative methods will be developed.

1.1.6 Exercises

1. Let $\mathbf{a}^T = [2\ 5\ 7\ 1\ -1]$, $\mathbf{b}^T = [0\ 1\ 2\ 7\ -4]$ and $\mathbf{c}^T = [1\ 4\ 2\ 1\ -3]$.

   (a). Find $3\mathbf{a} - 4\mathbf{b} + 2\mathbf{c}$.

   (b). Find the augmentations of $\mathbf{a}$, $\mathbf{b}$ and $\mathbf{c}$.

   (c). Find $\mathbf{a} \bullet \mathbf{b}$.

   (d). Find $\|\mathbf{a}\|$.

2. Consider the above problem.

   (a). Verify the $\mathbf{a} + (\mathbf{b} + \mathbf{c}) = (\mathbf{a} + \mathbf{b}) + \mathbf{c}$.

   (b). Prove this is true for all choices of the three vectors.

   (c). Verify the $\mathbf{a} \bullet (\mathbf{b} + \mathbf{c}) = \mathbf{a} \bullet \mathbf{b} + \mathbf{a} \bullet \mathbf{c}$.

   (d). Prove this is true for all choices of the three vectors.

3. Consider Example 1.1.2.

   (a). Find $A.\hat{\ }(1/2)$.

   (b). Find $\mathbf{b}. * \mathbf{c}$ where $\mathbf{b}$ and $\mathbf{c}$ are the second and third columns of A.

   (c). Find $\mathbf{b} \bullet \mathbf{c}$.

4. Consider the surface $z = f(x, y) = 400 - 2x^2 - y^2$.

   (a). By hand compute the 2×3 matrix

   $$F = \begin{bmatrix} f(1,1) & f(1,2) & f(1,3) \\ f(2,1) & f(2,2) & f(2,3) \end{bmatrix}.$$

   (b). Use MATLAB® and mesh(X,Y,F) to compute F and graph.

5. Consider the minimum cost of the box with no top, volume equals 200, bottom costs 4 \$/ft^2 and sides costs 1 \$/ft^2.

   (a). Find the cost function.

   (b). Use MATLAB® and mesh(X,Y,C) to approximate the minimum cost.

6. Consider the two-bar truss with $w = 100$ and variable $\theta = \pi/12$, $2\pi/12$, $3\pi/12$ and $4\pi/12$. Find the forces on the two bars.

7. Consider the two-loop circuit with $E_1 = 10$, $E_2 = 20$, $R_1 = 1$, $R_2 = 2$ and $R_3 = 2$. Find the three currents.

1.2   Matrix Products

In the previous sections matrices were used to describe systems of equations with two or three unknowns. Most realistic models have many more unknowns, and the resulting matrices are much larger. The general matrix-vector and matrix-matrix products will be defined and their properties will described.

An algebraic system for three unknowns x_1, x_2 and x_3 can be listed either as three scalar equations or as a vector equation

$$a_{11}x_1 + a_{12}x_2 + a_{13}x_3 = d_1,$$
$$a_{21}x_1 + a_{22}x_2 + a_{23}x_3 = d_2 \text{ and}$$
$$a_{31}x_1 + a_{32}x_2 + a_{33}x_3 = d_3.$$

The vector equation has the form

$$\begin{bmatrix} a_{11} & a_{12} & a_{13} \\ a_{21} & a_{22} & a_{23} \\ a_{31} & a_{32} & a_{33} \end{bmatrix} \begin{bmatrix} x_1 \\ x_2 \\ x_3 \end{bmatrix} = \begin{bmatrix} d_1 \\ d_2 \\ d_3 \end{bmatrix} \text{ or}$$

$A\mathbf{x} = \mathbf{d}$ where A is 3×3.

If there were two equations with four unknowns, then

$$a_{11}x_1 + a_{12}x_2 + a_{13}x_3 + a_{14}x_4 = d_1 \text{ and}$$
$$a_{21}x_1 + a_{22}x_2 + a_{23}x_3 + a_{24}x_4 = d_2.$$

The vector form is

$$\begin{bmatrix} a_{11} & a_{12} & a_{13} & a_{14} \\ a_{21} & a_{22} & a_{23} & a_{24} \end{bmatrix} \begin{bmatrix} x_1 \\ x_2 \\ x_3 \\ x_4 \end{bmatrix} = \begin{bmatrix} d_1 \\ d_2 \end{bmatrix} \text{ or}$$

$A\mathbf{x} = \mathbf{d}$ where A is 2×4.

1.2.1 Matrix-vector products

The product of a 3×3 matrix times a 3×1 column vector can be done by either products of rows in the matrix and the column vector, or by linear combinations of the columns of the matrix. The row version is

$$\begin{bmatrix} a_{11} & a_{12} & a_{13} \\ a_{21} & a_{22} & a_{23} \\ a_{31} & a_{32} & a_{33} \end{bmatrix} \begin{bmatrix} x_1 \\ x_2 \\ x_3 \end{bmatrix} = \begin{bmatrix} a_{11}x_1 + a_{12}x_2 + a_{13}x_3 \\ a_{21}x_1 + a_{22}x_2 + a_{23}x_3 \\ a_{31}x_1 + a_{32}x_2 + a_{33}x_3 \end{bmatrix}.$$

The column version is

$$\begin{bmatrix} a_{11} & a_{12} & a_{13} \\ a_{21} & a_{22} & a_{23} \\ a_{31} & a_{32} & a_{33} \end{bmatrix} \begin{bmatrix} x_1 \\ x_2 \\ x_3 \end{bmatrix} = x_1 \begin{bmatrix} a_{11} \\ a_{21} \\ a_{31} \end{bmatrix} + x_2 \begin{bmatrix} a_{12} \\ a_{22} \\ a_{32} \end{bmatrix} + x_3 \begin{bmatrix} a_{13} \\ a_{23} \\ a_{33} \end{bmatrix}.$$

This important observation also holds for general matrix-vector products.

Definition. *Let A and B be $m \times n$ matrices, $\mathbf{a}$ and $\mathbf{x}$ be $n \times 1$ column vectors and let s be a real number. The ij-component of A is a_{ij} where i is the row number and j is the column number. Two matrices are equal if they have the same dimension and equal components.*

Row vector $\mathbf{a}^T$ times a column vector $\mathbf{x}$ is a real number equal to the sum of the products of the components (also called the dot product of $\mathbf{a}$ and $\mathbf{x}$)

$$\mathbf{a}^T \mathbf{x} \equiv a_1 x_1 + \cdots + a_n x_n.$$

Matrix A times a column vector $\mathbf{x}$ is a $n \times 1$ column vector whose i^{th} component is row i of A times the column vector x

$$A\mathbf{x} \equiv [a_{i1}x_1 + \cdots + a_{in}x_n] \text{ where } i = 1, \cdots, m.$$

The scalar s times a matrix A is another $m \times n$ matrix whose ij-component is s times the ij-component of A

$$sA \equiv [sa_{ij}].$$

Notation: $sA = [sa_{ij}] = [a_{ij}s] = As$.
The addition of two matrices A and B is another $m \times n$ matrix whose ij-component is the sum of the ij-components of A and B

$$A + B \equiv [a_{ij} + b_{ij}].$$

Example 1.2.1. Let A and B be 4×2 and $\mathbf{x}$ be a 2×1 column vector

$$A = \begin{bmatrix} 1 & 2 \\ -1 & 0 \\ 7 & 4 \\ 2 & 3 \end{bmatrix}, \ B = \begin{bmatrix} -1 & 3 \\ 2 & 1 \\ 0 & 5 \\ -2 & 4 \end{bmatrix} \text{ and } \mathbf{x} = \begin{bmatrix} 3 \\ 4 \end{bmatrix}.$$

Then $A\mathbf{x}$ is defined to be a 4×1 column vector

$$A\mathbf{x} = \begin{bmatrix} 1(3) + 2(4) \\ -1(3) + 0(4) \\ 7(3) + 4(4) \\ 2(3) + 3(4) \end{bmatrix} = \begin{bmatrix} 11 \\ -3 \\ 37 \\ 18 \end{bmatrix}.$$

The matrix vector product could also be computed by a linear combination of the two column vectors of A

$$A\mathbf{x} = 3 \begin{bmatrix} 1 \\ -1 \\ 7 \\ 2 \end{bmatrix} + 4 \begin{bmatrix} 2 \\ 0 \\ 4 \\ 3 \end{bmatrix} = \begin{bmatrix} 3 \\ -3 \\ 21 \\ 6 \end{bmatrix} + \begin{bmatrix} 8 \\ 0 \\ 16 \\ 12 \end{bmatrix} = \begin{bmatrix} 11 \\ -3 \\ 37 \\ 18 \end{bmatrix}.$$

If

$$C = \begin{bmatrix} 2 & 4 \\ -1 & 2 \end{bmatrix},$$

then $A + C$ is not defined because their row and column numbers are not equal. If $s = 7$, then sA is defined to be the 4×2 matrix

$$sA = \begin{bmatrix} 7(1) & 7(2) \\ 7(-1) & 7(0) \\ 7(7) & 7(4) \\ 7(2) & 7(3) \end{bmatrix} = \begin{bmatrix} 7 & 14 \\ -7 & 0 \\ 49 & 28 \\ 14 & 21 \end{bmatrix} \text{ and}$$

$$A + B = \begin{bmatrix} 1-1 & 2+3 \\ -1+2 & 0+1 \\ 7+0 & 4+5 \\ 2-2 & 3+4 \end{bmatrix} = \begin{bmatrix} 0 & 5 \\ 1 & 1 \\ 7 & 9 \\ 0 & 7 \end{bmatrix}.$$

Theorem 1.2.1. *(Matrix-vector Products Properties) Let A, B and C be $m \times n$ matrices. Let $Z = 0_{m \times n}$ be a $m \times n$ matrix with all components equal to zero, called the zero matrix. Let $\mathbf{x}$ and $\mathbf{y}$ be $n \times 1$ column vectors. Then the following are true:*

(i). $A + (B + C) = (A + B) + C$, $A + B = B + A$ and $A + Z = A$.

(ii). $A(s\mathbf{x}) = (sA)\mathbf{x}$ where s is a real number, $A(\mathbf{x} + \mathbf{y}) = A\mathbf{x} + A\mathbf{y}$.

(iii). The matrix-vector product can be computed as a linear combination of the columns. Let column j of A be denoted by $\mathbf{a}_{:j}$

$$A\mathbf{x} = x_1 \mathbf{a}_{:1} + \cdots + x_n \mathbf{a}_{:n}.$$

.

The proofs of the above six properties are routine. The last property is the column version of a matrix-vector product whose proof is

$$A\mathbf{x} = \begin{bmatrix} a_{11}x_1 + \cdots + a_{1m}x_m \\ \vdots \\ a_{n1}x_1 + \cdots + a_{nm}x_m \end{bmatrix}$$

$$= \begin{bmatrix} a_{11}x_1 \\ \vdots \\ a_{n1}x_1 \end{bmatrix} + \cdots + \begin{bmatrix} a_{1m}x_n \\ \vdots \\ a_{nm}x_m \end{bmatrix}$$

$$= x_1 \begin{bmatrix} a_{11} \\ \vdots \\ a_{m1} \end{bmatrix} + \cdots + x_n \begin{bmatrix} a_{1n} \\ \vdots \\ a_{mn} \end{bmatrix}.$$

Example 1.2.2. Verify the matrix addition properties in the above theorem for

$$A = \begin{bmatrix} 1 & 2 \\ 2 & 0 \\ 4 & 6 \end{bmatrix}, \; B = \begin{bmatrix} -1 & 3 \\ 2 & 1 \\ 2 & 5 \end{bmatrix} \text{ and } C = \begin{bmatrix} 1 & -5 \\ 2 & 3 \\ 4 & 3 \end{bmatrix}.$$

$$A + (B + C) = \begin{bmatrix} 1 & 2 \\ 2 & 0 \\ 4 & 6 \end{bmatrix} + (\begin{bmatrix} -1 & 3 \\ 2 & 1 \\ 2 & 5 \end{bmatrix} + \begin{bmatrix} 1 & -5 \\ 2 & 3 \\ 4 & 3 \end{bmatrix})$$

$$= \begin{bmatrix} 1 + (-1 + 1) & 2 + (3 - 5) \\ 2 + (2 + 2) & 0 + (1 + 3) \\ 4 + (2 + 4) & 6 + (5 + 3) \end{bmatrix}$$

$$= \begin{bmatrix} (1 - 1) + 1 & (2 + 3) - 5 \\ (2 + 2) + 2 & (0 + 1) + 3 \\ (4 + 2) + 4 & (6 + 5) + 3 \end{bmatrix} = (A + B) + C.$$

$$A + B = \begin{bmatrix} 1 & 2 \\ 2 & 0 \\ 4 & 6 \end{bmatrix} + \begin{bmatrix} -1 & 3 \\ 2 & 1 \\ 2 & 5 \end{bmatrix}$$

$$= \begin{bmatrix} 1 - 1 & 2 + 3 \\ 2 + 2 & 0 + 1 \\ 4 + 2 & 6 + 5 \end{bmatrix}$$

$$= \begin{bmatrix} 1 - 1 & 3 + 2 \\ 2 + 2 & 1 + 0 \\ 2 + 4 & 5 + 6 \end{bmatrix} = B + A.$$

$$A + Z = \begin{bmatrix} 1 & 2 \\ 2 & 0 \\ 4 & 6 \end{bmatrix} + \begin{bmatrix} 0 & 0 \\ 0 & 0 \\ 0 & 0 \end{bmatrix}$$

$$= \begin{bmatrix} 1+0 & 2+0 \\ 2+0 & 0+0 \\ 4+0 & 6+0 \end{bmatrix} = A.$$

1.2.2 Matrix-matrix products

Matrix-vector products may be extended to matrix-matrix product as long as the number of columns in the left matrix is the same as the number of rows in the right matrix. Another important fact is that the order of matrix-matrix product is important, that is, AB may not be the same as BA! Like the matrix-vector product there is a row version and a column version. For example, consider the product of two 2×2 matrices

$$\begin{bmatrix} 2 & 3 \\ 4 & 5 \end{bmatrix} \begin{bmatrix} 6 & 7 \\ 8 & 9 \end{bmatrix} = \begin{bmatrix} 2(6)+3(8) & 2(7)+3(9) \\ 4(6)+5(8) & 4(7)+5(9) \end{bmatrix}$$

$$= \begin{bmatrix} 2 & 3 \\ 4 & 5 \end{bmatrix} \begin{bmatrix} 6 \\ 8 \end{bmatrix} + \begin{bmatrix} 2 & 3 \\ 4 & 5 \end{bmatrix} \begin{bmatrix} 7 \\ 9 \end{bmatrix}.$$

This product is either computed by rows times columns or by matrix times columns. Initially, we use the rows times columns approach for the general definition of a matrix-matrix product.

Definition. *Let A be an $m \times n$ and let B be an $n \times p$ matrices. The product AB is an $m \times p$ matrix whose ij-component is the product of row i of A times column j of B*

$$AB \equiv [a_{i1}b_{1j} + \cdots + a_{in}b_{nj}].$$

Example 1.2.3. First, let

$$A = \begin{bmatrix} 1 & 2 \\ 3 & 4 \end{bmatrix} \text{ and } B = \begin{bmatrix} 0 & 1 \\ 1 & 0 \end{bmatrix} \text{ and note}$$

$$AB = \begin{bmatrix} 2 & 1 \\ 4 & 3 \end{bmatrix} \neq \begin{bmatrix} 3 & 4 \\ 1 & 2 \end{bmatrix} = BA.$$

Second, note AB may be defined while BA is not defined. Let A be 2×3 and B be 3×3

$$A = \begin{bmatrix} 2 & 4 & 7 \\ 8 & 9 & 10 \end{bmatrix} \text{ and } B = \begin{bmatrix} 2 & 1 & 0 \\ 1 & 2 & 3 \\ 0 & 1 & 1 \end{bmatrix}.$$

$$AB = \begin{bmatrix} 2(2) + 4(1) + 7(0) & 2(1) + 4(2) + 7(1) & 2(0) + 4(3) + 7(1) \\ 8(2) + 9(1) + 10(0) & 8(1) + 9(2) + 10(1) & 8(0) + 9(3) + 10(1) \end{bmatrix}$$
$$= \begin{bmatrix} 8 & 17 & 19 \\ 25 & 36 & 37 \end{bmatrix}.$$

This product also can be written as columns of the product given by A times the column of B

$$AB = \begin{bmatrix} A\mathbf{b}_1 & A\mathbf{b}_2 & A\mathbf{b}_3 \end{bmatrix} \text{ where } B = \begin{bmatrix} \mathbf{b}_1 & \mathbf{b}_2 & \mathbf{b}_3 \end{bmatrix}$$

$$A\mathbf{b}_1 = \begin{bmatrix} 2 & 4 & 7 \\ 8 & 9 & 10 \end{bmatrix} \begin{bmatrix} 2 \\ 1 \\ 0 \end{bmatrix} = \begin{bmatrix} 8 \\ 25 \end{bmatrix}$$

$$A\mathbf{b}_2 = \begin{bmatrix} 2 & 4 & 7 \\ 8 & 9 & 10 \end{bmatrix} \begin{bmatrix} 1 \\ 2 \\ 1 \end{bmatrix} = \begin{bmatrix} 17 \\ 36 \end{bmatrix}$$

$$A\mathbf{b}_3 = \begin{bmatrix} 2 & 4 & 7 \\ 8 & 9 & 10 \end{bmatrix} \begin{bmatrix} 0 \\ 3 \\ 1 \end{bmatrix} = \begin{bmatrix} 19 \\ 37 \end{bmatrix}.$$

Theorem 1.2.2. *(Matrix-matrix Products Properties) Let $A, \widehat{A}$ be $m \times n$, B be $n \times p$, C be $p \times q$ and s be a real number. Then the following are true:*

(i). $A(BC) = (AB)C$, $s(AB) = (sA)B = A(sB)$ and

$$AB = A \begin{bmatrix} \mathbf{b}_1 & \cdots & \mathbf{b}_p \end{bmatrix} = \begin{bmatrix} A\mathbf{b}_1 & \cdots & A\mathbf{b}_p \end{bmatrix}.$$

(ii). $(A + \widehat{A})B = AB + \widehat{A}B.$

The proof of $(A + \widehat{A})B = AB + \widehat{A}B$ follows from the distributive property of real numbers. Let $1 \leq i \leq m$ and $1 \leq j \leq p$

$$(A + \widehat{A})B = [(a_{i1} + \widehat{a}_{i1})b_{1j} + \cdots + (a_{in} + \widehat{a}_{in})b_{nj}]$$
$$= [(a_{i1}b_{1j} + \widehat{a}_{i1}b_{1j}) + \cdots + (a_{in}b_{nj} + \widehat{a}_{in}b_{nj})]$$
$$= [a_{i1}b_{1j} + \cdots + a_{in}b_{nj}] + [\widehat{a}_{i1}b_{1j} + \cdots + \widehat{a}_{in}b_{nj}]$$
$$= AB + \widehat{A}B.$$

Example 1.2.4. Verify the associative rule $A(BC) = (AB)C$ for the following 3×3 matrices

$$A = \begin{bmatrix} 1 & 0 & 0 \\ 0 & 1 & 0 \\ 0 & 2/3 & 1 \end{bmatrix}, \ B = \begin{bmatrix} 1 & 0 & 0 \\ 1/2 & 1 & 0 \\ 0 & 0 & 1 \end{bmatrix}, \text{ and } C = \begin{bmatrix} 2 & -1 & 0 \\ -1 & 2 & -1 \\ 0 & -1 & 2 \end{bmatrix}.$$

$$A(BC) = \begin{bmatrix} 1 & 0 & 0 \\ 0 & 1 & 0 \\ 0 & 2/3 & 1 \end{bmatrix} \left(\begin{bmatrix} 1 & 0 & 0 \\ 1/2 & 1 & 0 \\ 0 & 0 & 1 \end{bmatrix} \begin{bmatrix} 2 & -1 & 0 \\ -1 & 2 & -1 \\ 0 & -1 & 2 \end{bmatrix} \right)$$

$$= \begin{bmatrix} 1 & 0 & 0 \\ 0 & 1 & 0 \\ 0 & 2/3 & 1 \end{bmatrix} \left(\begin{bmatrix} 2 & -1 & 0 \\ 0 & 3/2 & -1 \\ 0 & -1 & 2 \end{bmatrix} \right) = \begin{bmatrix} 2 & -1 & 0 \\ 0 & 3/2 & -1 \\ 0 & 0 & 4/3 \end{bmatrix}.$$

$$(AB)C = \left(\begin{bmatrix} 1 & 0 & 0 \\ 0 & 1 & 0 \\ 0 & 2/3 & 1 \end{bmatrix} \begin{bmatrix} 1 & 0 & 0 \\ 1/2 & 1 & 0 \\ 0 & 0 & 1 \end{bmatrix} \right) \begin{bmatrix} 2 & -1 & 0 \\ -1 & 2 & -1 \\ 0 & -1 & 2 \end{bmatrix}$$

$$= \left(\begin{bmatrix} 1 & 0 & 0 \\ 1/2 & 1 & 0 \\ 1/3 & 2/3 & 1 \end{bmatrix} \right) \begin{bmatrix} 2 & -1 & 0 \\ -1 & 2 & -1 \\ 0 & -1 & 2 \end{bmatrix} = \begin{bmatrix} 2 & -1 & 0 \\ 0 & 3/2 & -1 \\ 0 & 0 & 4/3 \end{bmatrix}.$$

1.2.3 Application to heat conduction

Consider a hot mass, which must be cooled by transferring heat from the mass to a cooler surrounding region. Examples include computer chips, electrical amplifiers, a transformer on a power line, or a gasoline engine. One way to do this is to attach *cooling fins* to this mass so that the surface area that transmits the heat will be larger. We wish to be able to model heat flow in order to determine whether or not a particular configuration will sufficiently cool the mass.

In order to start the modeling process, we will make some assumptions that will simplify the model. Later we will return to this model and reconsider some of these assumptions. First, assume no time dependence and the temperature is approximated by a function of only the distance from the mass to be cooled. Thus, there is diffusion in only one direction. This is depicted in Figure 1.2.1 where x is the direction perpendicular to the hot mass.

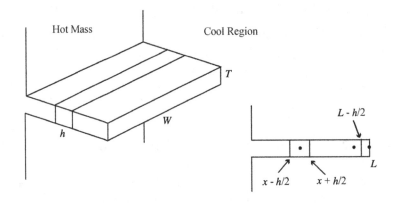

FIGURE 1.2.1
Thin cooling fin.

Second, assume the heat lost through the surface of the fin is similar to Newton's law of cooling so that for a slice of the lateral surface

$$\text{heat loss through a slice} = (\text{area})(\text{time interval})c(u_{sur} - u)$$
$$= h(2W + 2T)\ \Delta t\ c(u_{sur} - u).$$

Here u_{sur} is the surrounding temperature, and the c , the *heat transfer coefficient*, reflects the ability of the fin's surface to transmit heat to the surrounding region. If c is near zero, then little heat is lost. If c is large, then a larger amount of heat is lost through the lateral surface.

Third, assume heat diffuses in the x direction according to *Fourier's heat law* where K is the thermal conductivity. For interior volume elements with $x < L = 1$,

$$0 \approx (\text{heat through lateral surface })$$
$$+(\text{heat diffusing through front})$$
$$-(\text{heat diffusing through back})$$
$$= h\ (2W + 2T)\ \Delta t\ c(u_{sur} - u(x))$$
$$+TW\ \Delta t\ Ku_x(x + h/2)$$
$$-TW\ \Delta t\ Ku_x(x - h/2). \tag{1.2.1}$$

For the tip of the fin with $x = L$, we use $Ku_x(L) = c(u_{sur} - u(L))$ and

$$0 \approx (\text{heat through lateral surface of tip})$$
$$+(\text{heat diffusing through front})$$
$$-(\text{heat diffusing through back})$$
$$= (h/2)(2W + 2T)\ \Delta t\ c(u_{sur} - u(L))$$
$$+TW\ \Delta t\ c(u_{sur} - u(L))$$
$$-TW\ \Delta t\ Ku_x(L - h/2). \tag{1.2.2}$$

Note, the volume element near the tip of the fin is one half of the volume of the interior elements.

These are only approximations because the temperature changes continuously with space. In order to make these approximations in (1.2.1) and (1.2.2) more accurate, we divide by $h\ \Delta t\ TW$ and let h go to zero

$$0 = (2W + 2T)/(TW)\ c(u_{sur} - u) + (Ku_x)_x. \tag{1.2.3}$$

Let $C \equiv ((2W + 2T)/(TW))\ c$ and $f \equiv Cu_{sur}$. The *continuous* model is given by the following differential equation and two boundary conditions

$$-(Ku_x)_x + Cu = f, \tag{1.2.4}$$
$$u(0) = \text{given and} \tag{1.2.5}$$
$$Ku_x(L) = c(u_{sur} - u(L)). \tag{1.2.6}$$

The boundary condition in (1.2.6) is often called a *derivative or flux or Robin* boundary condition. *If c = 0, then no heat is allowed to pass through the* right boundary, and this type of boundary condition is often called a *Neumann* boundary condition. If c approaches infinity and the derivative remains bounded, then (1.2.6) implies $u_{sur} = u(L)$. When the value of the function is given at the boundary, this is often called the *Dirichlet* boundary condition.

The above derivation is useful because (1.2.1) and (1.2.2) suggest a way to discretize the continuous model. Let u_i be an approximation of $u(ih)$ where $h = L/n$. Approximate the derivative $u_x(ih + h/2)$ by $(u_{i+1} - u_i)/h$. Then, Equations (1.2.2) and (1.2.3) yield the *finite difference* approximation, a discrete model, of the continuum model (1.2.4)–(1.2.6).

Let u_0 be given and let $1 \leq i < n$:

$$-[K(u_{i+1} - u_i)/h - K(u_i - u_{i-1})/h] + hCu_i = hf(ih). \qquad (1.2.7)$$

Let $i = n$:

$$-[c(u_{sur} - u_n) - K(u_n - u_{n-1})/h] + (h/2)Cu_n = (h/2)f(nh). \qquad (1.2.8)$$

The discrete system (1.2.7) and (1.2.8) may be written in matrix form. For ease of notation we let $n = 4$, multiply (1.2.7) by h and (1.2.8) by $2h$, $B \equiv 2K + h^2C$ so that there are 4 equations and 4 unknowns:

$$Bu_1 - Ku_2 = h^2 f_1 + Ku_0,$$
$$-Ku_1 + Bu_2 - Ku_3 = h^2 f_2,$$
$$-Ku_2 + Bu_3 - Ku_4 = h^2 f_3 \text{ and}$$
$$-2Ku_3 + (B + 2hc)u_4 = h^2 f_4 + 2chu_{sur}.$$

The matrix form of this is $AU = F$ where A is, in general, $n \times n$ matrix and U and F are $n \times 1$ column vectors. For $n = 4$ we have

$$A = \begin{bmatrix} B & -K & 0 & 0 \\ -K & B & -K & 0 \\ 0 & -K & B & -K \\ 0 & 0 & -2K & B+2ch \end{bmatrix}$$

$$\text{where } U = \begin{bmatrix} u_1 \\ u_2 \\ u_3 \\ u_4 \end{bmatrix} \text{ and } F = \begin{bmatrix} h^2 f_1 + Ku_0 \\ h^2 f_2 \\ h^2 f_3 \\ h^2 f_4 + 2chu_{sur} \end{bmatrix}.$$

Example 1.2.5. Consider the above with $L = 4$, $K = 1$, $h = 4/4$, $C = 1$ and $c = 0$. The *steady state solution* is approximated by the solution of $A\mathbf{u} = \mathbf{d}$

$$\begin{bmatrix} 3 & -1 & 0 & 0 \\ -1 & 3 & -1 & 0 \\ 0 & -1 & 3 & -1 \\ 0 & 0 & -2 & 3 \end{bmatrix} \begin{bmatrix} u_1 \\ u_2 \\ u_3 \\ u_4 \end{bmatrix} = \begin{bmatrix} f_1 + u_0 \\ f_2 \\ f_3 \\ f_4 \end{bmatrix}.$$

1.2.4 Matrix computations using MATLAB®

The calculations for the heat conduction application can easily be done by MATLAB®. Here, we have computed the steady state solution. The interested reader should examine the MATLAB® codes fin1d_13.m and trid_13.m to see how this can be done using for-loops.

```
>> A = [3 -1 0 0; -1 3 -1 0; 0 -1 3 -1; 0 0 -2 3]
    A =
          3 -1 0 0
         -1 3 -1 0
          0 -1 3 -1
          0 0 -2 3
>> d = [1+10 1 1 1]'
    d =
         11
          1
          1
          1
>> u = A\d
    u =
        4.4468
        2.3404
        1.5745
        1.3830
```

1.2.5 Exercises

1. Compute $A\mathbf{x}$ (both row and column versions), $A+B$, $A+C$ and sA where

$$A = \begin{bmatrix} 1 & 5 & 3 \\ 8 & -1 & 2 \end{bmatrix}, \quad C = \begin{bmatrix} 0 & 1 & 8 \\ -4 & 2 & 5 \end{bmatrix}, \quad s = -7,$$

$$B = \begin{bmatrix} 1 & 0 & 10 \\ 2 & -3 & 1 \end{bmatrix} \text{ and } \mathbf{x} = \begin{bmatrix} 2 \\ 11 \\ 5 \end{bmatrix}.$$

2. (a). Verify $A + (B + C) = (A + B) + C$ where A, B and C are the same as in exercise one.

   (b). Prove this is true for all choices of the three matrices with the same dimension.

3. (a). Verify $A(2\mathbf{x} + 3\mathbf{y}) = 2A\mathbf{x} + 3A\mathbf{y}$ where A and $\mathbf{x}$ are the same as in exercise one and $\mathbf{y}^T = [1 \ 3 \ 2 \ -2]$.

   (b). Prove this is true for all choices of $m \times n$ matrices A and $n \times 1$ column vectors $\mathbf{x}$ and $\mathbf{y}$.

4. Compute AB (use both row and column versions) where

$$A = \begin{bmatrix} 1 & 5 & 3 \\ 8 & -1 & 2 \end{bmatrix} \text{ and } B = \begin{bmatrix} 1 & 0 & 2 & -1 \\ 5 & 4 & 7 & 1 \\ -1 & 2 & 0 & 1 \end{bmatrix}.$$

5. Verify $A(BC) = (AB)C$ where

$$A = \begin{bmatrix} 1 & 5 & 3 \\ 8 & -1 & 2 \end{bmatrix}, \ B = \begin{bmatrix} 1 & 2 \\ -3 & 0 \\ 1 & 3 \end{bmatrix} \text{ and } C = \begin{bmatrix} 3 & 7 \\ 0 & 2 \end{bmatrix}.$$

6. Verify $(A + \widehat{A})B = AB + \widehat{A}B$ where

$$A = \begin{bmatrix} 1 & 5 & 3 \\ 8 & -1 & 2 \end{bmatrix}, \ \widehat{A} = \begin{bmatrix} -1 & 0 & 1 \\ 3 & -11 & 0 \end{bmatrix} \text{ and } B = \begin{bmatrix} 1 & 2 \\ -3 & 0 \\ 1 & 3 \end{bmatrix}.$$

7. Consider Example 1.2.5 of heat conduction. Change the number of segments from $n = 4$ to $n = 5$ and leave the other parameters the same. Compute the steady state solution. Note there are now five unknowns, the vectors have five components, the matrix is 5×5.

1.3   Special Cases of $Ax = d$

In the previous sections, we have solved algebraic systems with two and three unknowns with $n = 2$ or 3. These could be written as a vector equation $Ax = d$ where A is an $n \times n$ matrix, d is a given $n \times 1$ column vector and x is an unknown $n \times 1$ column vector. The overall objective is be able to systematically solve problems with much larger n. In this section some important special matrices will be considered and an application to heat conduction in a wire will continue to be studied.

1.3.1   Five possible classes of "solutions"

Algebraic systems may have a solution, no solution or multiple solutions. Moreover, any solution may be very sensitive to the data in A or d. These will be illustrated by four simple 2×2 systems. One can gain insight to these examples by viewing them from both an algebraic and geometric perspective. Consider two equations

$$\begin{aligned} a_{11}x_1 + a_{12}x_2 &= d_1 \text{ and} \\ a_{21}x_1 + a_{22}x_2 &= d_2. \end{aligned}$$

Or, view this as a single vector equation

$$\begin{bmatrix} a_{11} & a_{12} \\ a_{21} & a_{22} \end{bmatrix} \begin{bmatrix} x_1 \\ x_2 \end{bmatrix} = \begin{bmatrix} d_1 \\ d_2 \end{bmatrix}.$$

Plot x_1 on the horizontal axis and x_2 on the vertical axis. The first scalar equation may be written in slope-intercept form of a line

$$x_2 = (-a_{11}/a_{12})x_1 + d_1/a_{12}, \quad \text{if } a_{12} \neq 0.$$

Do the same for the second scalar equation. The two lines are either not parallel, parallel and do not overlap, parallel and overlap, or they are "nearly" parallel. If they are not parallel, then there is exactly one solution. If they are parallel and do not overlap, then there is no solution. If they are parallel and overlap, then there are multiple solutions given by the points on the single line. If the two lines are "nearly" parallel, then small changes in the data can cause large changes in the intersection, that is, the solution. In the last case the small changes in the data can be caused by measurement, roundoff or human errors. Finally, there may be more equations than variables.

Case One. If the lines are not parallel, then

$$\det(A) = \det(\begin{bmatrix} a_{11} & a_{12} \\ a_{21} & a_{22} \end{bmatrix}) \neq 0.$$

Cramer's rule can be used to find the unique solution

$$x_1 = \det(\begin{bmatrix} d_1 & a_{12} \\ d_2 & a_{22} \end{bmatrix})/\det(A) \text{ and } x_2 = \det(\begin{bmatrix} a_{11} & d_1 \\ a_{21} & d_2 \end{bmatrix})/\det(A).$$

Case Two. The lines are parallel and do not overlap. For example,

$$x_1 + x_2 = 1 \text{ and}$$
$$2x_1 + 2x_2 = 4.$$

The second equation can be divided by 2 to get $x_1 + x_2 = 2$, which contradicts the first equation! So, there is no solution.

Case Three. The lines are parallel and do overlap. For example,

$$x_1 + x_2 = 1 \text{ and}$$
$$2x_1 + 2x_2 = 2.$$

The second equation can be divided by 2 so that $x_1 + x_2 = 1$, which is the same as the first equation. Any point on this line will be a solution of algebraic system.

Case Four. This example should be of great concern because it raises the issue of sensitivity of the solution to the data. Let ϵ represent a small variation in the data for $d_1 = 1$ and consider the system

$$x_1 + x_2 = 1 + \epsilon \text{ and}$$
$$(1 - 10^{-6})x_1 + x_2 = 1.$$

If $\epsilon = 0$, then the unique solution is $x_1 = 0$ and $x_2 = 1$. If $\epsilon = 10^{-3}$, then the unique solution is extremely different and is $x_1 = 1000$ and $x_2 = -998.999$. Moreover, for small variations in the right side the residuals, $\mathbf{r} = \mathbf{d} - A\mathbf{x}$, appear to be small

$$1 - (x_1 + x_2) = 1 - (1000 - 998.999) = -.001 \text{ and}$$
$$1 - ((1 - 10^{-6})x_1 + x_2) = 1 - ((1 - 10^{-6})1000 - 998.999) = 0!$$

Problems whose solutions are very sensitive to small changes in the data are called *ill-conditioned*.

Case Five. Consider finding a line (slope and intercept) that is "closest" to three data points in the plane. Unless the data point are on some line, there is no solution. But, one should be able to draw a line that is near the data!

The challenge is to detect these possibilities for large systems. This and next chapter are restricted to the unique solution case. Chapter 3 considers more equations than unknowns, and Chapter 4 studies the multiple solution case.

1.3.2   Triangular matrices

The simplest system to solve is the diagonal system where the equations are for $i = 1, \cdots, n$

$$a_{ii}x_i = d_i.$$

The matrix version of this uses the *diagonal* matrix $A = D$

$$\begin{bmatrix} a_{11} & 0 & \cdots & 0 \\ 0 & a_{22} & \ddots & \vdots \\ \vdots & \ddots & \ddots & 0 \\ 0 & \cdots & 0 & a_{nn} \end{bmatrix} \begin{bmatrix} x_1 \\ x_2 \\ \vdots \\ x_n \end{bmatrix} = \begin{bmatrix} d_1 \\ d_2 \\ \vdots \\ d_n \end{bmatrix}$$

$$D\mathbf{x} = \mathbf{d}.$$

If each component on the diagonal is not zero ($a_{ii} \neq 0$ for all i), then the solution is $x_i = d_i/a_{ii}$. The solution of a diagonal system requires order n operations.

More general $n \times n$ matrices are *lower triangular* $A = L$ and *upper triangular* $A = U$

$$\begin{bmatrix} a_{11} & 0 & \cdots & 0 \\ a_{21} & a_{22} & \ddots & \vdots \\ \vdots & \ddots & \ddots & 0 \\ a_{n1} & \cdots & a_{n,n-1} & a_{nn} \end{bmatrix} \begin{bmatrix} x_1 \\ x_2 \\ \vdots \\ x_n \end{bmatrix} = \begin{bmatrix} d_1 \\ d_2 \\ \vdots \\ d_n \end{bmatrix}$$

$$L\mathbf{x} = \mathbf{d} \text{ and}$$

$$\begin{bmatrix} a_{11} & a_{12} & \cdots & a_{1n} \\ 0 & a_{22} & \ddots & \vdots \\ \vdots & \ddots & \ddots & a_{n-1,n} \\ 0 & \cdots & 0 & a_{nn} \end{bmatrix} \begin{bmatrix} x_1 \\ x_2 \\ \vdots \\ x_n \end{bmatrix} = \begin{bmatrix} d_1 \\ d_2 \\ \vdots \\ d_n \end{bmatrix}$$

$$U\mathbf{x} = \mathbf{d}.$$

Both of these are relatively easy to solve provided the all the diagonal components a_{ii} are not zero. The lower triangular system is solved by finding first finding x_1 and then x_2 and so on until x_n is found; this is called *forward substitution*. The upper triangular system is solved in reverse order by first finding x_n and then x_{n-1} and so on until x_1 is determined; this is called *backward substitution*. The solves for triangular systems requires order n^2 operations.

Example 1.3.1. Let $n = 3$ and solve the following lower triangular system

$$\begin{bmatrix} 1 & 0 & 0 \\ 2 & 3 & 0 \\ 4 & 5 & 6 \end{bmatrix} \begin{bmatrix} x_1 \\ x_2 \\ x_3 \end{bmatrix} = \begin{bmatrix} 11 \\ 28 \\ 60 \end{bmatrix}.$$

The system is equivalent to the following three scalar equations

$$1x_1 = 11,$$
$$2x_1 + 3x_2 = 28 \text{ and}$$
$$4x_1 + 5x_2 + 6x_3 = 60.$$

The first equation gives $x_1 = 11$. Put this into the second equation $2(11) + 3x_2 = 28$ to compute $x_2 = (28 - 2(11))/3 = 2$. Finally, put $x_1 = 11$ and $x_2 = 2$ into the third equation and find $x_3 = (60 - 4(11) - 5(2))/6 = 1$.

Another way to solve this is by using the column version of the matrix-vector product

$$x_1 \begin{bmatrix} 1 \\ 2 \\ 4 \end{bmatrix} + x_2 \begin{bmatrix} 0 \\ 3 \\ 5 \end{bmatrix} + x_3 \begin{bmatrix} 0 \\ 0 \\ 6 \end{bmatrix} = \begin{bmatrix} 11 \\ 28 \\ 60 \end{bmatrix}.$$

The first equation gives $x_1 = 11$. Now, put this into the left column on the left side and move it to the right side

$$
x_2 \begin{bmatrix} 0 \\ 3 \\ 5 \end{bmatrix} + x_3 \begin{bmatrix} 0 \\ 0 \\ 6 \end{bmatrix} = \begin{bmatrix} 11 \\ 28 \\ 60 \end{bmatrix} - 11 \begin{bmatrix} 1 \\ 2 \\ 4 \end{bmatrix} = \begin{bmatrix} 0 \\ 6 \\ 16 \end{bmatrix}.
$$

The second equation gives $x_2 = 2$. The last step is to put $x_2 = 2$ into the left column and move it to the right side

$$
x_3 \begin{bmatrix} 0 \\ 0 \\ 6 \end{bmatrix} = \begin{bmatrix} 0 \\ 6 \\ 16 \end{bmatrix} - 2 \begin{bmatrix} 0 \\ 3 \\ 5 \end{bmatrix} = \begin{bmatrix} 0 \\ 0 \\ 6 \end{bmatrix}.
$$

The third equation yields $x_3 = 1$.

Example 1.3.2. Let $n = 4$ and solve the following upper triangular system

$$
\begin{bmatrix} 2 & 3 & 0 & 1 \\ 0 & 1 & -2 & 3 \\ 0 & 0 & 3 & -1 \\ 0 & 0 & 0 & 2 \end{bmatrix} \begin{bmatrix} x_1 \\ x_2 \\ x_3 \\ x_4 \end{bmatrix} = \begin{bmatrix} 25 \\ 1 \\ 9 \\ 6 \end{bmatrix}.
$$

The four equivalent scalar equations are

$$
2x_1 + 3x_2 + 0x_3 + 1x_4 = 25,
$$
$$
1x_2 - 2x_3 + 3x_4 = 1,
$$
$$
3x_3 - 1x_4 = 9 \text{ and}
$$
$$
2x_4 = 6.
$$

The solution of the last equation is $x_4 = 6/2 = 3$. Put this into the third equation and solve for $x_3 = (9 + 1(3))/3 = 4$. The second equation becomes $1x_2 - 2(4) + 3(3) = 1$ so that $x_2 = 0$. The first equation gives $2x_1 + 3(0) + 0(4) + 1(3) = 25$ so that $x_1 = 11$.

This can also be solved by using the column version of the matrix-vector product

$$
x_1 \begin{bmatrix} 2 \\ 0 \\ 0 \\ 0 \end{bmatrix} + x_2 \begin{bmatrix} 3 \\ 1 \\ 0 \\ 0 \end{bmatrix} + x_3 \begin{bmatrix} 0 \\ -2 \\ 3 \\ 0 \end{bmatrix} + x_4 \begin{bmatrix} 1 \\ 3 \\ -1 \\ 2 \end{bmatrix} = \begin{bmatrix} 25 \\ 1 \\ 9 \\ 6 \end{bmatrix}.
$$

The last equation gives $x_4 = 3$. Put this into the right column on the left side and move it to the right side

$$
x_1 \begin{bmatrix} 2 \\ 0 \\ 0 \\ 0 \end{bmatrix} + x_2 \begin{bmatrix} 3 \\ 1 \\ 0 \\ 0 \end{bmatrix} + x_3 \begin{bmatrix} 0 \\ -2 \\ 3 \\ 0 \end{bmatrix} = \begin{bmatrix} 25 \\ 1 \\ 9 \\ 6 \end{bmatrix} - 3 \begin{bmatrix} 1 \\ 3 \\ -1 \\ 2 \end{bmatrix} = \begin{bmatrix} 22 \\ -8 \\ 12 \\ 0 \end{bmatrix}.
$$

The third equation gives $x_3 = 4$. Put $x_3 = 4$ into the right column on the left side and move it to the right side

$$
x_1 \begin{bmatrix} 2 \\ 0 \\ 0 \\ 0 \end{bmatrix} + x_2 \begin{bmatrix} 3 \\ 1 \\ 0 \\ 0 \end{bmatrix} = \begin{bmatrix} 22 \\ -8 \\ 12 \\ 0 \end{bmatrix} - 4 \begin{bmatrix} 0 \\ -2 \\ 3 \\ 0 \end{bmatrix} = \begin{bmatrix} 22 \\ 0 \\ 0 \\ 0 \end{bmatrix}.
$$

The second and first equations give $x_2 = 0$ and then $x_1 = 11$.

The solution process in all of the above cases requires that the diagonal components not be zero. In this case, there is one and only one solution, which can be found by using either the row or column versions of the matrix-vector product. For large systems there may be some advantages to using one of the two matrix-vector products. The computing effort is related to storage size and location and the size of the matrix. If the matrix is diagonal $n \times n$, the solve step requires n divisions. If the matrix is triangular, the solve step requires n^2 operations.

Theorem 1.3.1. *(Triangular Solves)   Consider a lower or an upper triangular matrix $A = L$ or U. If all the diagonal component of the matrix are not zero, then the triangular systems $L\mathbf{x} = \mathbf{d}$ or $U\mathbf{x} = \mathbf{d}$ have one and only one solution.*

If the given matrix can be written as a product of a lower and upper triangular matrices, then one may be able to solve $A\mathbf{x} = \mathbf{d}$. Suppose $A = LU$ and both L and U have nonzero diagonal components. By the associative property

$$
A\mathbf{x} = \mathbf{d}
$$
$$
(LU)\mathbf{x} = \mathbf{d}
$$
$$
L(U\mathbf{x}) = \mathbf{d}.
$$

So, solve $L\mathbf{y} = \mathbf{d}$ by forward substitution, and then solve $U\mathbf{x} = \mathbf{y}$ by backward substitution.

Example 1.3.3. Consider the tridiagonal matrix in the following system

$$
\begin{bmatrix} 2 & -1 & 0 \\ -1 & 2 & -1 \\ 0 & -1 & 2 \end{bmatrix} \begin{bmatrix} x_1 \\ x_2 \\ x_3 \end{bmatrix} = \begin{bmatrix} 200 \\ 0 \\ 70 \end{bmatrix}.
$$

Observe the matrix can be factored as follows

$$
\begin{bmatrix} 2 & -1 & 0 \\ -1 & 2 & -1 \\ 0 & -1 & 2 \end{bmatrix} = \begin{bmatrix} 1 & 0 & 0 \\ -1/2 & 1 & 0 \\ 0 & -2/3 & 1 \end{bmatrix} \begin{bmatrix} 2 & -1 & 0 \\ 0 & 3/2 & -1 \\ 0 & 0 & 4/3 \end{bmatrix}
$$
$$
A = LU.
$$

First, solve $L\mathbf{y} = \mathbf{d}$

$$
\begin{bmatrix} 1 & 0 & 0 \\ -1/2 & 1 & 0 \\ 0 & -2/3 & 1 \end{bmatrix} \begin{bmatrix} y_1 \\ y_2 \\ y_3 \end{bmatrix} = \begin{bmatrix} 200 \\ 0 \\ 70 \end{bmatrix}.
$$

The equation associated with the first row gives $y_1 = 200$. The equation associated with the second row is $(-1/2)200 + 1y_2 = 0$ implies $y_2 = 100$. The equation associated with the last row is $(-2/3)100 + 1y_3 = 70$ implies $y_3 = 410/3$. Second, solve $U\mathbf{x} = \mathbf{y}$

$$
\begin{bmatrix} 2 & -1 & 0 \\ 0 & 3/2 & -1 \\ 0 & 0 & 4/3 \end{bmatrix} \begin{bmatrix} x_1 \\ x_2 \\ x_3 \end{bmatrix} = \begin{bmatrix} 200 \\ 100 \\ 410/3 \end{bmatrix}.
$$

The last row implies $x_3 = (410/3)/(4/3) = 102.5$. The second row gives $(3/2)x_2 - 102.5 = 100$ or $x_2 = 202.5/(3/2) = 135.0$. The top row gives $2x_1 - 135 = 200$ or $x_1 = 335/2 = 167.5$.

If the matrix is *tridiagonal*, a matrix whose only nonzero components are in the diagonal, subdiagonal and superdiagonal, then one may be able to find the LU factors by a process similar to the following. Let A be 3×3 tridiagonal matrix and assume the L and U factors have the special forms

$$
\begin{bmatrix} a_{11} & a_{12} & 0 \\ a_{21} & a_{22} & a_{23} \\ 0 & a_{32} & a_{33} \end{bmatrix} = \begin{bmatrix} 1 & 0 & 0 \\ \alpha_1 & 1 & 0 \\ 0 & \alpha_2 & 1 \end{bmatrix} \begin{bmatrix} \beta_1 & a_{12} & 0 \\ 0 & \beta_2 & a_{23} \\ 0 & 0 & \beta_3 \end{bmatrix} \tag{1.3.1}
$$

$$
= \begin{bmatrix} \beta_1 & a_{12} & 0 \\ \alpha_1\beta_1 & \alpha_1 a_{12} + \beta_2 & a_{23} \\ 0 & \alpha_2\beta_2 & \alpha_2 a_{23} + \beta_3 \end{bmatrix}.
$$

The matrices are equal if and only if they have the same number of rows and columns and their ij-components are all equal. This gives us five nontrivial scalar equations in the above 3×3 matrix equation: $a_{11} = \beta_1$, $a_{21} = \alpha_1\beta_1$, $a_{22} = \alpha_1 a_{12} + \beta_2$, $a_{32} = \alpha_2\beta_2$ and $a_{33} = \alpha_2 a_{23} + \beta_3$. If each of the β_1, β_2 and β_3 are nonzero, then we can solve for α_1 and α_2 as well as do the upper triangular solve. As an illustration consider Example 1.3.3 where the diagonal components are all equal to 2 and the other nonzero components are equal to -1 :

$$
\begin{aligned}
&2 = \beta_1, -1 = \alpha_1\beta_1 = \alpha_1 2 \text{ and so } \alpha_1 = -1/2, \\
&2 = -\alpha_1 + \beta_2 = -(-1/2) + \beta_2 \text{ and then } \beta_2 = 3/2, \\
&-1 = \alpha_2\beta_2 \text{ so that } \alpha_2 = -2/3 \text{ and} \\
&2 = -\alpha_2 + \beta_3 \text{ gives } \beta_3 = 4/3.
\end{aligned}
$$

The tridiagonal solve can be done in order $8n$ operations, see the MATLAB® code trid_13.m and the discussion at the end of Subsection 1.5.3.

1.3.3 Application to heat in wire with current

Consider heat conduction in a long thin wire. Assume there are five segments, not four, so that the unknown temperatures are $u_1, u_2, u_2,$ and u_4 with $u_0 = 70$ at the left end and $u_5 = 70$ at the right end. Also, assume there is an electrical current, which generates heat at a rate of f per unit volume, per unit time. The heat in a small i^{th} segment is now modeled or approximated by

$$\rho c(u_i^{new})(A\Delta x) = \rho c u_i (A\Delta x) - \Delta t\, A\, K \frac{u_i - u_{i-1}}{\Delta x} +$$
$$\Delta t\, A\, K \frac{u_{i+1} - u_i}{\Delta x} + f\,\Delta t\,(A\Delta x).$$

Divide by $\rho c(A\Delta x)$ and let $\alpha \equiv (\Delta t/\Delta x^2)(K/\rho c)$ to get for $i = 1, 2, 3$ and 4

$$u_i^{new} = u_i - \alpha(-u_{i-1} + 2u_i - u_{i+1}) + f\,\Delta t/(\rho c).$$

If there is no change, $u_i^{new} = u_i$, then this is a steady state model.

Consider the above with $L = 10$, $n = 5$, $\Delta x = 10/5$, $\rho = c = 1$, $K = 1/1000$, $\Delta t = 1000$ so that $\alpha \equiv (\Delta t/\Delta x^2)(K/\rho c) = 25/100 = 0.25$. If the electrical current in the wire generates heat at a rate equal to $f = 0.1$, then $f\,\Delta t/(\rho c) = 100$ and the above steady state model becomes

$$\begin{bmatrix} 2 & -1 & 0 & 0 \\ -1 & 2 & -1 & 0 \\ 0 & -1 & 2 & -1 \\ 0 & 0 & -1 & 2 \end{bmatrix} \begin{bmatrix} u_1 \\ u_2 \\ u_3 \\ u_3 \end{bmatrix} = \begin{bmatrix} 70 \\ 0 \\ 0 \\ 70 \end{bmatrix} + \begin{bmatrix} 400 \\ 400 \\ 400 \\ 400 \end{bmatrix}.$$

The reader should verify $A = LU$ where L and U are given below

$$\begin{bmatrix} 1 & 0 & 0 & 0 \\ -1/2 & 1 & 0 & 0 \\ 0 & -2/3 & 1 & 0 \\ 0 & 0 & -3/4 & 1 \end{bmatrix} \begin{bmatrix} 2 & -1 & 0 & 0 \\ 0 & 3/2 & -1 & 0 \\ 0 & 0 & 4/3 & -1 \\ 0 & 0 & 0 & 5/4 \end{bmatrix} \begin{bmatrix} u_1 \\ u_2 \\ u_3 \\ u_3 \end{bmatrix} = \begin{bmatrix} 470 \\ 400 \\ 400 \\ 470 \end{bmatrix}.$$

One must solve $A\mathbf{u} = L(U\mathbf{u}) = \mathbf{d}$. The solution of $L\mathbf{y} = \mathbf{d}$ is $y_1 = 470.0$, $y_2 = 635.0$, $y_3 = 823.3$ and $y_4 = 1087.5$. The solution of $U\mathbf{u} = \mathbf{y}$ is $u_1 = 870.0$, $u_2 = 1270.0$, $u_3 = 1270.0$ and $u_4 = 870.0$.

1.3.4 Matrix computations using MATLAB®

The calculations for the heat conduction with electrical current application can easily be done by MATLAB®. The steady state solution has been computed several ways.

```
>> A = [ 2 -1  0  0;
        -1  2 -1  0;
         0 -1  2 -1;
```

```
              0  0 -1  2];
>> d = [470 400 400 470]';
>> A\d                              % solves Au = d
      ans =
              1.0e+003 *
                  0.8700
                  1.2700
                  1.2700
                  0.8700
>> [L U] = lu(A)                    % finds the LU factorization of A
      L =
              1.0000  0.0000  0.0000  0.0000
             -0.5000  1.0000  0.0000  0.0000
              0.0000 -0.6667  1.0000  0.0000
              0.0000  0.0000 -0.7500  1.0000
      U =
              2.0000 -1.0000  0.0000  0.0000
              0.0000  1.5000 -1.0000  0.0000
              0.0000  0.0000  1.3333 -1.0000
              0.0000  0.0000  0.0000  1.2500
>> y = L\d                          % first step using LU
      y =
              1.0e+003 *
                  0.4700
                  0.6350
                  0.8233
                  1.0875
>> u = U\y                          % second step using LU
      u =
              1.0e+003 *
                  0.8700
                  1.2700
                  1.2700
                  0.8700
>> A3 = [A(:,1) A(:,2) d A(:,4)]    % use Cramer's Rule to find u3
      A3 =
              2 -1  470  0
             -1  2  400  0
              0 -1  400 -1
              0  0  470  2
>> u3 = det(A3)/det(A)
      u3 =
              1270
```

1.3.5 Exercises

1. Find the solution of

$$\begin{bmatrix} 2 & 0 & 0 \\ 1 & 3 & 0 \\ 3 & 4 & -1 \end{bmatrix} \begin{bmatrix} x_1 \\ x_2 \\ x_3 \end{bmatrix} = \begin{bmatrix} 4 \\ 14 \\ 19 \end{bmatrix}.$$

2. Find the solution of

$$\begin{bmatrix} 1 & 7 & 8 & 6 \\ 0 & 2 & 3 & 4 \\ 0 & 0 & 1 & 2 \\ 0 & 0 & 0 & 3 \end{bmatrix} \begin{bmatrix} x_1 \\ x_2 \\ x_3 \\ x_4 \end{bmatrix} = \begin{bmatrix} 10 \\ 23 \\ 7 \\ 3 \end{bmatrix}.$$

5. Use the LU factors in Example 1.3.3 to solve

$$\begin{bmatrix} 2 & -1 & 0 \\ -1 & 2 & -1 \\ 0 & -1 & 2 \end{bmatrix} \begin{bmatrix} x_1 \\ x_2 \\ x_3 \end{bmatrix} = \begin{bmatrix} 210 \\ 10 \\ 80 \end{bmatrix}.$$

6. Solve

$$\begin{bmatrix} 1 & 0 & 0 \\ -1 & 2 & 0 \\ 2 & 0 & 3 \end{bmatrix} \begin{bmatrix} 2 & 7 & 8 \\ 0 & 9 & 10 \\ 0 & 0 & 1 \end{bmatrix} \begin{bmatrix} x_1 \\ x_2 \\ x_3 \end{bmatrix} = \begin{bmatrix} 1 \\ 2 \\ 3 \end{bmatrix}.$$

7. Consider a 4×4 tridiagonal matrix similar to 3×3 tridiagonal matrix, see line (1.3.1).

   (a). Find the seven equations for the components in the LU factors.

   (b). Find the LU factors for

   $$A = \begin{bmatrix} 2 & -1 & 0 & 0 \\ -1 & 2 & -1 & 0 \\ 0 & -1 & 2 & -1 \\ 0 & 0 & -1 & 2 \end{bmatrix}.$$

1.4 Row Operations and Gauss Elimination

The objective of this section is to formulate a systematic way of transforming $A\mathbf{x} = \mathbf{d}$ to an equivalent upper triangular system $U\mathbf{x} = \widehat{\mathbf{d}}$. The solution of the upper triangular system can be found provided the diagonal components of the upper triangular matrix are all not zero. This method is called *Gauss elimination* and requires about n^2 storage and $n^3/3$ operations for an $n \times n$ matrix A. This systematic approach can be coded so that a computer can do these operations. An application to a six-bar truss will be given.

1.4.1   Introductory illustration

Consider the following algebraic system for three unknowns

$$
\begin{aligned}
x_1 - x_2 + x_3 &= 5 \\
-x_1 + x_2 + x_3 &= 1 \\
4x_2 + x_3 &= -1.
\end{aligned}
$$

The matrix version $A\mathbf{x} = \mathbf{d}$ is not in upper triangular form

$$
\begin{bmatrix} 1 & -1 & 1 \\ -1 & 1 & 1 \\ 0 & 4 & 1 \end{bmatrix} \begin{bmatrix} x_1 \\ x_2 \\ x_3 \end{bmatrix} = \begin{bmatrix} 5 \\ 1 \\ -1 \end{bmatrix}.
$$

If one adds equation one (row one) to equation two (row two), then the new algebraic system is

$$
\begin{aligned}
x_1 - x_2 + x_3 &= 5 \\
2x_3 &= 6 \\
4x_2 + x_3 &= -1.
\end{aligned}
$$

This is equivalent to multiplying the matrix equation by an elementary matrix $E_{21}(1)$

$$
E_{21}(1)\, A\mathbf{x} = E_{21}(1)\, \mathbf{d}
$$

$$
\begin{bmatrix} 1 & 0 & 0 \\ 1 & 1 & 0 \\ 0 & 0 & 1 \end{bmatrix} \begin{bmatrix} 1 & -1 & 1 \\ -1 & 1 & 1 \\ 0 & 4 & 1 \end{bmatrix} \begin{bmatrix} x_1 \\ x_2 \\ x_3 \end{bmatrix} = \begin{bmatrix} 1 & 0 & 0 \\ 1 & 1 & 0 \\ 0 & 0 & 1 \end{bmatrix} \begin{bmatrix} 5 \\ 1 \\ -1 \end{bmatrix}
$$

$$
\begin{bmatrix} 1 & -1 & 1 \\ 0 & 0 & 2 \\ 0 & 4 & 1 \end{bmatrix} \begin{bmatrix} x_1 \\ x_2 \\ x_3 \end{bmatrix} = \begin{bmatrix} 5 \\ 6 \\ -1 \end{bmatrix}.
$$

Next interchange equation two (row two) and equation three (row three)

$$
\begin{aligned}
x_1 - x_2 + x_3 &= 5 \\
4x_2 + x_3 &= -1 \\
2x_3 &= 6.
\end{aligned}
$$

The interchange is equivalent to multiplying the matrix equation by an elementary permutation matrix P_{23}

$$
P_{23}\, E_{21}(1)\, A\mathbf{x} = P_{23}\, E_{21}(1)\, \mathbf{d}
$$

$$
\begin{bmatrix} 1 & 0 & 0 \\ 0 & 0 & 1 \\ 0 & 1 & 0 \end{bmatrix} \begin{bmatrix} 1 & -1 & 1 \\ 0 & 0 & 2 \\ 0 & 4 & 1 \end{bmatrix} \begin{bmatrix} x_1 \\ x_2 \\ x_3 \end{bmatrix} = \begin{bmatrix} 1 & 0 & 0 \\ 0 & 0 & 1 \\ 0 & 1 & 0 \end{bmatrix} \begin{bmatrix} 5 \\ 6 \\ -1 \end{bmatrix}
$$

$$
\begin{bmatrix} 1 & -1 & 1 \\ 0 & 4 & 1 \\ 0 & 0 & 2 \end{bmatrix} \begin{bmatrix} x_1 \\ x_2 \\ x_3 \end{bmatrix} = \begin{bmatrix} 5 \\ -1 \\ 6 \end{bmatrix}.
$$

Note this is in upper triangular form $U\mathbf{x} = \widehat{\mathbf{d}}$ where $U = P_{23} E_{21}(1) A$ and $\widehat{\mathbf{d}} = P_{23} E_{21}(1)\,\mathbf{d}$. Since adding and interchanging equations can be reversed, the solution of the upper triangular system must also be a solution of the original system. The solution is $x_3 = 6/2 = 3$, $x_2 = (-1 - 3)/4 = -1$ and $x_1 = (5 - 3 + (-1)) = 1$.

1.4.2 Three types of row operations

In order to minimize the repeated writing of the equations, the augmented matrix notation will be used. Let A be an $n \times n$ matrix and let $\mathbf{d}$ be an $n \times 1$ column vector. The *augmented matrix* $[A\ \ \mathbf{d}]$ is an $n \times (n+1)$ matrix with the column vector augmented to the matrix. In the above example

$$[A\ \ \mathbf{d}] = \begin{bmatrix} 1 & -1 & 1 & 5 \\ -1 & 1 & 1 & 1 \\ 0 & 4 & 1 & -1 \end{bmatrix}.$$

The two row operations can be written as

$$P_{23} E_{21}(1)\,[A\ \ \mathbf{d}] = [U\ \ \widehat{\mathbf{d}}].$$

There are three *row operations*, which are used to transform matrices to upper triangular matrices. They can be used to solve algebraic systems, find the LU factors, evaluate determinants, find multiple solutions and eigenvectors. The row operations can be represented by *elementary matrices* $E_{ij}(a)$, P_{ij} and $E_i(c)$.

Add a **times** (row_j) **to** row_i. This can be represented by an $n \times n$ elementary matrix $E_{ij}(a)$ with the only nonzero components being ones on the diagonal and a equal to the ij-component. For example, for $n = 3$, $a = -2$, $i = 3$ and $j = 2$

$$E_{32}(-2) = \begin{bmatrix} 1 & 0 & 0 \\ 0 & 1 & 0 \\ 0 & -2 & 1 \end{bmatrix}.$$

Note, $E_{ij}(-a)$ is the inverse operation to $E_{ij}(a)$, that is, $E_{ij}(-a)\,E_{ij}(a) = I$. For example,

$$E_{32}(2)\,E_{32}(-2) = \begin{bmatrix} 1 & 0 & 0 \\ 0 & 1 & 0 \\ 0 & 2 & 1 \end{bmatrix} \begin{bmatrix} 1 & 0 & 0 \\ 0 & 1 & 0 \\ 0 & -2 & 1 \end{bmatrix} = \begin{bmatrix} 1 & 0 & 0 \\ 0 & 1 & 0 \\ 0 & 0 & 1 \end{bmatrix}.$$

Interchange row_i **and** row_j. This can be represented by an $n \times n$ matrix P_{ij} with the only nonzero components being ones on the diagonal except for row_i and row_j where the ij-component and ji-component equal one, respectively. For example, for $n = 4$, $i = 2$ and $j = 3$

$$P_{23} = \begin{bmatrix} 1 & 0 & 0 & 0 \\ 0 & 0 & 1 & 0 \\ 0 & 1 & 0 & 0 \\ 0 & 0 & 0 & 1 \end{bmatrix}.$$

Note, that $P_{ij} P_{ij} = I$.

Multiply *row_i* **by** c. The is represented by an $n \times n$ matrix $E_i(c)$ with the only nonzero components being ones on the diagonal except for *row_i* where the diagonal component is c. For example, for $n = 4$, $c = 9$ and $i = 3$

$$E_3(9) = \begin{bmatrix} 1 & 0 & 0 & 0 \\ 0 & 1 & 0 & 0 \\ 0 & 0 & 9 & 0 \\ 0 & 0 & 0 & 1 \end{bmatrix}.$$

It is easy to verify that $E_i(1/c) E_i(c) = I$.

Row operations are used, starting with the left columns in the augmented matrix, to transform it to an upper triangular matrix. Once the upper triangular matrix is found, the solution may be computed by backward substitution as long as the diagonal components of the upper triangular matrix are not zero. The following 2×2 example illustrates this may not always be possible

$$x_1 + x_2 = 1 \text{ and}$$
$$2x_1 + 2x_2 = 3.$$

The augmented matrix is

$$[A \ \mathbf{d}] = \begin{bmatrix} 1 & 1 & 1 \\ 2 & 2 & 3 \end{bmatrix}.$$

In order to obtain a zero in the 21-component add (-2) times *row_1* to *row_2*

$$E_{21}(-2) [A \ \mathbf{d}] = \begin{bmatrix} 1 & 0 \\ -2 & 1 \end{bmatrix} \begin{bmatrix} 1 & 1 & 1 \\ 2 & 2 & 3 \end{bmatrix} = \begin{bmatrix} 1 & 1 & 1 \\ 0 & 0 & 1 \end{bmatrix} = [U \ \widehat{\mathbf{d}}].$$

The 22-component of U is zero! Moreover, the second row is a shorthand way of writing $0x_1 + 0x_2 = 1$, which is impossible. Such algebraic systems are called *inconsistent* and have no solution.

1.4.3   Gauss elimination for solving $A\mathbf{x} = \mathbf{d}$

Assume $A\mathbf{x} = \mathbf{d}$ has a solution and it is unique. The Gauss elimination method is defined so that it can be implemented for larger systems and using computers. The reader will observe the by-hand calculations can become a little more than tiresome, but the objective here is to illustrate how the method can be used to solve larger systems.

Definition. *Consider the $n \times n$ algebraic system $Ax = d$. The Gauss elimination method for solving this system has of two stages:*

Stage 1. Transform $Ax = \mathbf{d}$ to upper triangular form $U\mathbf{x} = \widehat{\mathbf{d}}$, that is, transform the augmented matrix $[A \ \mathbf{d}]$ to an upper triangular matrix $[U \ \widehat{\mathbf{d}}]$.

(a). start with the left column, column $j = 1$, and use row operations to transform column $j = 1$ to zeros below row $i = 1$,

(b). move to the next column, column $j = 2$, and use row operations to transform column $j = 2$ to zeros below row $i = 2$ and

(c). repeat this until column $j = n - 1$ has been done.

Stage 2. Solve the upper triangular system $U\mathbf{x} = \widehat{\mathbf{d}}$ by backward substitution.

(a). start with the bottom row, row $i = n$, and solve the corresponding equation for x_n,

(b). put x_n into the equations corresponding to rows $i = n - 1, \cdots, 1$,

(c). solve equation $i = n - 1$ for x_{n-1} and

(d). repeat this until x_1 has been computed.

Example 1.4.1. Consider the 3×3 algebraic system

$$\begin{aligned} 5x_2 + 6x_3 &= 6 \\ x_1 + 3x_2 + x_3 &= 2 \\ 2x_1 + x_2 + x_3 &= 3. \end{aligned}$$

The augmented matrix is

$$[A \ \mathbf{d}] = \begin{bmatrix} 0 & 5 & 6 & 6 \\ 1 & 3 & 1 & 2 \\ 2 & 1 & 1 & 3 \end{bmatrix}.$$

Stage 1. In order to get zeros in column $j = 1$, the 11-component must not be zero. So, first interchange row_1 and row_2

$$P_{12} [A \ \mathbf{d}] = \begin{bmatrix} 0 & 1 & 0 \\ 1 & 0 & 0 \\ 0 & 0 & 1 \end{bmatrix} \begin{bmatrix} 0 & 5 & 6 & 6 \\ 1 & 3 & 1 & 2 \\ 2 & 1 & 1 & 3 \end{bmatrix} = \begin{bmatrix} 1 & 3 & 1 & 2 \\ 0 & 5 & 6 & 6 \\ 2 & 1 & 1 & 3 \end{bmatrix}.$$

Add (-2) times row_1 to row_3

$$E_{31}(-2) \, P_{12} \, [A \ \mathbf{d}] = \begin{bmatrix} 1 & 0 & 0 \\ 0 & 1 & 0 \\ -2 & 0 & 1 \end{bmatrix} \begin{bmatrix} 1 & 3 & 1 & 2 \\ 0 & 5 & 6 & 6 \\ 2 & 1 & 1 & 3 \end{bmatrix} = \begin{bmatrix} 1 & 3 & 1 & 2 \\ 0 & 5 & 6 & 6 \\ 0 & -5 & -1 & -1 \end{bmatrix}.$$

Move to the second column and add *row_2* to *row_3*

$$E_{32}(1)\, E_{31}(-2)\, P_{12}\, [A\ \mathbf{d}] = \begin{bmatrix} 1 & 0 & 0 \\ 0 & 1 & 0 \\ 0 & 1 & 1 \end{bmatrix} \begin{bmatrix} 1 & 3 & 1 & 2 \\ 0 & 5 & 6 & 6 \\ 0 & -5 & -1 & -1 \end{bmatrix}$$

$$= \begin{bmatrix} 1 & 3 & 1 & 2 \\ 0 & 5 & 6 & 6 \\ 0 & 0 & 5 & 5 \end{bmatrix}.$$

Stage 2. The last row corresponds to the equation $5x_3 = 5$ and so $x_3 = 1$. Put this into the equations corresponding the rows 1 and 2 to get

$$x_1 + 3x_2 + 1(1) = 2 \text{ and}$$
$$5x_2 + 6(1) = 6.$$

Then $x_2 = 0$ and $x_1 = 1$.

Example 1.4.2. Consider the two-loop circuit problem

$$i_1 - i_2 + i_3 = 0,$$
$$R_1 i_1 - R_3 i_3 = E_1 \text{ and}$$
$$-R_2 i_2 - R_3 i_3 = E_2.$$

Let $R_1 = 1$, $R_2 = 2$, $R_3 = 3$, $E_1 = 10$ and $E_2 = 20$ so that the augmented matrix is

$$[A\ \mathbf{d}] = \begin{bmatrix} 1 & -1 & 1 & 0 \\ 1 & 0 & -3 & 10 \\ 0 & -2 & -3 & 20 \end{bmatrix}.$$

Stage 1. Obtain zeros in the 21-component and 32-component by using

$$E_{32}(2)\, E_{21}(-1)\, [A\ \mathbf{d}] = \begin{bmatrix} 1 & -1 & 1 & 0 \\ 0 & 1 & -4 & 10 \\ 0 & 0 & -11 & 40 \end{bmatrix}.$$

Stage 2. The last row corresponds to the equation $-11i_3 = 40$ and so $i_3 = 40/(-11)$. Put this into the equations corresponding the rows 1 and 2 to get

$$i_1 - i_2 + 40/(-11) = 0 \text{ and}$$
$$i_2 - 4(40/(-11)) = 10.$$

This gives $i_2 = 50/(-11)$ and $i_1 = 10/(-11)$.

Example 1.4.3. Consider the steady state heat conduction in a wire where there were five segments and electrical current generating heat within the wire. The algebraic problem for the temperature in each interior segment is

$$\begin{bmatrix} 2 & -1 & 0 & 0 \\ -1 & 2 & -1 & 0 \\ 0 & -1 & 2 & -1 \\ 0 & 0 & -1 & 2 \end{bmatrix} \begin{bmatrix} u_1 \\ u_2 \\ u_3 \\ u_3 \end{bmatrix} = \begin{bmatrix} 470 \\ 400 \\ 400 \\ 470 \end{bmatrix}$$

and the augmented matrix is

$$[A \ d] = \begin{bmatrix} 2 & -1 & 0 & 0 & 470 \\ -1 & 2 & -1 & 0 & 400 \\ 0 & -1 & 2 & -1 & 400 \\ 0 & 0 & -1 & 2 & 470 \end{bmatrix}.$$

Stage 1. Obtain zeros in the subdiagonal by using

$$E_{43}(3/4) \ E_{32}(2/3) \ E_{21}(1/2) \ [A \ d] = \begin{bmatrix} 2 & -1 & 0 & 0 & 470.00 \\ 0 & 3/2 & -1 & 0 & 635.00 \\ 0 & 0 & 4/3 & -1 & 823.33 \\ 0 & 0 & 0 & 5/4 & 1087.50 \end{bmatrix}.$$

Stage 2. The last row corresponds to the equation $(5/4)u_4 = 1087.5$ and so $u_4 = 1087.5/1.25. = 870$. Put this into the equations corresponding the rows 1, 2 and 3 to get $u_3 = u_2 = 1270$ and $u_1 = 870$. The reader should note the similarities of these calculations with the LU factorization method as given in Subsection 1.3.3.

1.4.4 Application to six-bar truss

In Subsection 1.1.4, an application to a two-bar truss was presented. There was one stationary node where a balance of 2D forces gave two scalar equations for the forces within the two bars. If a bar is under tension, then the force within the bar will be positive. If the bar is under compression, then the force will be negative. The two-bar truss will be enhanced to a six-bar truss with three stationary nodes, see Figure 1.4.1.

Unlike the two-bar truss, it is not possible to assume the bars are under compression or tension forces. Hence, all the unknown bar forces in Figure 1.4.1 are initially displayed as under tension.

In order to derive an algebraic system for the six forces, use at each node the equilibrium condition that the sum of the force vectors must be the zero force vector. Let the angle between bars one and two be equal to θ and use

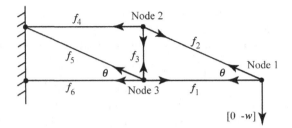

FIGURE 1.4.1
Six-bar truss.

the short notation for sine and cosine $s = \sin(\theta)$ and $c = \cos(\theta)$. Assume the
mass at the right end of the support structure has a weight equal to 10.

Node 1: $[-f_1 \ 0] + [-f_2 c \ f_2 s] + [0 \ -10] = [0 \ 0],$

Node 2: $[f_2 c \ -f_2 s] + [0 \ -f_3] + [-f_4 \ 0] = [0 \ 0]$ and

Node 3: $[f_1 \ 0] + [0 \ f_3] + [-f_5 c \ f_5 s] + [-f_6 \ 0] = [0 \ 0].$

Now equate the first and second components in each vector equation to obtain
a system with six scalar equations

$$\begin{aligned}
-f_1 - f_2 c &= 0 \\
f_2 s &= 10 \\
f_2 c - f_4 &= 0 \\
-f_2 s - f_3 &= 0 \\
f_1 - f_5 c - f_6 &= 0 \\
f_3 + f_5 s &= 0.
\end{aligned}$$

The vector version of these scalar equations is

$$\begin{bmatrix}
-1 & -c & 0 & 0 & 0 & 0 \\
0 & s & 0 & 0 & 0 & 0 \\
0 & c & 0 & -1 & 0 & 0 \\
0 & -s & -1 & 0 & 0 & 0 \\
1 & 0 & 0 & 0 & -c & -1 \\
0 & 0 & 1 & 0 & s & 0
\end{bmatrix}
\begin{bmatrix}
f_1 \\ f_2 \\ f_3 \\ f_4 \\ f_5 \\ f_6
\end{bmatrix}
=
\begin{bmatrix}
0 \\ 10 \\ 0 \\ 0 \\ 0 \\ 0
\end{bmatrix}$$

and the augmented matrix is

$$[A \ \mathbf{d}] =
\begin{bmatrix}
-1 & -c & 0 & 0 & 0 & 0 & 0 \\
0 & s & 0 & 0 & 0 & 0 & 10 \\
0 & c & 0 & -1 & 0 & 0 & 0 \\
0 & -s & -1 & 0 & 0 & 0 & 0 \\
1 & 0 & 0 & 0 & -c & -1 & 0 \\
0 & 0 & 1 & 0 & s & 0 & 0
\end{bmatrix}.$$

In order to do the first stage of Gauss elimination, one must use row operation to transform the lower portion of the augmented matrix to zero components. Fortunately, this matrix has many zero components. The required row operations are indicated by groups of operations on columns:

$$E_{65}(s/c)\ E_{63}(1)\ P_{43}\ E_{52}(c/s)\ E_{42}(1)\ E_{32}(-c/s)\ E_{51}(1)\ [A\ \mathbf{d}] =$$

$$\begin{bmatrix} -1 & -c & 0 & 0 & 0 & 0 & 0 \\ 0 & s & 0 & 0 & 0 & 0 & 10 \\ 0 & 0 & -1 & 0 & 0 & 0 & 10 \\ 0 & 0 & 0 & -1 & 0 & 0 & -10c/s \\ 0 & 0 & 0 & 0 & -c & -1 & 10c/s \\ 0 & 0 & 0 & 0 & 0 & -s/c & 20 \end{bmatrix}$$

The second stage is a backward substitution:

$$f_6 = -20c/s = -20\cos(\theta)/\sin(\theta),$$
$$f_5 = (10c/s - 20c/s)/(-c) = 10/s = 10/\sin(\theta),$$
$$f_4 = 10c/s = 10\cos(\theta)/\sin(\theta),$$
$$f_3 = -10,$$
$$f_2 = 10/s = 10/\sin(\theta) \text{ and}$$
$$f_1 = -c(10/s) = -10c/s = -10\cos(\theta)/\sin(\theta).$$

The negative signs indicate bars 1, 3 and 6 are under compression, and the other bars 2, 4 and 5 are under tension. As the angle θ decreases, the magnitudes of the forces increase, except for bar 3.

1.4.5 Gauss elimination using MATLAB®

Row operations using MATLAB® are illustrated in the MATLAB® code gauss_el.m, and for a more general matrix gauss_elim_13.m. Both stages of Gauss elimination are combined in the single MATLAB® command A\d, which is recommended for full matrices. Consider the six-beam support structure with $\theta = \pi/6$.

```
>>c = cos(pi/6)
>> s = sin(pi/6);
>> A = [-1 -c 0 0 0 0 ;
         0 s 0 0 0 0 ;
         0 c 0 -1 0 0 ;
         0 -s -1 0 0 0 ;
         1 0 0 0 -c -1 ;
         0 0 1 0 s 0];
>> d = [0 10 0 0 0 0]';
>> A\d
```

$$\text{ans} =$$

-17.3205	$\% = f_1$
20.0000	$\% = f_2$
-10.0000	$\% = f_3$
17.3205	$\% = f_4$
20.0000	$\% = f_5$
-34.6410	$\% = f_6$

The reader will find it interesting to experiment with different angles and weights, which can easily be done using the MATLAB® code support.m.

1.4.6 Exercises

1. Find the augmented matrix for

$$2x_1 - 2x_2 + x_3 = 7$$
$$-4x_1 + x_2 + x_3 = 0$$
$$3x_2 + x_3 = -2.$$

2. Find the augmented matrix for

$$x_1 - 2x_2 + x_3 - 5x_4 = 1$$
$$x_1 - 2x_2 + 5x_4 = 1$$
$$-3x_1 - 2x_2 + x_3 + 2x_4 = 0$$
$$4x_1 + x_3 - x_4 = 5.$$

3. Let $n = 4$ and find the elementary matrices $E_{43}(-7)$, P_{23} and $E_4(10)$.

4. Let $n = 3$ and find the elementary matrices $E_{21}(2)$, P_{13} and $E_3(\pi)$.

5. Consider the algebraic system

$$x_1 + 7x_2 + x_3 = 1$$
$$2x_1 + 0x_2 + 7x_3 = 2$$
$$14x_2 + x_3 = 6.$$

(a). Find the augmented matrix.

(b). Use row operations to transform the augmented matrix to an upper triangular matrix.

(c). Find the solution by using the upper triangular matrix.

6. Consider the algebraic system

$$x_1 - x_2 + x_4 = 6$$
$$x_1 - x_3 - x_4 = 7$$
$$2x_2 + 7x_3 = 4$$
$$2x_1 - x_4 = 8.$$

(a). Find the augmented matrix.

(b). Use row operations to transform the augmented matrix to an upper triangular matrix.

(c). Find the solution by using the upper triangular matrix.

7. Consider the steady state heat conduction within a fin in Example 1.2.5. Verify the row operations to find the indicated upper triangular matrix.

8. Consider the steady state heat conduction within a fin in Example 1.2.5. Use Gauss elimination to solve the steady state problem

$$
\begin{bmatrix}
3 & -1 & 0 & 0 \\
-1 & 3 & -1 & 0 \\
0 & -1 & 3 & -1 \\
0 & 0 & -2 & 3
\end{bmatrix}
\begin{bmatrix}
u_1 \\
u_2 \\
u_3 \\
u_3
\end{bmatrix}
=
\begin{bmatrix}
100 \\
10 \\
10 \\
10
\end{bmatrix}.
$$

9. Use MATLAB® to confirm any of your calculations in Exercises 5–8.

1.5 Inverse Matrices

The three types of elementary row operations can be reversed and be represented by elementary matrices. In terms of matrix operations this means for each elementary matrix there is another elementary matrix such that the product is the identity matrix (ones on the diagonal and zeros off diagonal):

$$
E_{ij}(-a) \; E_{ij}(a) = I
$$
$$
P_{ij} \; P_{ij} = I \text{ and}
$$
$$
E_i(1/c) \; E_i(c) = I.
$$

The objective of this section is to be able to find an $n \times n$ matrix B such that for a given $n \times n$ matrix A we have $AB = I$. An important use is to find the solution $Ax = d$ given by $x = A^{-1}d$ because

$$
Ax = A(A^{-1}d) = (AA^{-1})d = Id = d.
$$

Unfortunately, one cannot always find such matrices! We shall see in the last section A with a nonzero determinant implies that an inverse matrix exists.

1.5.1 Examples of inverse matrices

The identity $n \times n$ matrix I has all zero components except on the diagonal where the components are equal to one. The *identity matrix* can be written as a matrix whose j^{th} column vector is the jth unit column vector $\mathbf{e}_j$

$$I = \begin{bmatrix} \mathbf{e}_1 & \mathbf{e}_2 & \cdots & \mathbf{e}_n \end{bmatrix}$$

$$= \begin{bmatrix} 1 & 0 & \cdots & 0 \\ 0 & 1 & \cdots & 0 \\ \vdots & \vdots & \ddots & \vdots \\ 0 & 0 & \cdots & 1 \end{bmatrix}.$$

Definition. *Let A be an $n \times n$ matrix. An $n \times n$ matrix B is called an inverse matrix of A if and only if $AB = BA = I$.*

Notation. The inverse matrix is usually denoted by A^{-1}. Here one should be careful not to write this as $1/A$ or to have components equal to $1/a_{ij}$!

Example 1.5.1.

1. Elementary Matrices.

   If $A = E_{ij}(a)$, then $A^{-1} = E_{ij}(-a)$. If $A = P_{ij}$, then $A^{-1} = P_{ij}$. If $A = E_i(c)$, then $A^{-1} = E_i(1/c)$.

2. Diagonal Matrices.

   If A is a diagonal matrix with nonzero diagonal components $a_{ii} \neq 0$, then A^{-1} is a diagonal matrix with diagonal components equal to $1/a_{ii}$. For example, if $n = 3$ and

   $$A = \begin{bmatrix} 4 & 0 & 0 \\ 0 & 2 & 0 \\ 0 & 0 & 3 \end{bmatrix},$$

   then because $AA^{-1} = I$

   $$A^{-1} = \begin{bmatrix} 1/4 & 0 & 0 \\ 0 & 1/2 & 0 \\ 0 & 0 & 1/3 \end{bmatrix}.$$

3. Block Diagonal Matrices.

   If A is a block diagonal matrix with diagonal blocks that have inverses, then A has an inverse matrix which is also a block diagonal matrix. For example, consider a 5×5 matrix with two diagonal blocks

   $$A = \begin{bmatrix} A_{11} & \\ & A_{22} \end{bmatrix} = \begin{bmatrix} 1 & 0 & 0 & 0 & 0 \\ 0 & 1 & 0 & 0 & 0 \\ 0 & 4 & 1 & 0 & 0 \\ 0 & 0 & 0 & 2 & 0 \\ 0 & 0 & 0 & 0 & 3 \end{bmatrix} \quad \text{where}$$

   $$A_{11} = E_{32}(4) = \begin{bmatrix} 1 & 0 & 0 \\ 0 & 1 & 0 \\ 0 & 4 & 1 \end{bmatrix} \quad \text{and} \quad A_{22} = \begin{bmatrix} 2 & 0 \\ 0 & 3 \end{bmatrix}.$$

Both A_{11} and A_{22} have inverses and, hence, the inverse of A is

$$A^{-1} = \begin{bmatrix} A_{11}^{-1} & \\ & A_{22}^{-1} \end{bmatrix} = \begin{bmatrix} 1 & 0 & 0 & 0 & 0 \\ 0 & 1 & 0 & 0 & 0 \\ 0 & -4 & 1 & 0 & 0 \\ 0 & 0 & 0 & 1/2 & 0 \\ 0 & 0 & 0 & 0 & 1/3 \end{bmatrix}.$$

4. A is a 2×2 matrix with $det(A) \neq 0$.

The inverse matrix B must satisfy the matrix equation

$$AB = I$$
$$A \begin{bmatrix} \mathbf{b}_1 & \mathbf{b}_2 \end{bmatrix} = \begin{bmatrix} \mathbf{e}_1 & \mathbf{e}_2 \end{bmatrix}.$$

This is equivalent to two vector equations

$$A\mathbf{b}_1 = \mathbf{e}_1 \text{ and } A\mathbf{b}_2 = \mathbf{e}_2.$$

One can solve these by either Cramer's rule or Gauss elimination. For example, use Cramer's rule to find the inverse of

$$A = \begin{bmatrix} a_{11} & a_{12} \\ a_{21} & a_{22} \end{bmatrix}.$$

The first column $\mathbf{b}_1$ of the inverse matrix must satisfy

$$A\mathbf{b}_1 = \mathbf{e}_1$$
$$\begin{bmatrix} a_{11} & a_{12} \\ a_{21} & a_{22} \end{bmatrix} \begin{bmatrix} b_{11} \\ b_{21} \end{bmatrix} = \begin{bmatrix} 1 \\ 0 \end{bmatrix}.$$

Then

$$b_{11} = \det(\begin{bmatrix} 1 & a_{12} \\ 0 & a_{22} \end{bmatrix})/\det(A) = a_{22}/\det(A) \text{ and}$$

$$b_{21} = \det(\begin{bmatrix} a_{11} & 1 \\ a_{21} & 0 \end{bmatrix})/\det(A) = -a_{21}/\det(A).$$

The second column $\mathbf{b}_2$ of the inverse matrix must satisfy

$$A\mathbf{b}_2 = \mathbf{e}_2$$
$$\begin{bmatrix} a_{11} & a_{12} \\ a_{21} & a_{22} \end{bmatrix} \begin{bmatrix} b_{12} \\ b_{22} \end{bmatrix} = \begin{bmatrix} 0 \\ 1 \end{bmatrix}.$$

Then

$$b_{12} = \det(\begin{bmatrix} 0 & a_{12} \\ 1 & a_{22} \end{bmatrix})/\det(A) = -a_{12}/\det(A) \text{ and}$$

$$b_{22} = \det(\begin{bmatrix} a_{11} & 0 \\ a_{21} & 1 \end{bmatrix})/\det(A) = a_{11}/\det(A).$$

Therefore, the inverse of A is

$$A^{-1} = B = \begin{bmatrix} \mathbf{b}_1 & \mathbf{b}_2 \end{bmatrix} = \begin{bmatrix} a_{22} & -a_{12} \\ -a_{21} & a_{11} \end{bmatrix} / \det(A).$$

The alternative is to use row operations and Gauss elimination. The Gauss-Jordan method is a variation of Gauss elimination, which can be adapted to larger matrices and to computers.

1.5.2   Gauss–Jordan method to find inverse matrices

As motivation consider a special case of the above 2×2 example where

$$A = \begin{bmatrix} 2 & -1 \\ -1 & 2 \end{bmatrix} \text{ and } A^{-1} = \begin{bmatrix} 2/3 & 1/3 \\ 1/3 & 2/3 \end{bmatrix}.$$

First, use row operations and the augmented matrix to solve $A\mathbf{b}_1 = \mathbf{e}_1$

$$[A \ \mathbf{e}_1] = \begin{bmatrix} 2 & -1 & 1 \\ -1 & 2 & 0 \end{bmatrix}$$

$$E_{21}(1/2) \, [A \ \mathbf{e}_1] = \begin{bmatrix} 1 & 0 \\ 1/2 & 1 \end{bmatrix} \begin{bmatrix} 2 & -1 & 1 \\ -1 & 2 & 0 \end{bmatrix} = \begin{bmatrix} 2 & -1 & 1 \\ 0 & 3/2 & 1/2 \end{bmatrix}.$$

At this point either use backward substitution or more row operation to obtain an identity matrix in the left side of the augmented matrix. The row operations include two row multiples and then adding $(1/2)$ times row_2 to row_1 :

$$E_2(2/3) \, E_1(1/2) \, E_{21}(1/2) \, [A \ \mathbf{e}_1] = \begin{bmatrix} 1 & -1/2 & 1/2 \\ 0 & 1 & 1/3 \end{bmatrix}$$

$$E_{12}(1/2) \, E_2(2/3) \, E_1(1/2) \, E_{21}(1/2) \, [A \ \mathbf{e}_1] = \begin{bmatrix} 1 & 0 & 2/3 \\ 0 & 1 & 1/3 \end{bmatrix}.$$

The right column in the transform augmented matrix is the first column of the inverse matrix. Let E be the product of the four elementary matrices and write the above as

$$E \, [A \ \mathbf{e}_1] = [I \ \mathbf{b}_1].$$

In order to find the second column of the inverse matrix, the *same* row operations are needed to transform the augmented matrix from $[A \ \mathbf{e}_2]$ to $[I \ \mathbf{b}_2]$

$$E \, [A \ \mathbf{e}_2] = [I \ \mathbf{b}_2].$$

An efficient way to do both these solves is to combine the two columns of the inverse matrix into the right side of a larger augmented matrix. That is, use the row operations to transform $[A \ \mathbf{e}_1 \ \mathbf{e}_2] = [A \ I]$ to $[I \ \mathbf{b}_1 \ \mathbf{b}_2] = [I \ B]$

$$E \, [A \ \mathbf{e}_1 \ \mathbf{e}_2] = [I \ \mathbf{b}_1 \ \mathbf{b}_2].$$

In order to find the inverse of an $n \times n$ matrix A, one must solve n vector equations

$$AB = I$$

$$A \begin{bmatrix} \mathbf{b}_1 & \mathbf{b}_2 & \cdots & \mathbf{b}_n \end{bmatrix} = \begin{bmatrix} \mathbf{e}_1 & \mathbf{e}_2 & \cdots & \mathbf{e}_n \end{bmatrix}$$

$$\begin{bmatrix} A\mathbf{b}_1 & A\mathbf{b}_2 & \cdots & A\mathbf{b}_n \end{bmatrix} = \begin{bmatrix} \mathbf{e}_1 & \mathbf{e}_2 & \cdots & \mathbf{e}_n \end{bmatrix} \text{ or}$$

$$A\mathbf{b}_j = \mathbf{e}_j \text{ for } j = 1, \ldots, n.$$

Use row operations and elementary matrices to transform the augmented matrix $[A \ \ I]$ to $[I \ \ A^{-1}]$.

Definition. *Consider the $n \times n$ matrix A and find the inverse matrix, if it exists. The Gauss–Jordan method for solving this system has of two stages:*

Stage 1. Transform the augmented matrix $[A \ \ I]$ to upper triangular form $[U \ \ L]$.

(a). start with the left column, column $j = 1$, and use row operations to transform column $j = 1$ to zeros below row $i = 1$,

(b). move to the next column, column $j = 2$, and use row operations to transform column $j = 2$ to zeros below row $i = 2$ and

(c). repeat this until column $j = n - 1$ has been done.

Stage 2. Transform the upper triangular matrix $[U \ \ L]$ to $[I \ \ A^{-1}]$

(a). use elementary matrices so that the diagonal of U is transformed to have ones on all of the diagonal components,

(b). start with the right column of U, column $j = n$, and use row operations to transform column $j = n$ to zeros above row $i = n$,

(c). move to the next left column, column $j = n - 1$, and use row operations to transform column $j = n - 1$ to zeros above row $i = n - 1$ and

(d). repeat this until column $j = 2$ has been done.

Example 1.5.2. Use the Gauss–Jordan method to find the inverse of

$$A = \begin{bmatrix} 1 & 2 & 1 \\ 1 & 3 & 2 \\ 1 & 0 & 1 \end{bmatrix}.$$

The augmented matrix is

$$[A \ \ I] = \begin{bmatrix} 1 & 2 & 1 & 1 & 0 & 0 \\ 1 & 3 & 2 & 0 & 1 & 0 \\ 1 & 0 & 1 & 0 & 0 & 1 \end{bmatrix}.$$

Stage 1. Subtract *row_1* from *row_2* and *row_3*

$$E_{31}(-1)\, E_{21}(-1)\, [A \ I] = \begin{bmatrix} 1 & 2 & 1 & 1 & 0 & 0 \\ 0 & 1 & 1 & -1 & 1 & 0 \\ 0 & -2 & 0 & -1 & 0 & 1 \end{bmatrix}.$$

Add (2) times *row_2* to *row_3*

$$E_{32}(2)\, E_{31}(-1)\, E_{21}(-1)\, [A \ I] = \begin{bmatrix} 1 & 2 & 1 & 1 & 0 & 0 \\ 0 & 1 & 1 & -1 & 1 & 0 \\ 0 & 0 & 2 & -3 & 2 & 1 \end{bmatrix}.$$

Stage 2. Divide *row_3* by 2 and then subtract it from *row_2* and *row_1*

$$E_{13}(-1)\, E_{23}(-1)\, E_{3}(1/2)\, E_{32}(2)\, E_{31}(-1)\, E_{21}(-1)\, [A \ I] =$$

$$\begin{bmatrix} 1 & 2 & 0 & 5/2 & -1 & -1/2 \\ 0 & 1 & 0 & 1/2 & 0 & -1/2 \\ 0 & 0 & 1 & -3/2 & 1 & 1/2 \end{bmatrix}.$$

Finally, subtract (2) times *row_2* from *row_1*

$$E_{12}(-2)\, E_{13}(-1)\, E_{23}(-1)\, E_{3}(1/2)\, E_{32}(2)\, E_{31}(-1)\, E_{21}(-1)\, [A \ I] =$$

$$\begin{bmatrix} 1 & 0 & 0 & 3/2 & -1 & 1/2 \\ 0 & 1 & 0 & 1/2 & 0 & -1/2 \\ 0 & 0 & 1 & -3/2 & 1 & 1/2 \end{bmatrix}.$$

One can easily verify the inverse is given by the right side of the transformed augmented matrix

$$A^{-1} = \begin{bmatrix} 3/2 & -1 & 1/2 \\ 1/2 & 0 & -1/2 \\ -3/2 & 1 & 1/2 \end{bmatrix}.$$

The solution of $A\mathbf{x} = \mathbf{d} = [1 \ 2 \ 3]^T$ is easily computed by using the inverse matrix

$$\mathbf{x} = A^{-1}\mathbf{d} = \begin{bmatrix} 3/2 & -1 & 1/2 \\ 1/2 & 0 & -1/2 \\ -3/2 & 1 & 1/2 \end{bmatrix} \begin{bmatrix} 1 \\ 2 \\ 3 \end{bmatrix} = \begin{bmatrix} 1 \\ -1 \\ 2 \end{bmatrix}.$$

Example 1.5.3. Use the Gauss-Jordan method to find the inverse of

$$A = \begin{bmatrix} 0 & 5 & 6 \\ 1 & 3 & 1 \\ 2 & 1 & 1 \end{bmatrix}.$$

The augmented matrix is

$$[A \ I] = \begin{bmatrix} 0 & 5 & 6 & 1 & 0 & 0 \\ 1 & 3 & 1 & 0 & 1 & 0 \\ 2 & 1 & 1 & 0 & 0 & 1 \end{bmatrix}.$$

Stage 1. Because the 11-component is zero, row_1 and row_2 (or row_3) must be interchanged

$$P_{12}\,[A\ I] = \begin{bmatrix} 1 & 3 & 1 & 0 & 1 & 0 \\ 0 & 5 & 6 & 1 & 0 & 0 \\ 2 & 1 & 1 & 0 & 0 & 1 \end{bmatrix}$$

Subtract (2) times row_1 from row_3

$$E_{31}(-2)\,P_{12}\,[A\ I] = \begin{bmatrix} 1 & 3 & 1 & 0 & 1 & 0 \\ 0 & 5 & 6 & 1 & 0 & 0 \\ 0 & -5 & -1 & 0 & -2 & 1 \end{bmatrix}.$$

Add row_2 to row_3

$$E_{32}(1)\,E_{31}(-2)\,P_{12}\,[A\ I] = \begin{bmatrix} 1 & 3 & 1 & 0 & 1 & 0 \\ 0 & 5 & 6 & 1 & 0 & 0 \\ 0 & 0 & 5 & 1 & -2 & 1 \end{bmatrix}.$$

Stage 2. Divide row_2 and row_3 by 5

$$E_3(1/5)\,E_2(1/5)\,E_{32}(1)\,E_{31}(-2)\,P_{12}\,[A\ I]\,[A\ I] =$$

$$\begin{bmatrix} 1 & 3 & 1 & 0 & 1 & 0 \\ 0 & 1 & 6/5 & 1/5 & 0 & 0 \\ 0 & 0 & 1 & 1/5 & -2/5 & 1/5 \end{bmatrix}.$$

In order to transform the left side to the identity, subtract $(6/5)$ times row_3 from row_2, subtract row_3 from row_1 and last subtract (3) times row_2 from row_1

$$E_{12}(-3)\,E_{13}(-1)\,E_{23}(-6/5)\,E_3(1/5)\,E_2(1/5)\,E_{32}(1)\,E_{31}(-2)$$

$$P_{12}\,[A\ I] = \begin{bmatrix} 1 & 0 & 0 & -2/25 & -1/25 & 13/25 \\ 0 & 1 & 0 & -1/25 & 12/25 & -6/25 \\ 0 & 0 & 1 & 5/25 & -10/25 & 5/25 \end{bmatrix}.$$

The inverse is the right side of the transformed augmented matrix

$$A^{-1} = (1/25)\begin{bmatrix} -2 & -1 & 13 \\ -1 & 12 & -6 \\ 5 & -10 & 5 \end{bmatrix}.$$

The solution of $Ax = d = [6\ 2\ 3]^T$ is computed by using the inverse matrix

$$x = A^{-1}d = (1/25)\begin{bmatrix} -2 & -1 & 13 \\ -1 & 12 & -6 \\ 5 & -10 & 5 \end{bmatrix}\begin{bmatrix} 6 \\ 2 \\ 3 \end{bmatrix} = \begin{bmatrix} 1 \\ 0 \\ 1 \end{bmatrix}.$$

Example 1.5.4. Consider the steady state heat wire problem where there were five segments and current generating heat within the wire. The algebraic problem for the temperature in each interior segment is

$$\begin{bmatrix} 2 & -1 & 0 & 0 \\ -1 & 2 & -1 & 0 \\ 0 & -1 & 2 & -1 \\ 0 & 0 & -1 & 2 \end{bmatrix} \begin{bmatrix} u_1 \\ u_2 \\ u_3 \\ u_3 \end{bmatrix} = \begin{bmatrix} 470 \\ 400 \\ 400 \\ 470 \end{bmatrix}.$$

The augmented matrix for finding the inverse matrix is

$$[A \; I] = \begin{bmatrix} 2 & -1 & 0 & 0 & 1 & 0 & 0 & 0 \\ -1 & 2 & -1 & 0 & 0 & 1 & 0 & 0 \\ 0 & -1 & 2 & -1 & 0 & 0 & 1 & 0 \\ 0 & 0 & -1 & 2 & 0 & 0 & 0 & 1 \end{bmatrix}.$$

Stage 1. Obtain zeros in the subdiagonal by using

$$E_{43}(3/4) \, E_{32}(2/3) \, E_{21}(1/2) \, [A \; I] =$$

$$\begin{bmatrix} 2 & -1 & 0 & 0 & 1 & 0 & 0 & 0 \\ 0 & 3/2 & -1 & 0 & 1/2 & 1 & 0 & 0 \\ 0 & 0 & 4/3 & -1 & 1/3 & 2/3 & 1 & 0 \\ 0 & 0 & 0 & 5/4 & 1/4 & 2/4 & 3/4 & 1 \end{bmatrix}.$$

Stage 2. Multiply rows 1-4 by $1/2$, $2/3$, $3/4$ and $4/5$, respectively, to get for $D \equiv E_4(4/5) \, E_3(3/4) \, E_2(2/3) \, E_1(1/2)$

$$D \, E_{43}(3/4) \, E_{32}(2/3) \, E_{21}(1/2) \, [A \; I] =$$

$$\begin{bmatrix} 1 & -1/2 & 0 & 0 & 1/2 & 0 & 0 & 0 \\ 0 & 1 & -2/3 & 0 & 1/3 & 2/3 & 0 & 0 \\ 0 & 0 & 1 & -3/4 & 1/4 & 2/4 & 3/4 & 0 \\ 0 & 0 & 0 & 1 & 1/5 & 2/5 & 3/5 & 4/5 \end{bmatrix}.$$

In order to obtain zeros in the upper triangular part of the left side, use elementary matrices

$$E_{12}(1/2) \, E_{23}(2/3) \, E_{34}(3/4) \, D \, E_{43}(3/4) \, E_{32}(2/3) \, E_{21}(1/2) \, [A \; I] =$$

$$\begin{bmatrix} 1 & 0 & 0 & 0 & 8/10 & 6/10 & 4/10 & 2/10 \\ 0 & 1 & 0 & 0 & 6/10 & 12/10 & 8/10 & 4/10 \\ 0 & 0 & 1 & 0 & 4/10 & 8/10 & 12/10 & 6/10 \\ 0 & 0 & 0 & 1 & 2/10 & 4/10 & 6/10 & 8/10 \end{bmatrix}.$$

So, the inverse matrix is

$$A^{-1} = (1/10) \begin{bmatrix} 8 & 6 & 4 & 2 \\ 6 & 12 & 8 & 4 \\ 4 & 8 & 12 & 6 \\ 2 & 4 & 6 & 8 \end{bmatrix},$$

and the solution of the above algebraic system is

$$\mathbf{u} = A^{-1}\mathbf{d} = (1/10) \begin{bmatrix} 8 & 6 & 4 & 2 \\ 6 & 12 & 8 & 4 \\ 4 & 8 & 12 & 6 \\ 2 & 4 & 6 & 8 \end{bmatrix} \begin{bmatrix} 470 \\ 400 \\ 400 \\ 470 \end{bmatrix} = \begin{bmatrix} 870 \\ 1270 \\ 1270 \\ 870 \end{bmatrix}.$$

1.5.3 Properties of inverse matrices

Not all matrices have inverses! Any 2×2 with zero determinant cannot have an inverse. For example, if

$$A = \begin{bmatrix} 1 & 1 \\ 2 & 2 \end{bmatrix},$$

then one cannot find the first column of the inverse matrix because of the following contradiction

$$A\mathbf{b}_1 = \mathbf{e}_1$$
$$\begin{bmatrix} 1 & 1 \\ 2 & 2 \end{bmatrix} \begin{bmatrix} b_{11} \\ b_{21} \end{bmatrix} = \begin{bmatrix} 1 \\ 0 \end{bmatrix}$$
$$b_{11} + b_{21} = 1 \text{ and } 2b_{11} + 2b_{21} = 0!$$

In the last section of this chapter determinants of $n \times n$ matrices will be defined, and any matrix with a nonzero determinant will be shown to have an inverse matrix.

One very important property of inverse matrices is the solution of $A\mathbf{x} = \mathbf{d}$ is $A^{-1}\mathbf{d}$. Other properties are summarized in the following theorem.

Theorem 1.5.1. *(Properties of Inverse Matrices) Let A, A_1 and A_2 be $n \times n$ matrices that have inverses. Then the following statements hold:*

(i). *If $A^{-1} = B$, then A times column j of B is e_j.*

(ii). *$A^{-1}d$ is the unique solution of $A\mathbf{x} = \mathbf{d}$.*

(iii). *$(A_1 A_2)^{-1} = A_2^{-1} A_1^{-1}$ (note the reverse order).*

(iv). *There is only one inverse matrix, $(A^{-1})^{-1} = A$ and $(A^T)^{-1} = (A^{-1})^T$.*

(v). *Let $n = n_1 + n_2$ and A_{ii} be $n_i \times n_i$. The square matrices A_{11}, A_{22} have inverses if and only if the block diagonal has an inverse. Moreover,*

$$\begin{bmatrix} A_{11} & 0_{n_1 \times n_2} \\ 0_{n_2 \times n_1} & A_{22} \end{bmatrix}^{-1} = \begin{bmatrix} (A_{11})^{-1} & 0_{n_1 \times n_2} \\ 0_{n_2 \times n_1} & (A_{22})^{-1} \end{bmatrix}.$$

Solution of Ax = d

The proof of $(A_1 A_2)^{-1} = A_2^{-1} A_1^{-1}$ follows from the associative property of matrix products

$$\begin{aligned}
(A_1 A_2)^{-1}(A_1 A_2) &= (A_2^{-1} A_1^{-1})(A_1 A_2) \\
&= A_2^{-1}(A_1^{-1}(A_1 A_2)) \\
&= A_2^{-1}((A_1^{-1} A_1) A_2) \\
&= A_2^{-1}((I) A_2) = I.
\end{aligned}$$

The proof of item five is easy in one direction. Assume the block diagonal has an inverse and show the blocks must have an inverse. Let the inverse be given in B and require the blocks in B be chosen so that $AB = I$ and $BA = I$

$$\begin{bmatrix} A_{11} & 0_{n_1 \times n_2} \\ 0_{n_2 \times n_1} & A_{22} \end{bmatrix} \begin{bmatrix} B_{11} & B_{12} \\ B_{21} & B_{22} \end{bmatrix} = \begin{bmatrix} I_{n_1} & 0_{n_1 \times n_2} \\ 0_{n_2 \times n_1} & I_{n_2} \end{bmatrix} \text{ and}$$

$$\begin{bmatrix} B_{11} & B_{12} \\ B_{21} & B_{22} \end{bmatrix} \begin{bmatrix} A_{11} & 0_{n_1 \times n_2} \\ 0_{n_2 \times n_1} & A_{22} \end{bmatrix} = \begin{bmatrix} I_{n_1} & 0_{n_1 \times n_2} \\ 0_{n_2 \times n_1} & I_{n_2} \end{bmatrix}.$$

This gives $A_{11} B_{11} = I_{n_1}$ and $B_{11} A_{11} = I_{n_1}$, and so B_{11} is the inverse of A_{11}. Also, $A_{11} B_{12} = 0_{n_1 \times n_2}$ and therefore B_{12} must be a zero matrix. The proof for $B_{22} = (A_{22})^{-1}$ and $B_{21} = 0_{n_2 \times n_1}$ is similar.

Example 1.5.5. This illustrates the third property in the above theorem. Find the inverse matrix of $A = A_1 A_2$ where

$$A = \begin{bmatrix} 4 & 0 & 0 \\ -4 & 2 & 0 \\ 0 & 0 & 1 \end{bmatrix} = \begin{bmatrix} 4 & 0 & 0 \\ 0 & 2 & 0 \\ 0 & 0 & 1 \end{bmatrix} \begin{bmatrix} 1 & 0 & 0 \\ -2 & 1 & 0 \\ 0 & 0 & 1 \end{bmatrix} = A_1 A_2.$$

A_1 is a diagonal matrix and A_2 is an elementary matrix with inverses

$$A_1^{-1} = \begin{bmatrix} 1/4 & 0 & 0 \\ 0 & 1/2 & 0 \\ 0 & 0 & 1 \end{bmatrix} \text{ and } A_2^{-1} = \begin{bmatrix} 1 & 0 & 0 \\ 2 & 1 & 0 \\ 0 & 0 & 1 \end{bmatrix}.$$

The inverse of A is

$$A^{-1} = A_2^{-1} A_1^{-1} = \begin{bmatrix} 1 & 0 & 0 \\ 2 & 1 & 0 \\ 0 & 0 & 1 \end{bmatrix} \begin{bmatrix} 1/4 & 0 & 0 \\ 0 & 1/2 & 0 \\ 0 & 0 & 1 \end{bmatrix} = \begin{bmatrix} 1/4 & 0 & 0 \\ 2/4 & 1/2 & 0 \\ 0 & 0 & 1 \end{bmatrix}.$$

In the Gauss elimination method, the augmented matrix was transformed by a product of elementary matrices, which have inverses. Therefore, the product of the elementary matrices, E, has an inverse and

$$E[A \ \mathbf{d}] = [U \ \widehat{\mathbf{d}}].$$

If all the diagonal components of U are not zero, the solutions of $A\mathbf{x} = \mathbf{d}$ and $U\mathbf{x} = EA\mathbf{x} = E\mathbf{d} = \widehat{\mathbf{d}}$ are the same. If the diagonal components are not zero, then $A = E^{-1}U$ is a product matrices with inverse and so $A^{-1} = U^{-1}E$ exists.

Assume A has an inverse and show the the Gauss elimination process gives $EA = U$ and has an inverse. So the first column of A must have at least one nonzero component. There is permutation matrix, P_1, so that

$$P_1 A = \begin{bmatrix} b & e^T \\ f & C \end{bmatrix}$$

where $b \neq 0$ is a single number, f and e are $(n-1) \times 1$ vectors and C is $(n-1) \times (n-1)$ matrix. The product of the next $n-1$ row operations can be done by a block elementary matrix, often called a *Gauss transform*,

$$G_1 = \begin{bmatrix} 1 & 0_{1 \times (n-1)} \\ -f/b & I_{n-1} \end{bmatrix}.$$

This zeros out the vector f and changes the matrix C

$$G_1 P_1 A = \begin{bmatrix} b & e^T \\ 0_{(n-1) \times 1} & \widehat{C} \end{bmatrix}$$

where $\widehat{C} \equiv C - (f/b)e^T$. Now, move to the second column and repeat this on the smaller matrix $\widehat{C}$. We need $\widehat{C}$ to have an inverse, and this follows by multiplying on the right by another block elementary matrix

$$G_1 P_1 A \begin{bmatrix} 1 & -e^T/b \\ 0_{(n-1) \times 1} & I_{n-1} \end{bmatrix} = \begin{bmatrix} b & e^T \\ 0_{(n-1) \times 1} & \widehat{C} \end{bmatrix} \begin{bmatrix} 1 & -e^T/b \\ 0_{(n-1) \times 1} & I_{n-1} \end{bmatrix}$$

$$= \begin{bmatrix} b & 0_{1 \times (n-1)} \\ 0_{(n-1) \times 1} & \widehat{C} \end{bmatrix}.$$

The diagonal block matrix on the right is a product of matrices with inverses. So, we may continue this elimination procedure on the remaining columns to get

$$(G_{n-1}P_{n-1}) \cdots (G_1 P_1) A = \begin{bmatrix} b & e^T \\ 0_{(n-1) \times 1} & \widehat{U} \end{bmatrix} = U.$$

The matrix $\widehat{U}$ is upper triangular. Since the left side is a product of matrices with inverses, the right side must have an inverse and the diagonal components of $\widehat{U}$ must not be zero.

The last two paragraphs give a constructive proof of the following equivalent statements:

(i) A has an inverse

(ii) $Ax = 0_{n \times 1}$ implies $x = 0_{n \times 1}$, and

(iii) $EA = U$

    where U is upper triangular with nonzero diagonal components and E is a product of elementary matrices.

We proved (i) implies (ii), (ii) implies (iii). The proof that (iii) implies (i) follows from $EA = U$ because both E and U have inverses giving $A^{-1} = U^{-1}E$.

    The second criteria is often used to the determine if an inverse of matrix exists. One class of matrices is the *strictly diagonally dominant matrices*:

$$|a_{ii}| > \sum_{j \neq i} |a_{ij}| \text{ for all } i. \tag{1.5.1}$$

The cooling fin application is an example.

Theorem 1.5.2. *(Diagonal Dominant Matrix) If A is strictly diagonally dominant, then A has an inverse.*

Proof. Suppose there is a nonzero vector, x, and $Ax = 0_{n \times 1}$. This leads to a contradiction of the assumption. Let i be such that $|x_i| = \max |x_j| > 0$.

$$a_{ii}x_i = -\sum_{j \neq i} a_{ij}x_j$$

$$|a_{ii}x_i| \leq \sum_{j \neq i} |a_{ij}x_j|$$

$$|a_{ii}| \leq \sum_{j \neq i} |a_{ij}| |x_j| / |x_i| \leq \sum_{j \neq i} |a_{ij}|.$$

∎

    A related result requires the strict inequality for at least one component, and greater than or equal to for the other components in line (1.5.1). If there is no subset of the indices, S, such that

$$a_{ij} = 0 \text{ for } i \in S \text{ and } j \notin S,$$

then one can also argue by contradiction to show such a matrix has an inverse. More details on these matrices (irreducible) can be found in [20, Section 2.3]

    In the definition of the inverse matrix two conditions are required, $AB = I$ and $BA = I$. As long as the matrices are both $n \times n$, the second condition follows from the first. Since $AB = I$, B must satisfy property (ii) and hence, has an inverse. For if not, then there is nonzero x with $Bx = 0$. Then $AB(x) = A(Bx) = 0$ and this is a contradiction to $AB = I$. Next, multiply $AB = I$ from the left by B and then from the right by the inverse of B

$$B(AB) = BI$$
$$(BA)B = B \text{ and}$$
$$BA = I.$$

The elementary matrix products are not directly used in computations, but the less costly row operations are executed. The first step requires $n - 1$ row operation with vectors of length $n - 1$. The operations count is order $2(n - 1)^2$. The next column requires $2(n - 2)^2$. The total operations for all $n - 1$ columns is of order $2n^3/3$. See Exercise 11 on these summations.

A special case for A is the tridiagonal matrix with nonzero entries the subdiagonal, diagonal and superdiagonal. Because of the many zero entries, the operation count for a tridiagonal matrix solve is of order $8n$, and therefore, the full Gauss elimination method should not be used. In order see this, consider the 3×3 solve $Ax = d$. The row operations are

$$E_{21}(-a_{21}/a_{11})[A \ \ d] = \begin{bmatrix} a_{11} & a_{12} & 0 & d_1 \\ 0 & \widehat{a}_{22} & a_{23} & \widehat{d}_2 \\ 0 & a_{32} & a_{33} & d_3 \end{bmatrix}$$

where $\widehat{a}_{22} = a_{22} - (a_{21}/a_{11})a_{12}$ and $\widehat{d}_2 = d_2 - (a_{21}/a_{11})d_1$. Next zero the 32-component

$$E_{32}(-a_{32}/\widehat{a}_{22})E_{21}(-a_{21}/a_{11})[A \ \ d] = \begin{bmatrix} a_{11} & a_{12} & 0 & d_1 \\ 0 & \widehat{a}_{22} & a_{23} & \widehat{d}_2 \\ 0 & 0 & \widehat{a}_{33} & \widehat{d}_3 \end{bmatrix}$$

where $\widehat{a}_{33} = a_{33} - (a_{32}/\widehat{a}_{22})a_{23}$ and $\widehat{d}_3 = d_3 - (a_{32}/\widehat{a}_{22})\widehat{d}_2$. If this is extended to $n \times n$, operation count is $5n$ and an another $3n$ for the upper triangular solve step, see the MATLAB® code trid_13.m.

1.5.4 Inverse matrices and MATLAB®

The by-hand computation of inverse matrices can be an extremely tedious process. Even using computing tools, the number of calculations is of order $2n^3$ arithmetic operations for an $n \times n$ matrix! The MATLAB® command inv(A) will compute the inverse of a matrix A. Some illustrations of this and row operations are given in the MATLAB®code inv_mat.m. Example 1.5.2 done using MATLAB® is

```
>> A = [0 5 6;1 3 1;2 1 1]
    A =
         0 5 6
         1 3 1
         2 1 1
>> B = inv(A)              % computes the inverse matrix
    B =
        -0.0800 -0.0400  0.5200
        -0.0400  0.4800 -0.2400
         0.2000 -0.4000  0.2000
```

```
>> A*B
        ans =
                1.0000  0.0000  0.0000
               -0.0000  1.0000 -0.0000
                0.0000  0.0000  1.0000
>> d = [6 2 3]'
        d =
                6
                2
                3
>> solution1 = inv(A)*d        % uses inverse matrix
        solution1 =
                1.0000
               -0.0000
                1.0000
>> solutions2 = A\d            % uses Gauss elimination
        solutions2 =
                1
                0
                1
```

1.5.5   Exercises

1. Let $A = \begin{bmatrix} 3 & -1 \\ 1 & 3 \end{bmatrix}$.

   (a). Use Cramer's rule to find the inverse of A.

   (b). Use Gauss–Jordan method to find the inverse.

   (c). Find the inverse of

   $$\begin{bmatrix} 3 & -1 & 0 & 0 \\ 1 & 3 & 0 & 0 \\ 0 & 0 & 3 & 0 \\ 0 & 0 & 0 & 4 \end{bmatrix}.$$

2. Let $A = \begin{bmatrix} 1 & -1 & 1 \\ 1 & 0 & -3 \\ 0 & -2 & -3 \end{bmatrix}$.

   (a). Use the Gauss–Jordan method to find the inverse matrix.

   (b). Use the inverse to solve $A\mathbf{x} = \mathbf{d} = [0 \ 10 \ 20]^T$.

3. Verify the row operations used in Example 1.5.4.

4. Let $A = \begin{bmatrix} 1 & 2 & 0 & 1 \\ 2 & 1 & 0 & 0 \\ 0 & 0 & 2 & 0 \\ 1 & 2 & 0 & 2 \end{bmatrix}$.

(a). Use the Gauss–Jordan method to find the inverse matrix.

(b). Use the inverse to solve $A\mathbf{x} = \mathbf{d} = [2 \ 1 \ 4 \ -1]^T$.

5. Let $A_1 = \begin{bmatrix} 1 & 0 & 0 \\ 0 & 1 & 0 \\ -4 & 0 & 1 \end{bmatrix}$ and $A_2 = \begin{bmatrix} 1 & 2 & 0 \\ 0 & 1 & 0 \\ 0 & 0 & 1 \end{bmatrix}$.

(a). Find the inverses of both matrices.

(b). Find the inverses of $A_1 A_2$ and $A_2 A_1$.

6. Use MATLAB® to confirm any by-hand calculations done in Exercises 2–5.

7. Prove: if A has an inverse, then there is only one inverse matrix.

8. Prove: $(A^{-1})^{-1} = A$ and $(A^{-1})^T = (A^T)^{-1}$.

9. Let $\mathbf{u}$ and $\mathbf{v}$ be column vectors and A be an $n \times n$ real matrix.

(a). If $1 + \mathbf{v}^T \mathbf{u} \neq 0$, then show

$$(I + \mathbf{u}\mathbf{v}^T)^{-1} = I - \frac{1}{(1 + \mathbf{v}^T\mathbf{u})}\mathbf{u}\mathbf{v}^T.$$

(b). Use the above to show

$$(A + \mathbf{u}\mathbf{v}^T)^{-1} = A^{-1} - \frac{1}{(1 + \mathbf{v}^T A^{-1}\mathbf{u})}A^{-1}\mathbf{u}\mathbf{v}^T A^{-1}.$$

(c). Let U and V be $n \times k$ matrices and assume A and $I + V^T A^{-1} U$ have inverses. Show the *Sherman–Morrison–Woodbury* identity holds

$$(A + UV^T)^{-1} = A^{-1} - A^{-1}U(I + V^T A^{-1}U)^{-1}V^T A^{-1}.$$

10. Consider the five-bar truss in Figure 1.5.1, see the MATLAB® code bridge.m

(a). Formulate an algebraic system for the eight forces.

(b). Use the MATLAB® commands A\d and inv(A) to solve this when $w = -10000$ and variable $\theta = \pi/6, \pi/4$ and $2\pi/6$.

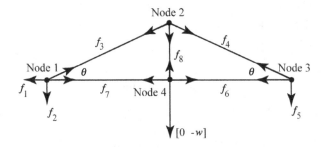

FIGURE 1.5.1
Five-bar truss.

11. The operation counts for triangular solves and Gauss elimination used the
following:

$$1 + 2 + \cdots + n = n(n+1)/2 \text{ and}$$
$$1 + 2^2 + \cdots + n^2 = n(n+1)(2n+1)/6.$$

The evolution of the first formula can be done by noting

$$S_n = 1 + 2 + \cdots + (n-1) + n = n + (n-1) + \cdots + 2 + 1.$$

Add the two version of the sum to get

$$2S_n = n(n+1).$$

The second formula is not evident, but it appears the summation may
have the form

$$S(n) = a_3 n^3 + a_2 n^2 + a_1 n + a_0.$$

First, choose the coefficients so that $S(1) = 1, S(2) = 5, S(3) = 14$ and
$S(4) = 30$:

$$\begin{bmatrix} 1 & 1 & 1 & 1 \\ 1 & 2 & 4 & 8 \\ 1 & 3 & 9 & 27 \\ 1 & 4 & 16 & 64 \end{bmatrix} \begin{bmatrix} a_0 \\ a_1 \\ a_2 \\ a_3 \end{bmatrix} = \begin{bmatrix} 1 \\ 5 \\ 14 \\ 30 \end{bmatrix}.$$

Second, these particular coefficients may or may not work for other choices
of n! In order to confirm that the proposed formula works, one can use
mathematical induction.

(a). Use Gauss elimination to solve the above system.

(b). Use mathematical induction to show the proposed formula is valid
for all n.

1.6 Determinants and Cramer's Rule

Determinants of 2×2 and 3×3 matrices were introduced as natural ways to compute areas and volumes between vectors, which are columns in the matrix. Also, the unique solution of algebraic systems with two or three unknowns is characterized by the determinant of the 2×2 or 3×3 coefficient matrix not being zero. In this section these topics will be reviewed and extended to $n \times n$ matrices. Basic properties of determinants will be presented and inverse matrix computation via determinants will be formulated via Cramer's rule.

1.6.1 Determinants for 2×2 and 3×3 matrices

Consider a 2×2 matrix with column vectors **a** and **b**

$$A = \begin{bmatrix} a_{11} & a_{12} \\ a_{21} & a_{22} \end{bmatrix} = [\mathbf{a} \ \mathbf{b}].$$

The area of the parallelogram formed by the vectors, which are not parallel, is plus or minus the determinant of A

$$\det(A) \equiv a_{11}a_{22} - a_{12}a_{21}.$$

Determinants can be used to solve 2×2 algebraic systems $A\mathbf{x} = \mathbf{d}$ by Cramer's rule when $\det(A) \neq 0$

$$x_1 = \det(\begin{bmatrix} d_1 & a_{12} \\ d_2 & a_{22} \end{bmatrix})/\det(A) \text{ and } x_2 = \det(\begin{bmatrix} a_{11} & d_1 \\ a_{21} & d_2 \end{bmatrix})/\det(A).$$

Notice the determinant does not change if the rows and columns are interchanged

$$\det(\begin{bmatrix} a_{11} & a_{21} \\ a_{12} & a_{22} \end{bmatrix}) = \det(A^T) = \det(A).$$

And, $\det(A)$ changes sign if the rows are interchanged

$$\det(\begin{bmatrix} a_{21} & a_{22} \\ a_{11} & a_{12} \end{bmatrix}) = -\det(A).$$

The extension of determinants to 3×3 matrices can be done in an inductive way. Consider a 3×3 matrix with column vectors **a**, **b** and **c**

$$A = \begin{bmatrix} a_{11} & a_{12} & a_{13} \\ a_{21} & a_{22} & a_{23} \\ a_{31} & a_{32} & a_{33} \end{bmatrix} = [\mathbf{a} \ \mathbf{b} \ \mathbf{c}].$$

The volume of the parallelepiped formed be the three vectors is plus or minus the dot product of **a** with **b** × **c**, which can be written as the determinant of A

$$\det(A) \equiv a_{11}(a_{22}a_{33} - a_{23}a_{32})$$
$$-a_{12}(a_{21}a_{33} - a_{23}a_{31})$$
$$+a_{13}(a_{21}a_{32} - a_{22}a_{31}), \quad \text{expansion by row one}$$
$$= a_{11}(a_{22}a_{33} - a_{23}a_{32})$$
$$-a_{21}(a_{12}a_{33} - a_{13}a_{32})$$
$$+a_{31}(a_{12}a_{23} - a_{13}a_{22}), \quad \text{expansion by column one}$$
$$= \mathbf{a} \bullet (\mathbf{b} \times \mathbf{c}).$$

Here we have interchanged rows and columns and the determinant does not change. That is, $\det(A) = \det(A^T)$ where A^T is called the *transpose of A* and is defined as

$$A^T \equiv \begin{bmatrix} a_{11} & a_{21} & a_{31} \\ a_{12} & a_{22} & a_{32} \\ a_{13} & a_{23} & a_{33} \end{bmatrix} = \begin{bmatrix} \mathbf{a}^T \\ \mathbf{b}^T \\ \mathbf{c}^T \end{bmatrix}.$$

The full expansion of the 3 × 3 matrix can be reorganized

$$\det(A) \equiv a_{11}a_{22}a_{33} - a_{11}a_{23}a_{32}$$
$$-a_{12}a_{21}a_{33} + a_{12}a_{23}a_{31}$$
$$+a_{13}a_{21}a_{32} - a_{13}a_{22}a_{31}.$$

There are $n! = 3! = 6$ permutations of the $n = 3$ indices. The sign changes are in pairs and the following table input into 6 × 4 matrix $\sigma(k, indices)$

$$\begin{bmatrix} 1 & 1 & 2 & 3 \\ 2 & 1 & 3 & 2 \\ 4 & 2 & 1 & 3 \\ 3 & 2 & 3 & 1 \\ 5 & 3 & 1 & 2 \\ 6 & 3 & 2 & 1 \end{bmatrix}$$

The above full expansion can be expressed as

$$\det(A) = \sum_{k=1}^{k=3!} (-1)^{k+1} a_{1,\sigma(k,1)} a_{2,,\sigma(k,2)} a_{3,\sigma(k,3)}$$

Cramer's rule, for $\det(A) \neq 0$, allows one to easily solve the 3 × 3 algebraic systems $A\mathbf{x} = \mathbf{d}$

$$x_1 = \det([\mathbf{d} \ \ \mathbf{b} \ \ \mathbf{c}])/\det(A),$$
$$x_2 = \det([\mathbf{a} \ \ \mathbf{d} \ \ \mathbf{c}])/\det(A) \text{ and}$$
$$x_2 = \det([\mathbf{a} \ \ \mathbf{b} \ \ \mathbf{d}])/\det(A).$$

The justification of the formula for the first unknown follows from the properties of the cross product and the column version of a matrix-vector product

$$A\mathbf{x} = \mathbf{d}$$
$$x_1\mathbf{a} + x_2\mathbf{b} + x_3\mathbf{c} = \mathbf{d}$$
$$x_1\mathbf{a} = \mathbf{d} - x_2\mathbf{b} - x_3\mathbf{c}.$$

Recall the cross product $\mathbf{b} \times \mathbf{c}$ is perpendicular to both $\mathbf{b}$ and $\mathbf{c}$. Use this fact and take the dot product of both sides with $\mathbf{b} \times \mathbf{c}$

$$(x_1\mathbf{a}) \bullet (\mathbf{b} \times \mathbf{c}) = (\mathbf{d} - x_2\mathbf{b} - x_3\mathbf{c}) \bullet (\mathbf{b} \times \mathbf{c})$$
$$(\mathbf{a} \bullet (\mathbf{b} \times \mathbf{c}))x_1 = \mathbf{d} \bullet (\mathbf{b} \times \mathbf{c}) - \mathbf{b} \bullet (\mathbf{b} \times \mathbf{c})x_2 - \mathbf{c} \bullet (\mathbf{b} \times \mathbf{c})x_3)$$
$$\det(A)x_1 = \det([\mathbf{d} \quad \mathbf{b} \quad \mathbf{c}]) - 0x_2 - 0x_3.$$

The derivations of the equations for the other two unknowns are similar. Furthermore, these concepts generalize to algebraic systems with n unknowns.

1.6.2   Determinant of an $n \times n$ matrix

The determinant of an $n \times n$ matrix is defined inductively from determinant of $(n-1) \times (n-1)$ matrices in much the same way as for 3×3 matrices. Let M_{ij} be the determinant of the $(n-1) \times (n-1)$ matrix formed by deleting row i and column j of A. The determinant of A may be computed by expanding about any row or column. Here, we use the first row of the matrix. Any of the first three properties (listed below) can be used as the definition of a determinant of a matrix. This is the result of associative, commutative and distributive arithmetic rules as was previously illustrated for 3×3 matrices.

The elementary matrices are the matrices in properties 5-7 corresponding to the three row operations in Gauss elimination. Note, $\det(P_{ij}) = -1, \det(E_{ij}(a)) = 1$ and $\det(E_i(c)) = c$. Let $\mathbf{E}_k$ be any of the three elementary matrices, and then properties 5-7 can be written as

$$\det(\mathbf{E}_1 B) = \det(\mathbf{E}_1)\det(B).$$

Replace B by $\mathbf{E}_1 B$ and use the associative property for matrix products

$$\det(\mathbf{E}_2(\mathbf{E}_1 B)) = \det(\mathbf{E}_2)\det(\mathbf{E}_1 B)$$
$$\det((\mathbf{E}_2\mathbf{E}_1)B) = \det(\mathbf{E}_2)\det(\mathbf{E}_1)\det(B).$$

Continue this to get

$$\det((\mathbf{E}_K \cdots \mathbf{E}_1)B) = \det(\mathbf{E}_K) \cdots \det(\mathbf{E}_1)\det(B).$$

For the special case $B = I$ use $\det(B) = \det(I) = 1$ so that

$$\det(A) = \det(\mathbf{E}_K \cdots \mathbf{E}_1) = \det(\mathbf{E}_K) \cdots \det(\mathbf{E}_1).$$

This gives property 8.

The last property follows from matrices with inverses and their inverse can be represented as a product of elementary matrices. Also, $EA = U$ with nonzero diagonal components gives another product of elementary matrices, F, such that $FEA = FU = I$. Then $\det(EA) = \det(E)\det(A) = \det(U)$. Since $\det(E) \neq 0$ and $\det(U) \neq 0$, $\det(E)\det(A) = \det(U)$ implies $\det(A) \neq 0$. Moreover, $\det(FE)\det(A) = \det(I) = 1$ and $A^{-1} = FE$.

Definition. *The row expansion of a determinant of an $n \times n$ matrix A is along the first row*

$$\det(A) \equiv a_{11}(-1)^{1+1}M_{11} + a_{12}(-1)^{1+2}M_{12} + \cdots + a_{1n}(-1)^{1+n}M_{1n}.$$

The coefficients of a_{ij}, $(-1)^{i+j}M_{ij}$, are called cofactors.

Determinant Properties:

1. $\det(A) = a_{i1}(-1)^{i+1}M_{i1} + a_{i2}(-1)^{i+2}M_{i2} + \cdots + a_{in}(-1)^{i+n}M_{in}$ (row i expansion).

2. $\det(A) = a_{1j}(-1)^{1+j}M_{1j} + a_{2j}(-1)^{2+j}M_{2j} + \cdots + a_{nj}(-1)^{n+j}M_{nj}$ (column j expansion).

3. $\det(A) = \sum_{k=1}^{k=n!}(-1)^{k+1}a_{1,\sigma(k,1)}a_{2,,\sigma(k,2)} \cdots a_{n,\sigma(k,n)}$ (full expansion).

4. The determinant of the transpose matrix is $\det(A^T) = \det(A)$.

5. The determinant of the matrix formed by an interchange of two rows is $\det(P_{ij}A) = -\det(A)$.

6. The determinant of the matrix formed by adding a multiple of a row j to row i is unchanged $\det(E_{ij}(a)A) = \det(A)$.

7. The determinant of the matrix formed by multiplying a row by c is $\det(E_i(c)A) = c\det(A)$.

8. If A is a product of elementary matrices, then

$$\det(AB) = \det(A)\det(B).$$

9. If A has an inverse, then

$$1 = \det(I) = \det(A)\det(A^{-1}) \text{ and } \det(A) \neq 0.$$

The following example illustrates how they may be used to reduce the computational burden.

Example 1.6.1. First, the determinant is evaluated by expanding down column one, which has only two nonzero components,

$$\det(A) = \det(\begin{bmatrix} 1 & -1 & 1 \\ 1 & 0 & -3 \\ 0 & -2 & -3 \end{bmatrix})$$

$$= (-1)^2(1)\det(\begin{bmatrix} 0 & -3 \\ -2 & -3 \end{bmatrix}) + (-1)^{2+1}(1)\det(\begin{bmatrix} -1 & 1 \\ -2 & -3 \end{bmatrix})$$

$$= -6 - 5 = -11.$$

One can also use the following elementary row operations to transform the matrix to upper triangular form

$$E_{32}(2) \, E_{11}(-1) \, A = U = \begin{bmatrix} 1 & -1 & 1 \\ 0 & 1 & -4 \\ 0 & 0 & -11 \end{bmatrix}.$$

Note $\det(E_{32}(2)(E_{11}(-1) \, A)) = \det(E_{11}(-1) \, A) = \det(A) = \det(U)$. The determinant of an upper triangular matrix can be expanded by the left column so that the determinant is the product of the diagonal components. Therefore, $\det(A) = \det(U) = (1)(1)(-11) = -11$.

Example 1.6.2. Consider the 4×4 matrix for the heat conduction problem and expand it by column one

$$\det(A) = \det(\begin{bmatrix} 2 & -1 & 0 & 0 \\ -1 & 2 & -1 & 0 \\ 0 & -1 & 2 & -1 \\ 0 & 0 & -1 & 2 \end{bmatrix})$$

$$= 2\det(\begin{bmatrix} 2 & -1 & 0 \\ -1 & 2 & -1 \\ 0 & -1 & 2 \end{bmatrix}) - (-1)\det(\begin{bmatrix} -1 & 0 & 0 \\ -1 & 2 & -1 \\ 0 & -1 & 2 \end{bmatrix})$$

$$= 2(4) - 3 = 5.$$

This determinant may also be computed by elementary matrix transformation to obtain an upper triangular form

$$E_{43}(3/4) \, E_{32}(2/3) \, E_{21}(1/2) \, A = U = \begin{bmatrix} 2 & -1 & 0 & 0 \\ 0 & 3/2 & -1 & 0 \\ 0 & 0 & 4/3 & -1 \\ 0 & 0 & 0 & 5/4 \end{bmatrix}.$$

$$\det(E_{43}(3/4) \, E_{32}(2/3) \, E_{21}(1/2) \, A) = \det(A) = \det(U) = 2(3/2)(4/3)(5/4)$$
$$= 5.$$

1.6.3   Cramer's rule and inverses

In order to generalize Cramer's rule from 3×3 to $n \times n$ matrices, we must generalize perpendicular three dimensional vectors to n-dimensional vectors. Two n-dimensional column vectors $\mathbf{a} = [a_1 \cdots a_n]^T$ and $\mathbf{b} = [b_1 \cdots b_n]^T$ are called *perpendicular or orthogonal* if and only if the dot product is zero, that is, $\mathbf{a}^T\mathbf{b} = \mathbf{a} \bullet \mathbf{b} \equiv a_1 b_1 + \cdots + a_n b_n = 0$. Let A be an $n \times n$ matrix with components a_{ij}, and denote its column j by $\mathbf{a}_{:j}$ so that $A = [\mathbf{a}_{:1} \cdots \mathbf{a}_{:n}]$. The matrix-vector product $A\mathbf{x}$ can be written as a linear combination of the column vectors, see Theorem 1.2.1,

$$A\mathbf{x} = \mathbf{d}$$
$$x_1 \mathbf{a}_{:1} + \cdots + x_n \mathbf{a}_{:n} = \mathbf{d}.$$

In order find the first unknown x_1, we must solve

$$x_1 \mathbf{a}_{:1} = \mathbf{d} - (x_2 \mathbf{a}_{:2} + \cdots + x_n \mathbf{a}_{:n}).$$

In the case $n = 3$, we computed the dot product of both sides with the cross product of the second and third columns. Since the cross product of two vectors are perpendicular to both the vectors, the dot product on the right side simplified to the $\det([\mathbf{d} \quad \mathbf{a}_{:2} \quad \mathbf{a}_{:3}])$ and, consequently, $x_1 = \det([\mathbf{d} \quad \mathbf{a}_{:2} \quad \mathbf{a}_{:3}])/\det(A)$.

The generalization of the cross product to higher dimensions requires an n-dimensional vector that will be perpendicular to $n - 1$ given n-dimensional vectors. Let the $n - 1$ given column vectors $j = 2, \cdots, n$ in the matrix A be $\mathbf{a}_{:2}, \cdots, \mathbf{a}_{:n}$. We now claim the column vector $\mathbf{c}$ of cofactors with $j = 1$ will be perpendicular to each of these vectors

$$\mathbf{c} = \begin{bmatrix} (-1)^{1+1} M_{11} \\ (-1)^{2+1} M_{21} \\ \vdots \\ (-1)^{n+1} M_{n1} \end{bmatrix}.$$

To see how this is established, consider the reduced notational burden of the case $n = 4$. We must find x_1 in

$$x_1 \mathbf{a}_{:1} = \mathbf{d} - (x_2 \mathbf{a}_{:2} + x_3 \mathbf{a}_{:3} + x_4 \mathbf{a}_{:4}).$$

Compute the dot product of both sides with $\mathbf{c}$

$$(x_1 \mathbf{a}_{:1}) \bullet \mathbf{c} = (\mathbf{d} - (x_2 \mathbf{a}_{:2} + x_3 \mathbf{a}_{:3} + x_4 \mathbf{a}_{:4})) \bullet \mathbf{c} \qquad (1.6.1)$$
$$(\mathbf{a}_{:1} \bullet \mathbf{c})x_1 = \mathbf{d} \bullet \mathbf{c} - (\mathbf{a}_{:2} \bullet \mathbf{c})x_2 - (\mathbf{a}_{:3} \bullet \mathbf{c})x_3 - (\mathbf{a}_{:4} \bullet \mathbf{c})x_4.$$

All five of the dot products can be expressed as determinants.

$$\mathbf{a}_{:1} \bullet \mathbf{c} = a_{11}(-1)^{1+1} M_{11} + a_{21}(-1)^{2+1} M_{21} +$$
$$a_{31}(-1)^{3+1} M_{31} + a_{41}(-1)^{4+1} M_{41}$$
$$= \det(A) \quad \text{(expansion by column one) and similarly}$$
$$\mathbf{d} \bullet \mathbf{c} = \det([\mathbf{d} \quad \mathbf{a}_{:2} \quad \mathbf{a}_{:3} \quad \mathbf{a}_{:4}]).$$

The other three determinants are also expanded by column one and are equal to zero because they have two identical columns.

$$(\mathbf{a}_{:2} \bullet \mathbf{c}) = a_{12}(-1)^{1+1}M_{11} + a_{22}(-1)^{2+1}M_{21} +$$
$$a_{32}(-1)^{3+1}M_{31} + a_{42}(-1)^{4+1}M_{41}$$

$$= \det\left(\begin{bmatrix} a_{12} & a_{12} & a_{13} & a_{14} \\ a_{22} & a_{22} & a_{23} & a_{24} \\ a_{32} & a_{32} & a_{33} & a_{34} \\ a_{42} & a_{42} & a_{43} & a_{44} \end{bmatrix} \right) = \det([\mathbf{a}_{:2} \ \mathbf{a}_{:2} \ \mathbf{a}_{:3} \ \mathbf{a}_{:4}]) = 0,$$

$$(\mathbf{a}_{:3} \bullet \mathbf{c}) = \det([\mathbf{a}_{:3} \ \mathbf{a}_{:2} \ \mathbf{a}_{:3} \ \mathbf{a}_{:4}]) = 0 \text{ and}$$
$$(\mathbf{a}_{:4} \bullet \mathbf{c}) = \det([\mathbf{a}_{:4} \ \mathbf{a}_{:2} \ \mathbf{a}_{:3} \ \mathbf{a}_{:4}]) = 0.$$

Then, equation (1.6.1) becomes

$$\det(A)x_1 = \det([\mathbf{d} \ \mathbf{a}_{:2} \ \mathbf{a}_{:3} \ \mathbf{a}_{:4}]).$$

By using other columns of cofactors one can solve for the other unknowns.

Theorem 1.6.1. *(Cramer's Rule) Let A be an $n \times n$ matrix, and let A_i be an $n \times n$ matrix with the same columns of A except column i is replaced by column $\mathbf{d}$. If $\det(A) \neq 0$, then the solution of $A\mathbf{x} = \mathbf{d}$ is given by*

$$x_i = \det(A_i)/\det(A).$$

Cramer's rule is useful for smaller dimensional matrices, special calculations where only a few of the unknown components are required and some theoretical considerations. However, it does allows one to give an explicit formula for the inverse matrix

$$AA^{-1} = I$$
$$A[\mathbf{b}_1 \cdots \mathbf{b}_n] = [\mathbf{e}_1 \cdots \mathbf{e}_n] \text{ or equivalently}$$
$$A\mathbf{b}_1 = \mathbf{e}_1, \cdots, A\mathbf{b}_n = \mathbf{e}_n.$$

In order to find the first column $\mathbf{b}_1$ of the inverse matrix, apply Cramer's rule with $\mathbf{d} = \mathbf{e}_1$

$$b_{i1} = \det(A_i)/\det(A) \text{ where}$$
$$b_{11} = \det([\mathbf{e}_1 \ \mathbf{a}_{:2} \cdots \mathbf{a}_{:n}])/\det(A) = (-1)^{1+1}M_{11}/\det(A)$$
$$b_{21} = \det([\mathbf{a}_{:1} \ \mathbf{e}_1 \cdots \mathbf{a}_{:n}])/\det(A) = (-1)^{1+2}M_{12}/\det(A)$$

$$\vdots$$

$$b_{n1} = \det([\mathbf{a}_{:1} \ \mathbf{a}_{:2} \cdots \mathbf{e}_1])/\det(A) = (-1)^{1+n}M_{1n}/\det(A).$$

The other columns of the inverse matrix are similar.

Theorem 1.6.2. *(Inverse Matrix via Determinants) Let A be an $n \times n$ matrix with nonzero determinant. Then the inverse matrix exists and column j in the inverse matrix is*

$$\mathbf{b}_j = \begin{bmatrix} b_{1j} \\ b_{2j} \\ \vdots \\ b_{nj} \end{bmatrix} = \begin{bmatrix} (-1)^{j+1} M_{j1}/\det(A) \\ (-1)^{j+2} M_{j2}/\det(A) \\ \vdots \\ (-1)^{j+n} M_{jn}/\det(A) \end{bmatrix}.$$

Example 1.6.3. Use determinants to find the inverse of

$$A = \begin{bmatrix} 1 & 2 & 1 \\ 1 & 3 & 2 \\ 1 & 0 & 1 \end{bmatrix}.$$

Since $\det(A) = 2 \neq 0$, the inverse matrix exists. The first column in the inverse matrix $A^{-1} = B = [\mathbf{b}_1 \ \mathbf{b}_2 \ \mathbf{b}_3]$ is

$$\mathbf{b}_1 = \begin{bmatrix} b_{11} \\ b_{21} \\ b_{31} \end{bmatrix} = \begin{bmatrix} (-1)^{1+1} M_{11}/2 \\ (-1)^{1+2} M_{12}/2 \\ (-1)^{1+3} M_{13}/2 \end{bmatrix} = \begin{bmatrix} \det\left(\begin{bmatrix} 3 & 2 \\ 0 & 1 \end{bmatrix}\right)/2 \\ -\det\left(\begin{bmatrix} 1 & 2 \\ 1 & 1 \end{bmatrix}\right)/2 \\ \det\left(\begin{bmatrix} 1 & 3 \\ 1 & 0 \end{bmatrix}\right)/2 \end{bmatrix} = \begin{bmatrix} 3/2 \\ 1/2 \\ -3/2 \end{bmatrix}.$$

The other two columns in the inverse matrix may be computed in a similar fashion to yield

$$A^{-1} = (1/2) \begin{bmatrix} 3 & -2 & 1 \\ 1 & 0 & -1 \\ -3 & 2 & 1 \end{bmatrix}.$$

If any column of an $n \times n$ matrix is a linear combination of the other columns, then the determinant of the matrix must be zero. Equivalently, if the determinant of the matrix is not zero, then no column of the matrix A is a linear combination of the other columns (the columns are *linearly independent*)

$$A\mathbf{x} = x_1 \mathbf{a}_{:1} + \cdots + x_n \mathbf{a}_{:n} = \mathbf{0} \text{ implies } \mathbf{x} = \mathbf{0}.$$

Each column of such matrices must have at least one nonzero component so that row operations can be done to transform the matrix to upper triangular form, $EA = U$. This eventually leads to the following theorem, which gives a characterization of matrices that have inverses. We have given a proof using row operations on the sequence of columns of the matrix. Here we given an alternate proof using mathematical induction on matrix size starting with 2×2, and showing if the equivalence is true for $n \times n$, then it must true for $(n + 1) \times (n + 1)$.

Theorem 1.6.3. *(Inverse Matrix Equivalence)* *Let A be an n × n matrix. The following are equivalent:*

(i). *A has an inverse,*

(ii). $A\mathbf{x} = 0_{n \times 1}$ *implies* $\mathbf{x} = 0_{n \times 1}$,

(iii). *diagonal components of* $EA = U$ *are not zero and*

(iv). $\det(A) \neq 0$.

Proof. First, consider a 2 × 2 matrix that has an inverse and permutation matrix such that
$$PA = \begin{bmatrix} b & e \\ f & c \end{bmatrix} \text{ and } b \neq 0.$$

Use a single row operation given by the elementary matrix
$$\begin{bmatrix} 1 & 0 \\ -f/b & 1 \end{bmatrix} \begin{bmatrix} b & e \\ f & c \end{bmatrix} = \begin{bmatrix} b & e \\ 0 & \widehat{c} \end{bmatrix}$$

where $\widehat{c} = c - (f/b)e$. The equivalence of the four is easy to see.

Second, assume the four conditions are equivalent for n and show they are equivalent for the matrices of dimension $n + 1$. Item (i) for $n + 1$ implies item (ii) for $n + 1$, which gives a permutation matrix, P, such that
$$P_1 A = \begin{bmatrix} b & e^T \\ f & C \end{bmatrix}$$

where b is a single number, f and e are $n \times 1$ vectors and C is $n \times n$ matrix. The product of the next n row operations can be done by a block elementary matrix, often called a *Gauss transform*,
$$G_1 = \begin{bmatrix} 1 & 0_{1 \times n} \\ -f/b & I_n \end{bmatrix}.$$

This zeros out the vector f and changes the matrix C
$$G_1 P_1 A = \begin{bmatrix} b & e^T \\ 0_{n \times 1} & \widehat{C} \end{bmatrix}$$

where $\widehat{C} = C - (f/b)e^T$. The $n \times n$ matrix $\widehat{C}$ has an inverse because
$$G_1 P_1 A \begin{bmatrix} 1 & -e^T/b \\ 0_{n \times 1} & I_n \end{bmatrix} = \begin{bmatrix} b & e^T \\ 0_{n \times 1} & \widehat{C} \end{bmatrix} \begin{bmatrix} 1 & -e^T/b \\ 0_{n \times 1} & I_n \end{bmatrix}$$
$$= \begin{bmatrix} b & 0_{1 \times n} \\ 0_{n \times 1} & \widehat{C} \end{bmatrix}.$$

Since the left side is a product the matrices with inverses, the right side $n \times n$ matrix $\widehat{C}$ must have an inverse. Now, use the induction assumption that all four items are true for the matrix $n \times n$ matrix $\widehat{C}$: $\widehat{C}^{-1}$ exists, $\widehat{C}\widehat{x} = 0_{n \times 1}$ implies $\widehat{x} = 0_{n \times 1}$, $\widehat{E}\widehat{C} = \widehat{U}$ and $\det(\widehat{C}) \neq 0$.

$$G_1 P_1 A = \begin{bmatrix} b & e^T \\ 0_{n \times 1} & \widehat{E}\widehat{U} \end{bmatrix}$$

$$= \begin{bmatrix} 1 & 0_{1 \times n} \\ 0_{n \times 1} & \widehat{E} \end{bmatrix} \begin{bmatrix} b & e^T \\ 0_{n \times 1} & \widehat{U} \end{bmatrix}.$$

The $\widehat{E}$ is a product the elementary matrices and the $\widehat{U}$ is an upper triangular matrix with nonzero diagonal components. This proves item (iii) for $n + 1$ because

$$\begin{bmatrix} 1 & 0_{1 \times n} \\ 0_{n \times 1} & \widehat{E}^{-1} \end{bmatrix} G_1 P_1 A = \begin{bmatrix} b & e^T \\ 0_{n \times 1} & \widehat{U} \end{bmatrix}.$$

Item (iv) follows from the above. The left side has the form EA where E is a product of elementary matrices. Apply property eight for matrix products to get

$$\det(EA) = \det(E)\det(A) = \det(\begin{bmatrix} b & e^T \\ 0_{n \times 1} & \widehat{U} \end{bmatrix}).$$

Because both $\det(E)$ and the right side $b\det(\widehat{U})$ are not zero, $\det(A)$ cannot be zero.

Item (i) for $n + 1$ follows from item (iv) for $n + 1$ from both $\det(E)$ and $\det(A)$ being not zero. This means the determinant of the above upper triangular matrix must not be zero. But, the determinant is a product of the diagonal components and so none of the diagonal components can be zero. The means $EA = U$ for $n + 1$ has invertible E and U and so A has an inverse. ∎

1.6.4   Determinants using MATLAB®

Return to Example 1.6.2 where the matrix is 4×4 and we want to solve $Au = d = [470 \ 400 \ 400 \ 470]^T$ using Cramer's rule. The first two unknowns are as follows.

```
>> A =[2 -1  0  0;
       -1  2 -1  0;
        0 -1  2 -1;
        0  0 -1  2];
>> d = [470 400 400 470]';
>> det(A)
    ans =
         5
>> A1 = [d A(:,2) A(:,3) A(:,4)]
```

A1 =
```
     470 -1  0  0
     400  2 -1  0
     400 -1  2 -1
     470  0 -1  2
```
>> u1 = det(A1)/5
u1 =
```
     870
```
>> A2 = [A(:,1) d A(:,3) A(:,4)]
A2 =
```
      2 470  0  0
     -1 400 -1  0
      0 400  2 -1
      0 470 -1  2
```
>> u2 = det(A2)/5
u2 =
```
    1270
```

1.6.5 Exercises

1. Consider Example 1.6.1 and verify

   (a). $\det(A^T) = \det(A)$ and

   (b). $\det(P_{23}A) = -\det(A)$, that is, interchange rows 2 and 3.

2. Consider Example 1.6.1 and verify

   (a). $\det(E_{32}(10)\ A) = \det(A)$, that is, multiply row 2 by 10 and add to row 3 and

   (b). $\det(E_2(4)A) = 4\det(A)$, that is, multiply row 2 by 4.

3. Consider Example 1.6.1 and compute the determinant

   (a). by expanding row $i = 3$ and

   (b). by expanding column $j = 2$.

4. Let A be the 4×4 matrix

$$A = \begin{bmatrix} 3 & -1 & 0 & 0 \\ -1 & 3 & -1 & 0 \\ 0 & -1 & 3 & -1 \\ 0 & 0 & -1 & 3 \end{bmatrix}.$$

   (a). Find the determinant by expanding row one.

   (b). Use elementary matrices to transform the matrix to upper triangular form and then compute the determinant.

5. Let A be the 4×4 matrix

$$A = \begin{bmatrix} 1 & 2 & 1 & 0 \\ 0 & 1 & 0 & 0 \\ 8 & 16 & 11 & 0 \\ 1 & 2 & 3 & 7 \end{bmatrix}.$$

(a). Find the determinant by expanding row two.

(b). Use elementary matrices to transform the matrix to upper triangular form and then compute the determinant.

6. In Example 1.6.3 use determinants to compute the other two columns of the inverse matrix.

7. In Example 1.6.1 use determinants to compute inverse matrix.

8. Consider the derivation of Cramer's rule for $n = 4$ (see Subsection 1.6.3) and justify the equation for x_4.

9. Use MATLAB® to confirm any calculations done in Exercises 4–7.

10. Consider Theorem 1.6.3. Prove the equivalence for $n = 2$.

2

Matrix Factorizations

There are various classes of matrices that have an inverse. Triangular matrices with invertible diagonal blocks have inverses. Strictly diagonally dominant matrices also have inverses. In this chapter four additional classes will be introduced: Schur complement, nonsingular and LU factors, M-matrix and symmetric positive definite (SPD).

2.1   The Schur Complement

The goal is to extend the Gauss transform in Gauss elimination to larger blocks

$$G_1 P_1 A = \begin{bmatrix} 1 & 0 \\ -f/b & I_{n-1} \end{bmatrix} \begin{bmatrix} b & e^T \\ f & C \end{bmatrix} = \begin{bmatrix} b & e^T \\ 0 & \widehat{C} \end{bmatrix}.$$

In order to extend this to $n \times n$ matrices, consider a 2×2 block matrix where the diagonal blocks are square but may not have the same dimension

$$A = \begin{bmatrix} B & E^T \\ F & C \end{bmatrix}. \tag{2.1.1}$$

In general A is $n \times n$ with $n = k + m$, B is $k \times k$, C is $m \times m$, E^T is $k \times m$ and F is $m \times k$. If B has an inverse, then we can multiply block row one by FB^{-1} and subtract it from block row two. This is equivalent to multiplication of A by a *block elementary matrix* of the form

$$\begin{bmatrix} I_k & 0 \\ -FB^{-1} & I_m \end{bmatrix}.$$

If $Ax = d$ is viewed in block form, then

$$\begin{bmatrix} B & E^T \\ F & C \end{bmatrix} \begin{bmatrix} x_1 \\ x_2 \end{bmatrix} = \begin{bmatrix} d_1 \\ d_2 \end{bmatrix}. \tag{2.1.2}$$

The above block elementary matrix multiplication gives

$$\begin{bmatrix} B & E^T \\ 0 & C - FB^{-1}E^T \end{bmatrix} \begin{bmatrix} x_1 \\ x_2 \end{bmatrix} = \begin{bmatrix} d_1 \\ d_2 - FB^{-1}d_1 \end{bmatrix}. \tag{2.1.3}$$

DOI: 10.1201/9781003304128-2

So, if the block upper triangular matrix has an inverse, then this last block equation can be solved.

The following basic properties of square matrices play an important role in the solution of (2.1.2). These properties follow directly from the definition of an inverse matrix.

Theorem 2.1.1. *(Inverse Matrix Properties) Let B and C be square matrices with dimensions k and m, respectively, and with inverses. The off diagonal matrices, E and F must be $m \times k$. Then the following equalities hold:*

1. $\begin{bmatrix} B & 0 \\ 0 & C \end{bmatrix}^{-1} = \begin{bmatrix} B^{-1} & 0 \\ 0 & C^{-1} \end{bmatrix}$,

2. $\begin{bmatrix} I_k & 0 \\ F & I_m \end{bmatrix}^{-1} = \begin{bmatrix} I_k & 0 \\ -F & I_m \end{bmatrix}$,

3. $\begin{bmatrix} B & 0 \\ F & C \end{bmatrix} = \begin{bmatrix} B & 0 \\ 0 & C \end{bmatrix} \begin{bmatrix} I_k & 0 \\ C^{-1}F & I_m \end{bmatrix}$,

4. $\begin{bmatrix} B & 0 \\ 0 & C \end{bmatrix} = \begin{bmatrix} B & E^T \\ 0 & C \end{bmatrix} \begin{bmatrix} I_k & -B^{-1}E^T \\ 0 & I_m \end{bmatrix}$ and

5. $\begin{bmatrix} B & 0 \\ F & C \end{bmatrix}^{-1} = \begin{bmatrix} B^{-1} & 0 \\ -C^{-1}FB^{-1} & C^{-1} \end{bmatrix}$.

Definition. *Let A have the form in (2.1.1) and B be nonsingular. The Schur complement of B in A is $\widehat{C} \equiv C - FB^{-1}E^T$.*

Theorem 2.1.2. *(Schur Complement Existence) Consider A as in (2.1.1) and let B have an inverse. A has an inverse if and only if the Schur complement of B in A, $\widehat{C}$, has an inverse.*

Proof. The proof of the Schur complement theorem is a direct consequence of using a block elementary row operation to get a zero matrix in the block row 2 and column 1 position, the 21-block,

$$\begin{bmatrix} I_k & 0 \\ -FB^{-1} & I_m \end{bmatrix} \begin{bmatrix} B & E^T \\ F & C \end{bmatrix} = \begin{bmatrix} B & E^T \\ 0 & \widehat{C} \end{bmatrix}.$$

Assume that A has an inverse and show the Schur complement must have an inverse. Since the two matrices on the left side have inverses, the matrix on the right side has an inverse. Because B has an inverse, the Schur complement is defined. Multiply the above on the right by

$$\begin{bmatrix} I_k & -B^{-1}E^T \\ 0 & I_m \end{bmatrix}$$

to get

$$\begin{bmatrix} B & E^T \\ 0 & \widehat{C} \end{bmatrix} \begin{bmatrix} I_k & -B^{-1}E^T \\ 0 & I_m \end{bmatrix} = \begin{bmatrix} B & 0 \\ 0 & \widehat{C} \end{bmatrix}.$$

This is a product of three matrices that have inverses. Thus, the Schur complement in second diagonal block must have an inverse. Conversely, A may be factored as

$$\begin{bmatrix} B & E^T \\ F & C \end{bmatrix} = \begin{bmatrix} I_k & 0 \\ FB^{-1} & I_m \end{bmatrix} \begin{bmatrix} B & E^T \\ 0 & \widehat{C} \end{bmatrix}.$$

If both B and the Schur complement have inverses, then both matrices on the right side have inverses so that A also has an inverse. ∎

In the row operations for Gauss elimination and in the above proof, the following identity concerning row operations was used. It is used in the following special cases for matrices with inverses.

Fundamental Schur Identity. Assume the 11-block B has an inverse.

$$\begin{bmatrix} I_k & 0 \\ -FB^{-1} & I_m \end{bmatrix} \begin{bmatrix} B & E^T \\ F & C \end{bmatrix} \left(\begin{bmatrix} I_k & -B^{-1}E^T \\ 0 & I_m \end{bmatrix} \right) = \begin{bmatrix} B & 0 \\ 0 & \widehat{C} \end{bmatrix}.$$

The choice of the blocks B and C can play a very important role. Often the choice of the physical object, which is being modeled, does this. For example, consider the airflow over an aircraft. Here we might partition the aircraft into wing, rudder, fuselage and "connecting" components. Such partitions of the physical object or the matrix are called *domain decompositions*.

The solution of $Ax = d$ when using the above notation for Schur complement requires the solution of

$$\begin{bmatrix} B & E^T \\ 0 & \widehat{C} \end{bmatrix} \begin{bmatrix} x_1 \\ x_2 \end{bmatrix} = \begin{bmatrix} d_1 \\ \widehat{d_2} \end{bmatrix}. \tag{2.1.4}$$

The solution of $\widehat{C}x_2 = \widehat{d_2} \equiv d_2 - FB^{-1}d_1$ often involves a matrix with less zero components but smaller is size. When x_2 is computed, then solve $Bx_1 = d_1 - E^Tx_2$.

Example 2.1.1. Three-loop circuit.
Consider a three-loop circuit as depicted Figure 2.1.1. There are five unknown currents and two unknown potentials at nodes one and two.

The derivation of the five voltage equations and two current equations at nodes one and two assumes the potentials at nodes one and two are given by p_1 and p_2 and the potential at the bottoms nodes is set to zero. The voltage drops across resistors R_k for $k = 1, \cdots, 5$ are plus or minus $i_k R_k$, which depends on the direction of the current.

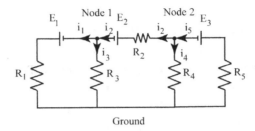

FIGURE 2.1.1
Three-loop circuit.

Voltage from node one to ground:

$$p_1 + E_1 - i_1 R_1 = 0 \text{ and}$$
$$p_1 - i_3 R_3 = 0.$$

Voltage from node one to node two:

$$p_1 + E_2 - (-i_2) R_2 = p_2.$$

Voltage from node two to ground:

$$p_2 + E_3 - (-i_5) R_5 = 0 \text{ and}$$
$$p_2 - i_4 R_4 = 0.$$

At each node the sum of the currents must be zero:

$$-i_1 + i_2 - i_3 = 0 \text{ and}$$
$$-i_2 - i_4 + i_5 = 0.$$

The matrix version of these seven equations is

$$
\begin{bmatrix}
R_1 & 0 & 0 & 0 & 0 & -1 & 0 \\
0 & R_2 & 0 & 0 & 0 & 1 & -1 \\
0 & 0 & R_3 & 0 & 0 & -1 & 0 \\
0 & 0 & 0 & R_4 & 0 & 0 & -1 \\
0 & 0 & 0 & 0 & R_5 & 0 & 1 \\
-1 & 1 & -1 & 0 & 0 & 0 & 0 \\
0 & -1 & 0 & -1 & 1 & 0 & 0
\end{bmatrix}
\begin{bmatrix}
i_1 \\ i_2 \\ i_3 \\ i_4 \\ i_5 \\ p_1 \\ p_2
\end{bmatrix}
=
\begin{bmatrix}
E_1 \\ -E_2 \\ 0 \\ 0 \\ -E_3 \\ 0 \\ 0
\end{bmatrix}.
$$

One can view the coefficient matrix as either a 7×7 matrix or as 2×2 block matrix where the 11-block is a 5×5 diagonal matrix with the resistances on the diagonal. The block LU factorization method reduces to solving a 2×2 matrix equation $\widehat{C} x_2 = \widehat{d}_2$ where the unknown vector has the two potentials, and solving a 5×5 diagonal matrix equation $B x_1 = d_1 - E^T x_2$. This reduction in the computational complexity illustrates the merits of using appropriate block structure. The MATLAB® code circuit3.m can be used to solve this circuit problem.

Example 2.1.2. Domain-decomposition.

Consider the steady state heat conduction problem with five unknowns and matrix system

$$
\begin{bmatrix}
2 & -1 & 0 & 0 & 0 \\
-1 & 2 & -1 & 0 & 0 \\
0 & -1 & 2 & -1 & 0 \\
0 & 0 & -1 & 2 & -1 \\
0 & 0 & 0 & -1 & 2
\end{bmatrix}
\begin{bmatrix}
u_1 \\ u_2 \\ u_3 \\ u_4 \\ u_2
\end{bmatrix}
=
\begin{bmatrix}
f_1 \\ f_2 \\ f_3 \\ f_4 \\ f_2
\end{bmatrix}.
$$

The unknowns are the approximations to the temperatures from left to right. Reorder these unknowns by listing u_3 last. This is a domain decomposition of the wire and the resulting system is

$$
\begin{bmatrix}
2 & -1 & 0 & 0 & 0 \\
-1 & 2 & 0 & 0 & -1 \\
0 & 0 & 2 & -1 & -1 \\
0 & 0 & -1 & 2 & 0 \\
0 & -1 & -1 & 0 & 2
\end{bmatrix}
\begin{bmatrix}
u_1 \\ u_2 \\ u_4 \\ u_5 \\ u_3
\end{bmatrix}
=
\begin{bmatrix}
f_1 \\ f_2 \\ f_4 \\ f_5 \\ f_3
\end{bmatrix}.
$$

The new matrix has 2×2 block structure with the 11-block having two 2×2 diagonal blocks and the 22-block a single number. The 11-block is easy to invert and concurrent calculations can be used.

Example 2.1.3. Gauss transform in block Gauss elimination.

Consider a matrix $k^2 \times k^2$ whose components are $k \times k$ blocks, and assume A_{11} has an inverse. The block Gauss transform is

$$
GA = \begin{bmatrix}
I_k & 0 \\
-FA_{11}^{-1} & I_{k(k-1)}
\end{bmatrix}.
$$

2.1.1 Heat diffusion in fin with two directions

Diffusion in 2D. Let $u = u(x, y) =$ temperature on a fin

$$0 = f + (Ku_x)_x + (Ku_y)_y + Cu \text{ and} \tag{2.1.5}$$
$$u = \text{given on the boundary.} \tag{2.1.6}$$

Diffusion in 2D. Let u_{ij} approximate $u(ih, jh)$ with $h = L/n = \Delta x = \Delta y$.

$$0 = f + \beta(u_{i+1,j} + u_{i-1,j}) - \beta 2 u_{i,j} +$$
$$\beta(u_{i,j+1} + u_{i,j-1}) - \beta 2 u_{i,j} + Cu_{ij} \tag{2.1.7}$$
$$\text{where } i, j = 1, ..., n - 1 \text{ and } \beta \equiv K/h^2 \text{ and}$$

$$u_{0,j}, u_{n,j}, u_{i,0}, u_{i,n} = \text{given.} \tag{2.1.8}$$

The matrix version of the discrete 2D model with $n = 6$ will have $5^2 = 25$ unknowns. Consequently, the matrix A will be 25×25. The location of its components will evolve from line (2.1.7) and will depend on the ordering of the unknowns u_{ij}. The classical method of ordering is to start with the bottom grid row ($j = 1$) and move from left ($i = 1$) to right ($i = n - 1$) so that

$$u = \begin{bmatrix} U_1^T & U_2^T & U_3^T & U_4^T & U_5^T \end{bmatrix}^T \text{ with } U_j = \begin{bmatrix} u_{1j} & u_{2j} & u_{3j} & u_{4j} & u_{5j} \end{bmatrix}^T$$

is a grid row j of unknowns. The final grid row corresponds to $j = n - 1$. So, it is reasonable to think of A as a 5×5 *block matrix* where each block is 5×5 and corresponds to a grid row. With careful writing of the equation (2.1.7) one can derive A as

$$\begin{bmatrix} B & -I & & & \\ -I & B & -I & & \\ & -I & B & -I & \\ & & -I & B & -I \\ & & & -I & B \end{bmatrix} \begin{bmatrix} U_1 \\ U_2 \\ U_3 \\ U_4 \\ U_5 \end{bmatrix} = (1/\beta) \begin{bmatrix} F_1 \\ F_2 \\ F_3 \\ F_4 \\ F_5 \end{bmatrix} \text{ where}$$

$$B = \begin{bmatrix} 4 + C/\beta & -1 & & & \\ -1 & 4 + C/\beta & -1 & & \\ & -1 & 4 + C/\beta & -1 & \\ & & -1 & 4 + C/\beta & -1 \\ & & & -1 & 4 + C/\beta \end{bmatrix} \text{ and}$$

$$I = \begin{bmatrix} 1 & & & & \\ & 1 & & & \\ & & 1 & & \\ & & & 1 & \\ & & & & 1 \end{bmatrix}.$$

There are three possible options. First, one can try to zero out the 21-block. The matrix B is strictly diagonally dominant and so it does have an inverse. Can one continue the block row operations? Second, use the block tridiagonal method, see tridblock_13.m. A third option is to use domain decomposition. List the third grid row last

$$\begin{bmatrix} B & -I & & & \\ -I & B & & & \_I \\ & & B & -I & -I \\ & & -I & B & \\ & & -I & -I & B \end{bmatrix} \begin{bmatrix} U_1 \\ U_2 \\ U_4 \\ U_5 \\ U_3 \end{bmatrix} = (1/\beta) \begin{bmatrix} F_1 \\ F_2 \\ F_4 \\ F_4 \\ F_3 \end{bmatrix}.$$

The big 11-block is 4×4 with two 2×2 diagonal blocks. So, one can take advantage of any concurrent computation tools.

2.1.2   Exercises

1. Let A be the 4×4 matrix

$$A = \begin{bmatrix} 2 & -1 & 0 & 0 \\ -1 & 2 & -1 & 0 \\ 0 & -1 & 2 & -1 \\ 0 & 0 & -1 & 2 \end{bmatrix}.$$

    (a). View this as a block 2×2 matrix where $n = 4 = 2 + 2 = n_1 + n_2$. Find the four blocks and the Schur complement matrix $\widehat{A}_{22}$.

    (b). Use these and line (2.1.4) to solve $Ax = d = [470 \ \ 400 \ \ 400 \ \ 470]^T$.

2. Let A be the 5×5 matrix

$$\begin{bmatrix} 2 & 0 & 1 & 0 & 1 \\ 0 & 3 & 0 & 2 & 0 \\ 1 & 0 & 3 & -1 & 0 \\ 0 & 1 & -1 & 3 & -1 \\ 1 & 1 & 0 & -1 & 3 \end{bmatrix}$$

    (a). View this as a block 2×2 matrix where $n = 5 = 2 + 3 = n_1 + n_2$. Find the four blocks and the Schur complement matrix $\widehat{A}_{22}$.

    (b). Use the resulting factors and line (2.1.4) to solve $Ax = d = [5 \ \ 6 \ \ 7 \ \ 8 \ \ 9]^T$.

3. Use MATLAB® to confirm any calculations in exercises 1 and 2.

4. Prove the identities in Theorem 2.1.1.

5. Prove the fundamental Schur identity.

6. Can one extend Gauss elimination to block Gauss elimination?

2.2 $PA = LU$ and A Nonsingular

The condition, $Az = 0$ implies $z = 0$, is very helpful in finding the factorization $PA = LU$. For example, if A is 2×2 and $Az = 0$ implies $z = 0$, then column one must have at least one nonzero component. If $a_{21} \neq 0$ and $a_{11} = 0$, then interchange the rows to get

$$PA = \begin{bmatrix} a_{21} & a_{22} \\ 0 & a_{12} \end{bmatrix} = U.$$

If $a_{11} \neq 0$, then use one row operation

$$\begin{bmatrix} 1 & 0 \\ -a_{21}/a_{11} & a_{12} \end{bmatrix} \begin{bmatrix} a_{11} & a_{12} \\ a_{21} & a_{22} \end{bmatrix} = \begin{bmatrix} a_{11} & a_{12} \\ 0 & \widehat{a}_{22} \end{bmatrix} = U.$$

Because A has an inverse, the triangular matrix must have an inverse. In the more general $n \times n$ a possible permutation may be required to compute the next Gauss transform. Can one group all the permutations and not have them interlaced with the Gauss transforms? Yes, this generalizes via mathematical induction when A is $n \times n$ and has an inverse.

Definition. *If there is a permutation matrix P such that $PA = LU$ where L and U are invertible lower and upper triangular matrices, respectively, then the matrix PA is said to have an LU factorization.*

Gaussian elimination method for the solution of $Ax = d$ uses permutations of the rows and elementary row operations to find the LU factorization so that

$$PAx = L(Ux) = Pd$$
$$\text{solve } Ly = Pd \text{ and}$$
$$\text{solve } Ux = y.$$

Example 2.2.1. Consider the 3×3 matrix

$$\begin{bmatrix} 1 & 2 & 0 \\ 1 & 2 & 1 \\ 0 & 1 & 3 \end{bmatrix}.$$

Since the component in the first row and column is not zero, no row interchange is necessary for the first column. The elementary row operation on the first column is

$$\begin{bmatrix} 1 & 0 & 0 \\ -1 & 1 & 0 \\ 0 & 0 & 1 \end{bmatrix} \begin{bmatrix} 1 & 2 & 0 \\ 1 & 2 & 1 \\ 0 & 1 & 3 \end{bmatrix} = \begin{bmatrix} 1 & 2 & 0 \\ 0 & 0 & 1 \\ 0 & 1 & 3 \end{bmatrix}.$$

For column two we must interchange rows two and three

$$\begin{bmatrix} 1 & 0 & 0 \\ 0 & 0 & 1 \\ 0 & 1 & 0 \end{bmatrix} \begin{bmatrix} 1 & 0 & 0 \\ -1 & 1 & 0 \\ 0 & 0 & 1 \end{bmatrix} \begin{bmatrix} 1 & 2 & 0 \\ 1 & 2 & 1 \\ 0 & 1 & 3 \end{bmatrix} = \begin{bmatrix} 1 & 2 & 0 \\ 0 & 1 & 3 \\ 0 & 0 & 1 \end{bmatrix}.$$

Note the first two factors on the left side can be rewritten as

$$\begin{bmatrix} 1 & 0 & 0 \\ 0 & 0 & 1 \\ 0 & 1 & 0 \end{bmatrix} \begin{bmatrix} 1 & 0 & 0 \\ -1 & 1 & 0 \\ 0 & 0 & 1 \end{bmatrix} = \begin{bmatrix} 1 & 0 & 0 \\ 0 & 1 & 0 \\ -1 & 0 & 1 \end{bmatrix} \begin{bmatrix} 1 & 0 & 0 \\ 0 & 0 & 1 \\ 0 & 1 & 0 \end{bmatrix}.$$

This gives $L^{-1}PA = U$

$$\begin{bmatrix} 1 & 0 & 0 \\ 0 & 1 & 0 \\ -1 & 0 & 1 \end{bmatrix} \begin{bmatrix} 1 & 0 & 0 \\ 0 & 0 & 1 \\ 0 & 1 & 0 \end{bmatrix} \begin{bmatrix} 1 & 2 & 0 \\ 1 & 2 & 1 \\ 0 & 1 & 3 \end{bmatrix} = \begin{bmatrix} 1 & 2 & 0 \\ 0 & 1 & 3 \\ 0 & 0 & 1 \end{bmatrix}$$

and the desired factorization $PA = LU$

$$\begin{bmatrix} 1 & 0 & 0 \\ 0 & 0 & 1 \\ 0 & 1 & 0 \end{bmatrix} \begin{bmatrix} 1 & 2 & 0 \\ 1 & 2 & 1 \\ 0 & 1 & 3 \end{bmatrix} = \begin{bmatrix} 1 & 0 & 0 \\ 0 & 1 & 0 \\ 1 & 0 & 1 \end{bmatrix} \begin{bmatrix} 1 & 2 & 0 \\ 0 & 1 & 3 \\ 0 & 0 & 1 \end{bmatrix}.$$

If the diagonal components of the triangular matrices are not zero, then the solution of the algebraic equation can be done in two triangular solves, which require order n^2 operations. This is particularly useful when the solution of $Ax = d$ required for multiple choices the right side and the same matrix.

Theorem 2.2.1. *(LU Factorization)* *If A has an inverse so that $Az = 0$ implies $z = 0$, then there exist permutation matrix P, and invertible lower and upper triangular matrices L and U, respectively, such that $PA = LU$.*

Proof. We have already proved the $n = 2$ case. Assume it is true for any $(n-1) \times (n-1)$ matrix. If A is invertible, then column one must have some nonzero component, say in row i. So, if the first component is not zero, interchange the first row with the row i. Let P be the permutation matrix

$$PA = \begin{bmatrix} b & e^T \\ f & C \end{bmatrix}$$

where $b \neq 0$ is 1×1, e^T is $1 \times (n-1)$, f is $(n-1) \times 1$, and C is $(n-1) \times (n-1)$. Apply the Schur complement analysis and the block elementary matrix operation

$$\begin{bmatrix} 1 & 0 \\ -fb^{-1} & I_{n-1} \end{bmatrix} \begin{bmatrix} b & e^T \\ f & C \end{bmatrix} = \begin{bmatrix} b & e^T \\ 0 & \widehat{C} \end{bmatrix}$$

where $\widehat{C} = C - fb^{-1}e^T$ is the $(n-1) \times (n-1)$ Schur complement. By the Schur complement theorem $\widehat{C}$ must have an inverse. Use the inductive assumption to write $\widehat{C} = \widehat{P}\widehat{L}\widehat{U}$.

$$\begin{bmatrix} b & e^T \\ 0 & \widehat{C} \end{bmatrix} = \begin{bmatrix} b & e^T \\ 0 & \widehat{P}\widehat{L}\widehat{U} \end{bmatrix}$$

$$= \begin{bmatrix} 1 & 0 \\ 0 & \widehat{P} \end{bmatrix} \begin{bmatrix} 1 & 0 \\ 0 & \widehat{L} \end{bmatrix} \begin{bmatrix} b & e^T \\ 0 & \widehat{U} \end{bmatrix}.$$

Since $\widehat{P}$ is a permutation matrix,

$$\begin{bmatrix} 1 & 0 \\ 0 & \widehat{P} \end{bmatrix} \begin{bmatrix} 1 & 0 \\ -fb^{-1} & I_{n-1} \end{bmatrix} PA = \begin{bmatrix} 1 & 0 \\ 0 & \widehat{L} \end{bmatrix} \begin{bmatrix} b & e^T \\ 0 & \widehat{U} \end{bmatrix}.$$

Note the change in order of the products

$$\begin{bmatrix} 1 & 0 \\ 0 & \widehat{P} \end{bmatrix} \begin{bmatrix} 1 & 0 \\ -fb^{-1} & I_{n-1} \end{bmatrix} = \begin{bmatrix} 1 & 0 \\ \widehat{P}(-fb^{-1}) & \widehat{P} \end{bmatrix}$$

$$= \begin{bmatrix} 1 & 0 \\ \widehat{P}(-fb^{-1}) & I_{n-1} \end{bmatrix} \begin{bmatrix} 1 & 0 \\ 0 & \widehat{P} \end{bmatrix}.$$

This gives

$$\begin{bmatrix} 1 & 0 \\ \widehat{P}(-fb^{-1}) & I_{n-1} \end{bmatrix} \begin{bmatrix} 1 & 0 \\ 0 & \widehat{P} \end{bmatrix} PA = \begin{bmatrix} 1 & 0 \\ 0 & \widehat{L} \end{bmatrix} \begin{bmatrix} b & e \\ 0 & \widehat{U} \end{bmatrix}.$$

Finally, multiply by the inverse of the left factor on the left side to get

$$\begin{bmatrix} 1 & 0 \\ 0 & \widehat{P} \end{bmatrix} PA = \begin{bmatrix} 1 & 0 \\ \widehat{P}fb^{-1} & I_{n-1} \end{bmatrix} \begin{bmatrix} 1 & 0 \\ 0 & \widehat{L} \end{bmatrix} \begin{bmatrix} b & e^T \\ 0 & \widehat{U} \end{bmatrix}$$

$$= \begin{bmatrix} 1 & 0 \\ \widehat{P}fb^{-1} & \widehat{L} \end{bmatrix} \begin{bmatrix} b & e^T \\ 0 & \widehat{U} \end{bmatrix}.$$

■

2.2.1   Exercises

1. Find the LU factors such that $PA = LU$ and solve $AX = d$

$$\begin{bmatrix} 0 & -1 & 2 \\ -1 & 2 & -1 \\ 1 & -1 & 0 \end{bmatrix} \begin{bmatrix} x_1 \\ x_2 \\ x_3 \end{bmatrix} = \begin{bmatrix} 100 \\ 10 \\ 70 \end{bmatrix}.$$

2. In the proof of the theorem explain why the triangular matrices have inverses.

2.3 $A = LU, A^{-1} \geq 0$ and M-Matrix

This a subset of the nonsingular matrices that has positive diagonal components and the inverse matrices have nonnegative components. There are a number of equivalent definition of an M-matrix. We will study two of these.

The notation with a $M \geq 0$ will be understood to mean M and 0 are matrices with the same dimension, $M = [m_{ij}]$, and 0 has all components equal to zero, and $m_{ij} \geq 0$.

Definition. *Let $A = [a_{ij}]$ have nonpositive off diagonal components $a_{ij} \leq 0$ for $i \neq j$. A is called an M-matrix if and only if A has an inverse and each of the components of the inverse are nonnegative, $A^{-1} \geq 0$.*

Example 2.3.1. The block lower triangular matrix with $g \leq 0$ is an M-matrix

$$A = \begin{bmatrix} 1 & 0 \\ g & I_{n-1} \end{bmatrix} \text{ and } A^{-1} = \begin{bmatrix} 1 & 0 \\ -g & I_{n-1} \end{bmatrix} \geq 0.$$

Example 2.3.2. The following matrix and its LU factors are M-matrices

$$\begin{bmatrix} 2 & -1 & 0 & 0 \\ -1 & 2 & -1 & 0 \\ 0 & -1 & 2 & -1 \\ 0 & 0 & -1 & 2 \end{bmatrix} = \begin{bmatrix} 1 & 0 & 0 & 0 \\ -1/2 & 1 & 0 & 0 \\ 0 & -2/3 & 1 & 0 \\ 0 & 0 & -3/4 & 1 \end{bmatrix} \begin{bmatrix} 2 & -1 & 0 & 0 \\ 0 & 3/2 & -1 & 0 \\ 0 & 0 & 4/3 & -1 \\ 0 & 0 & 0 & 5/4 \end{bmatrix}.$$

The next matrix is not symmetric, but it is also an M-matrix

$$A = \begin{bmatrix} 2 & -1 & 0 \\ -1 & 2 & -1 \\ 0 & -2 & 2 \end{bmatrix}.$$

Theorem 2.3.1. *(M-Matrix Properties) Let A have nonpositive off diagonal components.*

   (i). *The diagonal components of an M-matrix must be positive.*

   (ii). *If A is an M-matrix, $Ax = d$, $Ay = d + \Delta d$ and $\Delta d \geq 0$, then $x \leq y$.*

   (iii). *If $A = LU$ where L and U are lower and upper triangular matrices, respectively, have positive diagonals and nonpositive off diagonals components, then A is an M-matrix.*

Proof. If $a_{ii} \leq 0$, then column i of $A \leq 0$. But, $A^{-1}A = I$ so that

$$A^{-1}(\text{col } i \text{ of } A) = e_i \text{ which has one positive component.}$$

Thus, $a_{ii} \leq 0$ cannot be true.

   Note $A(y - x) = Ay - Ax = d + \Delta d - d = \Delta d \geq 0$. Multiply by the inverse to get $y - x = A^{-1}\Delta d \geq 0$.

   Since both L and U are triangular matrices with positive diagonal components, both L and U have inverses and $A^{-1} = U^{-1}L^{-1}$. Because the off diagonal components are nonpositive, the triangular solves for either $Ux = d \geq 0$ or $Ly = d \geq 0$ will give $x \geq 0$ or $y \geq 0$. Apply this to each column in the identity matrix with $d = e_i$. So, $U^{-1} \geq 0$, $L^{-1} \geq 0$ to conclude $A^{-1} \geq 0$. ∎

Since the diagonal components of an M-matrix are positive, Gauss elimination does not require permutation. So the Gauss elimination stage is just a product of Gauss transforms

$$G_{n-1} \cdots G_2 G_1 A = U \text{ and}$$
$$A = G_1^{-1} G_2^{-1} \cdots G_{n-1}^{-1} U.$$

The factors $G_1^{-1} G_2^{-1} \cdots G_{n-1}^{-1}$ have the form of a lower triangular matrix because any lower triangular matrix can be viewed as a special product

$$\begin{bmatrix} L_{11} & & \\ L_{21} & L_{22} & \\ L_{31} & L_{32} & L_{33} \end{bmatrix} = \begin{bmatrix} L_{11} & & \\ L_{21} & I & \\ L_{31} & 0 & I \end{bmatrix} \begin{bmatrix} I & & \\ 0 & L_{22} & \\ 0 & L_{32} & I \end{bmatrix} \begin{bmatrix} I & & \\ 0 & I & \\ 0 & 0 & L_{33} \end{bmatrix}.$$

In the above the first Gauss transform is

$$G_1 = \begin{bmatrix} 1 & \\ -f/a_{11} & I_{n-1} \end{bmatrix} \text{ and } G_1^{-1} = \begin{bmatrix} 1 & \\ +f/a_{11} & I_{n-1} \end{bmatrix}$$

The next theorem characterizes M-matrices in terms of the LU factors.

Theorem 2.3.2. *(M-Matrix and LU) Let A have nonpositive off diagonal components. A is an M-matrix if and only if $A = LU$ where L and U have positive diagonals and nonpositive off diagonals.*

Proof. The proof is by induction. The 2×2 is easy, but we give it here because it is a pattern for the inductive case. Use the fundamental Schur identity

$$\begin{bmatrix} 1 & 0 \\ -a_{21}/a_{11} & 1 \end{bmatrix} \begin{bmatrix} a_{11} & a_{12} \\ a_{21} & a_{22} \end{bmatrix} \begin{bmatrix} 1 & -a_{12}/a_{11} \\ 0 & 1 \end{bmatrix} = \begin{bmatrix} a_{11} & 0 \\ 0 & \widehat{a}_{22} \end{bmatrix}.$$

If A has an inverse, then the left side has an inverse, and therefore $\widehat{a}_{22} \neq 0$.

$$A^{-1} = \begin{bmatrix} a_{22} & -a_{12} \\ -a_{21} & a_{11} \end{bmatrix} / \det(A) = \begin{bmatrix} b_{11} & b_{12} \\ b_{21} & b_{22} \end{bmatrix}.$$

If $A^{-1} \geq 0$, then $b_{22} = a_{11}/\det(A) \geq 0$. Take the inverse of the fundamental Schur identity

$$\begin{bmatrix} 1 & a_{12}/a_{11} \\ 0 & 1 \end{bmatrix} \begin{bmatrix} b_{11} & b_{12} \\ b_{21} & b_{22} \end{bmatrix} \begin{bmatrix} 1 & 0 \\ a_{21}/a_{11} & 1 \end{bmatrix} = \begin{bmatrix} 1/a_{11} & 0 \\ 0 & 1/\widehat{a}_{22} \end{bmatrix}.$$

Next equate the 22-components on the left and right sides to get $\widehat{a}_{22} > 0$.

Assume A is an M-matrix and the characterization is true for $(n-1) \times (n-1)$ M-matrices. Since $a_{11} > 0$, the first Gauss transform is

$$G_1 = \begin{bmatrix} 1 & 0 \\ -f/a_{11} & I_{n-1} \end{bmatrix} \text{ and } f \geq 0.$$

The fundamental Schur identity is

$$\begin{bmatrix} 1 & 0 \\ -f/a_{11} & I_{n-1} \end{bmatrix} \begin{bmatrix} a_{11} & e^T \\ f & C \end{bmatrix} \begin{bmatrix} 1 & -e^T/a_{11} \\ 0 & I_{n-1} \end{bmatrix} = \begin{bmatrix} a_{11} & 0 \\ 0 & \widehat{C} \end{bmatrix}.$$

Since the left side factors have inverses, the right side has an inverse and $\widehat{C}$ has an inverse. We need to show $\widehat{C} = C - f/a_{11}e^T$ is an M-matrix. Note, the off diagonal components C are not positive, and the components of f and e^T are not positive. So, the off diagonal components $\widehat{C}$ are not positive. In order to show the inverse of $\widehat{C}$ must be greater than zero matrix, take the inverse of the above fundamental identity

$$\begin{bmatrix} 1 & e^T/a_{11} \\ 0 & I_{n-1} \end{bmatrix} \begin{bmatrix} b_{11} & B_{12} \\ B_{21} & B_{22} \end{bmatrix} \begin{bmatrix} 1 & 0 \\ f/a_{11} & I_{n-1} \end{bmatrix} = \begin{bmatrix} 1/a_{11} & 0 \\ 0 & (\widehat{C})^{-1} \end{bmatrix}$$

where the inverse of A is

$$\begin{bmatrix} b_{11} & B_{12} \\ B_{21} & B_{22} \end{bmatrix}.$$

Since $A^{-1} \geq 0$, $B_{22} \geq 0$. Equate the 22-blocks of the left and right side to get $(\widehat{C})^{-1} \geq 0$. Thus, $\widehat{C}$ is an M-matrix.

The next step is either to continue the Gauss transforms on the smaller M-matrix $\widehat{C}$ or to use the inductive assumption on $\widehat{C}$. The latter option gives $\widehat{C} = \widehat{L}\,\widehat{U}$ and

$$
\begin{bmatrix} 1 & 0 \\ -f/a_{11} & I_{n-1} \end{bmatrix} \begin{bmatrix} a_{11} & e^T \\ f & C \end{bmatrix} = \begin{bmatrix} a_{11} & e^T \\ 0 & \widehat{L}\widehat{U} \end{bmatrix}
$$
$$
= \begin{bmatrix} 1 & 0 \\ 0 & \widehat{L} \end{bmatrix} \begin{bmatrix} a_{11} & e^T \\ 0 & \widehat{U} \end{bmatrix}.
$$

Multiply from the left by the inverse Gauss transform to get the desired LU factors

$$
\begin{bmatrix} a_{11} & e^T \\ f & C \end{bmatrix} = \begin{bmatrix} 1 & 0 \\ f/a_{11} & I_{n-1} \end{bmatrix} \begin{bmatrix} 1 & 0 \\ 0 & \widehat{L} \end{bmatrix} \begin{bmatrix} a_{11} & e^T \\ 0 & \widehat{U} \end{bmatrix}
$$
$$
= \begin{bmatrix} 1 & 0 \\ f/a_{11} & \widehat{L} \end{bmatrix} \begin{bmatrix} a_{11} & e^T \\ 0 & \widehat{U} \end{bmatrix}.
$$

     ∎

Since M-matrices have inverses, the matrix may be characterized by a special Schur complement which must also be an M-matrix. Consider the M-matrix

$$
A = \begin{bmatrix} B & E^T \\ F & C \end{bmatrix} = \begin{bmatrix} L_{11} & 0 \\ L_{21} & L_{22} \end{bmatrix} \begin{bmatrix} U_{11} & U_{12} \\ 0 & U_{22} \end{bmatrix}.
$$

Equate the 11-blocks to conclude $B = L_{11}U_{11}$ and therefore must be an M-matrix.

Theorem 2.3.3. *(Schur Complement and M-Matrix) Let A have nonpositive off diagonal components. A is an M-matrix if and only if B and Schur complement $\widehat{C}$ are M-matrices.*

Proof. Use the fundamental Schur complement identity

$$
\begin{bmatrix} I_k & 0 \\ -FB^{-1} & I_m \end{bmatrix} \begin{bmatrix} B & E^T \\ F & C \end{bmatrix} \begin{bmatrix} I_k & -B^{-1}E^T \\ 0 & I_m \end{bmatrix} = \begin{bmatrix} B & 0 \\ 0 & \widehat{C} \end{bmatrix}.
$$

Assume A is an M-matrix.

Then B is an M-matrix and has an inverse. So, the above identity is valid. Also, the off diagonal components are not positive and $F \leq 0$ and $E \leq 0$. Then the off diagonal component in $\widehat{C} = C - FB^{-1}E^T$ must not be positive. The left side is a product of three matrices with inverses giving

$$
\begin{bmatrix} I_k & 0 \\ B^{-1}E^T & I_m \end{bmatrix} \begin{bmatrix} B_{11} & B_{12} \\ B_{21} & B_{22} \end{bmatrix} \begin{bmatrix} I_k & 0 \\ FB^{-1} & I_m \end{bmatrix} = \begin{bmatrix} B^{-1} & 0 \\ 0 & (\widehat{C})^{-1} \end{bmatrix}
$$

where B_{22} is the inverse of the 22-block of $A^{-1} \geq 0$. The 22-block of the left side product is $B_{22} = (\widehat{C})^{-1} \geq 0$ and $\widehat{C}$ must also be an M-matrix. ∎

2.3.1   Exercises

1. Find the LU factors of A and show A, L and U are M-matrices

$$A = \begin{bmatrix} 2 & -1 & 0 \\ -1 & 2 & -1 \\ 0 & -2 & 2 \end{bmatrix}.$$

2. In the proof of Theorem 2.3.2 complete the proof, in place of the inductive option, by continuing to use Gauss transforms.

3. In the proof of Theorem 2.3.3 show A is an M-matrix when both B and $\widehat{C}$ are M-matrices.

2.4 $A = GG^T$ and A SPD

In this section we will restrict the matrices to symmetric positive definite matrices. Although this restriction may seem a little severe, there are a number of important applications, which include some classes of partial differential equations and some classes of least squares problems. The advantage of this restriction is that the number of operations to do Gaussian elimination can be cut in half.

Definition.   *Let A be an $n \times n$ real matrix. A is a real symmetric positive definite matrix (SPD) if and only if $A = A^T$ and for all $x \neq 0$, $x^T A x > 0$.*

Example 2.4.1.   1. Consider the 2×2 matrix $\begin{bmatrix} 2 & -1 \\ -1 & 2 \end{bmatrix}$ and note

$$x^T A x = x_1^2 + (x_1 - x_2)^2 + x_2^2 > 0.$$

A similar $n \times n$ matrix is positive definite

$$\begin{bmatrix} 2 & -1 & & \\ -1 & 2 & \ddots & \\ & \ddots & \ddots & -1 \\ & & -1 & 2 \end{bmatrix}.$$

2. The matrix A for which $A = A^T$ with $a_{ii} > \sum_{j \neq i} |a_{ij}|$ is positive definite. The symmetry implies that the matrix is also column *strictly diagonally dominant*, that is, $a_{jj} > \sum_{i \neq j} |a_{ij}|$. Now use this and the inequality $|ab| \leq \frac{1}{2}(a^2 + b^2)$ to show for all $x \neq 0$, $x^T A x > 0$.

3. Consider the normal equation from the least squares problem where A is $m \times n$ where $m > n$. Assume A has *full column rank* ($Ax = 0$ implies $x = 0$). The normal equation $A^T Ax = A^T d$ is equivalent to finding the least squares solution of $Ax = d$. Here $A^T A$ is SPD because if $x \neq 0$, then $Ax \neq 0$, and $x^T (A^T A)x = (Ax)^T (Ax) > 0$.

Theorem 2.4.1. *(Basic Properties of SPD Matrices) If A is an $n \times n$ SPD matrix, then*

    (i). The diagonal components of A are positive, $a_{ii} > 0$,

    (ii). If $A = \begin{bmatrix} B & F^T \\ F & C \end{bmatrix}$, then B and C are SPD,

    (iii). $Ax = 0$ implies $x = 0$ so that A has an inverse and

    (iv). If S is $m \times n$ with $m \geq n$ and has full column rank ($Sx = 0$ implies $x = 0$), then $S^T AS$ is positive definite.

Proof.

    (i). Choose $x = e_i$, the unit vector with 1 in component i so that $x^T Ax = a_{ii} > 0$.

    (ii). Choose $x = \begin{bmatrix} X_1 \\ 0 \end{bmatrix}$ so that $x^T Ax = X_1^T B X_1 > 0$.

    (iii). If $Ax = 0$, then $x^T Ax = 0$. Since A is positive definite, $x = 0$.

    (iv). Let $x \neq 0$ so that by the full rank assumption $Sx \neq 0$. By the positive definite assumption on A

$$(Sx)^T A(Sx) = x^T (S^T AS)x > 0.$$

■

The next theorem uses the Schur complement to give a characterization of the block 2×2 SPD matrix

$$A = \begin{bmatrix} B & F^T \\ F & C \end{bmatrix}. \tag{2.4.1}$$

Theorem 2.4.2. *(Schur Complement and SPD) Let A as in (2.4.1) be symmetric. A is SPD if and only if B and the Schur complement of B in A, $\widehat{C} = C - FB^{-1}F^T$, are SPD.*

Proof. Assume A is SPD so that B is also SPD and has an inverse. Use the fundamental Schur identity

$$\begin{bmatrix} I_k & 0 \\ -FB^{-1} & I_m \end{bmatrix} \begin{bmatrix} B & F^T \\ F & C \end{bmatrix} \begin{bmatrix} I_k & -B^{-1}F^T \\ 0 & I_m \end{bmatrix} = \begin{bmatrix} B & 0 \\ 0 & \widehat{C} \end{bmatrix}.$$

Since B is SPD, $\left(B^{-1}\right)^T = \left(B^T\right)^{-1} = B^{-1}$ and thus

$$
S = \begin{bmatrix} I_k & -B^{-1}F^T \\ 0 & I_m \end{bmatrix} \text{ and } S^T = \begin{bmatrix} I_k & -B^{-1}F^T \\ 0 & I_m \end{bmatrix}^T = \begin{bmatrix} I_k & 0 \\ -FB^{-1} & I_m \end{bmatrix}.
$$

Then

$$
S^T A S = \begin{bmatrix} B & 0 \\ 0 & \widehat{C} \end{bmatrix}.
$$

Since S has an inverse, it has full rank and B and $\widehat{C}$ must be SPD. The converse is also true by reversing the above argument. ∎

Example 2.4.2. Consider the 3×3 matrix

$$
A = \begin{bmatrix} 2 & -1 & 0 \\ -1 & 2 & -1 \\ 0 & -1 & 2 \end{bmatrix}.
$$

The first elementary row operation is

$$
\begin{bmatrix} 1 & 0 & 0 \\ 1/2 & 1 & 0 \\ 0 & 0 & 1 \end{bmatrix} \begin{bmatrix} 2 & -1 & 0 \\ -1 & 2 & -1 \\ 0 & -1 & 2 \end{bmatrix} = \begin{bmatrix} 2 & -1 & 0 \\ 0 & 3/2 & -1 \\ 0 & -1 & 2 \end{bmatrix}.
$$

The Schur complement of the 1×1 matrix $B = [2]$ is

$$
\widehat{C} = C - FB^{-1}F^T = \begin{bmatrix} 3/2 & -1 \\ -1 & 2 \end{bmatrix}.
$$

Do another elementary row operation, but now on the second column

$$
\begin{bmatrix} 1 & 0 & 0 \\ 0 & 1 & 0 \\ 0 & 2/3 & 1 \end{bmatrix} \begin{bmatrix} 2 & -1 & 0 \\ 0 & 3/2 & -1 \\ 0 & -1 & 2 \end{bmatrix} = \begin{bmatrix} 2 & -1 & 0 \\ 0 & 3/2 & -1 \\ 0 & 0 & 4/3 \end{bmatrix}.
$$

Thus the matrix A can be factored as

$$
\begin{aligned}
A &= \begin{bmatrix} 1 & 0 & 0 \\ -1/2 & 1 & 0 \\ 0 & -2/3 & 1 \end{bmatrix} \begin{bmatrix} 2 & -1 & 0 \\ 0 & 3/2 & -1 \\ 0 & 0 & 4/3 \end{bmatrix} \\
&= \begin{bmatrix} 1 & 0 & 0 \\ -1/2 & 1 & 0 \\ 0 & -2/3 & 1 \end{bmatrix} \begin{bmatrix} 2 & 0 & 0 \\ 0 & 3/2 & 0 \\ 0 & 0 & 4/3 \end{bmatrix} \begin{bmatrix} 1 & -1/2 & 0 \\ 0 & 1 & -2/3 \\ 0 & 0 & 1 \end{bmatrix} \\
&= \begin{bmatrix} \sqrt{2} & 0 & 0 \\ -1/\sqrt{2} & \sqrt{3}/\sqrt{2} & 0 \\ 0 & -\sqrt{2}/\sqrt{3} & 2/\sqrt{3} \end{bmatrix} \begin{bmatrix} \sqrt{2} & 0 & 0 \\ -1/\sqrt{2} & \sqrt{3}/\sqrt{2} & 0 \\ 0 & -\sqrt{2}/\sqrt{3} & 2/\sqrt{3} \end{bmatrix}^T.
\end{aligned}
$$

Definition. *The Cholesky factorization of A is $A = GG^T$ where G is a lower triangular matrix with positive diagonal components.*

Any SPD matrix has a Cholesky factorization. The following proof is again by mathematical induction on the dimension of the matrix. The virtue of the Cholesky factorization is the construction of factors requires about half of operations as in the LU factors for a nonsingular matrix.

Theorem 2.4.3. *(Cholesky Factorization) A is SPD if and only if it has a Cholesky factorization.*

Proof. Assume A is SPD. The $n = 2$ case is clearly true. Let $b = a_{11} > 0$ and apply fundamental Schur identity to $A = \begin{bmatrix} b & f^T \\ f & C \end{bmatrix}$

$$\begin{bmatrix} 1 & 0 \\ -fb^{-1} & I \end{bmatrix} \begin{bmatrix} b & f^T \\ f & C \end{bmatrix} \begin{bmatrix} 1 & -b^{-1}f^T \\ 0 & I \end{bmatrix} = \begin{bmatrix} b & 0 \\ 0 & C - fb^{-1}f^T \end{bmatrix}.$$

The Schur complement $\widehat{C} = C - fb^{-1}f^T$ must be SPD and has dimension $n - 1$. Therefore, by the mathematical induction assumption it must have a Cholesky factorization $\widehat{C} = \widehat{G}\widehat{G}^T$. Then

$$\begin{bmatrix} b & 0 \\ 0 & C - fb^{-1}f^T \end{bmatrix} = \begin{bmatrix} b & 0 \\ 0 & \widehat{G}\widehat{G}^T \end{bmatrix}$$

$$= \begin{bmatrix} \sqrt{b} & 0 \\ 0 & \widehat{G} \end{bmatrix} \begin{bmatrix} \sqrt{b} & 0 \\ 0 & \widehat{G}^T \end{bmatrix}.$$

$$A = \begin{bmatrix} 1 & 0 \\ fb^{-1} & I \end{bmatrix} \begin{bmatrix} \sqrt{b} & 0 \\ 0 & \widehat{G} \end{bmatrix} \begin{bmatrix} \sqrt{b} & 0 \\ 0 & \widehat{G}^T \end{bmatrix} \begin{bmatrix} 1 & b^{-1}f^T \\ 0 & I \end{bmatrix}$$

$$= \begin{bmatrix} \sqrt{b} & 0 \\ f/\sqrt{b} & \widehat{G} \end{bmatrix} \begin{bmatrix} \sqrt{b} & 0 \\ f/\sqrt{b} & \widehat{G} \end{bmatrix}^T.$$

∎

The mathematical induction proofs are not fully constructive, but they do imply that the Schur complement is either invertible or is SPD. This allows one to continue with possible permutations and elementary column operations. This process can be done until the upper triangular matrix is obtained. In the case of the SPD matrix, the Schur complement is also SPD and the first pivot must be positive and so no row interchanges are required.

2.4.1   SPD and minimization

The one variable problem $ax = d$ with $a > 0$ is equivalent to the finding the minimum of

$$\frac{1}{2}ax^2 - xd.$$

This can be generalized provided a is replaced by a SPD matrix. One example is the model of the steady state membrane where the shape of the membrane should minimize the potential energy, see [28, section 3.5].

Theorem 2.4.4. *(Energy Equivalence) If A is $n \times n$ SPD matrix, then $Ax = d$ is equivalent to*

$$J(x) = \min_y J(y) \text{ where } J(y) \equiv \frac{1}{2}x^T A x - x^T d.$$

Proof. The next identity follows from $A = A^T$

$$
\begin{aligned}
J(x + ty) &= \frac{1}{2}(x + ty)^T A(x + ty) - (x + ty)^T d \\
&= \frac{1}{2}x^T A x + (ty)^T A x + \frac{1}{2}(ty)^T A(ty) \\
&\quad - x^T d - (ty)^T d \\
&= J(x) + ty^T(Ax - d) + \frac{1}{2}t^2 y^T A y \equiv f(t). \qquad (2.4.2)
\end{aligned}
$$

Since A is SPD, $y^T A y > 0$ and the right side is a quadratic function of t for any nonzero vector y. Note $f(0) = J(x)$, $f'(0) = y^T(Ax - d)$ and $f''(0) = y^T A y > 0$. If $Ax = d$, then line (2.4.2) gives

$$J(x + ty) = J(x) + \frac{1}{2}t^2 y^T A y \geq J(x).$$

If $J(x) = \min J(x + ty)$, then $f'(0) = y^T(Ax - d) = 0$. This is true for all y and choose $y = e_i$. Thus, $y^T(Ax - d) = [Ax - d]_i = 0$ and $Ax - d = 0_{n \times 1}$. ∎

The importance of general energy formulations is for iterative methods (see Section 11.4) and for boundary values problems (see Section 10.5). The steepest descent method can be applied to $J(x)$ and is outlined in Exercises 6-8 and in [28, section 3.5].

2.4.2   Exercises

1. By-hand calculations find the Cholesky factorization for

$$A = \begin{bmatrix} 3 & -1 & 0 \\ -1 & 3 & -1 \\ 0 & -1 & 3 \end{bmatrix}.$$

2. In Theorem 2.4.2 prove $\widehat{C}$ is SPD.

3. In Theorem 2.4.2 prove the converse part.

4. In Theorem 2.4.3 prove the $n = 2$ case.

5. In Theorem 2.4.3 prove the converse part.

6. Consider Theorem 2.4.4 and use line (2.4.2).

(a). For $t > 0$ and $t \downarrow 0$ and show

$$0 \le \frac{J(x + ty) - J(x)}{t} \text{ converges to } 0 \le y^T(Ax - d).$$

(b). This is true for both y and $-y$ and show

$$y^T(Ax - d) = 0 \text{ and } Ax = d.$$

7. Find the directional derivative and gradient of J.

(a). Let $g(t) \equiv J(x + tu)$ and $u^T u = 1$. Show $g'(t) = (\nabla J)^T u$.

(b). Use line (2.4.2) as in 6(a). Show $\nabla J = Ax - d = -r(x)$.

8. Show $c = r^T r / (r^T Ar)$ so that $J(x + cr)$ is a minimum.

The first step in the steepest descent method is $x^+ = x + cr$.

3

Least Squares and Normal Equations

The goal is to solve systems with more equations than unknowns where A is $m \times n$ and $m > n$. The least squares solution(s) of $Ax = d$ are equivalent to finding the solution(s) of the normal equations $A^T A x = A^T d$. If $A^T A$ is nonsingular, then there is a unique solution. An application to parameter identification is illustrated in the second section. The third and fourth sections construct multiple solutions. This is done by using the basis for the subspace $R(A)$ and the projection of d onto $R(A)$.

3.1 Normal Equations

As a motivating example consider finding the straight line "closest" to the data points

x	$y = mx + c$
1	10
2	9
3	7

The unknowns are $m = x_2$ and $c = x_1$. The matrix form of this is

$$\begin{bmatrix} 1 & 1 \\ 1 & 2 \\ 1 & 3 \end{bmatrix} \begin{bmatrix} x_1 \\ x_2 \end{bmatrix} = \begin{bmatrix} 10 \\ 9 \\ 7 \end{bmatrix}.$$

The solution of the normal equations will be shown to be equivalent to finding the minimum of the residual. More precisely, these terms are defined as follows.

Definition. *Let A be $m \times n$. The least squares solution, x, of $Ax = d$ if and only if for $r(x) \equiv d - Ax$ and all y*

$$r(x)^T r(x) \leq r(y)^T r(y).$$

Definition. *The normal equations are $A^T A x = A^T d$ or equivalently $A^T r(x) = 0_{n \times 1}$.*

DOI: 10.1201/9781003304128-3

Remark. This means the columns of A are perpendicular to $r(x)$.

Theorem 3.1.1. *(Least Squares via Normal Equations)* x *is a least squares solution of* $Ax = d$ *if and only if* x *is some solution of the normal equations.*

Proof. The proof follows from

$$r(y) = d - Ay$$
$$= d - A(x + y - x)$$
$$= r(x) - A(y - x) \text{ and}$$
$$r(y)^T r(y) = (r(x) - A(y - x))^T (r(x) - A(y - x))$$
$$= r(x)^T r(x) - 2(y - x)^T A^T r(x) +$$
$$(A(y - x))^T A(y - x) \tag{3.1.1}$$
$$\geq r(x)^T r(x) - 2(y - x)^T A^T r(x).$$

If $A^T r(x) = 0_{n \times 1}$, then $r(y)^T r(y) \geq r(x)^T r(x)$.

If x is a least squares solution, then in line (3.1.1) replace y with $x + ty$ where and $t > 0$. Use the least squares inequality and the above equality to get

$$0 \leq r(x + ty)^T r(x + ty) - r(x)^T r(x)$$
$$= -2ty^T A^T r(x) + t^2 (Ay)^T Ay.$$

Divide by $t > 0$

$$0 \leq \frac{r(x + ty)^T r(x + ty) - r(x)^T r(x)}{t}$$
$$= -2y^T A^T r(x) + t(Ay)^T Ay.$$

Let $t \downarrow 0$ and $y = e_i$ to get

$$0 \leq -2y^T A^T r(x) = -2[A^T r(x)]_i.$$

Likewise, choose $y = -e_i$ to get the other inequality

$$0 \leq 2e_i^T A^T r(x) = 2[A^T r(x)]_i.$$

So, $A^T r(x) = 0_{n \times 1}$. ∎

Remark. It is interesting to compare Theorem 2.4.4 and Theorem 3.1.1. The identities in line (2.4.2) and line (3.1.1) play similar roles. In the least squares problem Ay may be a zero vector and so $r(x + ty)^T r(x + ty)$ may not be a quadratic function of t.

When A is a square $n \times n$ matrix, that satisfies $Ax = 0_{n \times 1}$ implies $x = 0_{n \times 1}$, then the matrix has an inverse. If A is a rectangular $m \times n$ matrix with $m > n$ and satisfies $Ax = 0_{m \times 1}$ implies $x = 0_{n \times 1}$, then the matrix $A^T A$ will have an inverse (see the next theorem). Examples of such matrices are as follows.

Example 3.1.1. $Ax = \begin{bmatrix} 1 & 1 \\ 1 & 2 \\ 1 & 3 \end{bmatrix} \begin{bmatrix} x_1 \\ x_2 \end{bmatrix} = \begin{bmatrix} 0 \\ 0 \\ 0 \end{bmatrix}$ means

$$E_{31}(-1)E_{21}(-1)Ax = \begin{bmatrix} 1 & 1 \\ 0 & 1 \\ 0 & 2 \end{bmatrix} \begin{bmatrix} x_1 \\ x_1 \end{bmatrix} = \begin{bmatrix} 0 \\ 0 \\ 0 \end{bmatrix}.$$

Since the elementary matrices have inverses, both equations have the same solution which is the zero vector.

Example 3.1.2. $A = \begin{bmatrix} 1 & 1 & 2 \\ 1 & 2 & 3 \\ 1 & 3 & 4 \\ 1 & 4 & 5 \end{bmatrix}$. Since column three is the sum of columns one and two, $Ax = 0_{4 \times 1}$ does not imply $x = 0_{3 \times 1}$.

Example 3.1.3. $Ax = \begin{bmatrix} 1 & 1 & 1 \\ 1 & 2 & 4 \\ 1 & 3 & 9 \\ 1 & 4 & 16 \end{bmatrix} \begin{bmatrix} x_1 \\ x_2 \\ x_3 \end{bmatrix} = \begin{bmatrix} 0 \\ 0 \\ 0 \\ 0 \end{bmatrix}$. Use row operations indicated by E to get

$$EAx = \begin{bmatrix} 1 & 1 & 1 \\ 0 & 1 & 3 \\ 0 & 0 & 2 \\ 0 & 0 & 6 \end{bmatrix} \begin{bmatrix} x_1 \\ x_2 \\ x_3 \end{bmatrix} = \begin{bmatrix} 0 \\ 0 \\ 0 \\ 0 \end{bmatrix}.$$

The third and fourth rows give $x_3 = 0$. Then the first and second rows imply $x_2 = 0$ and $x_1 = 0$.

Theorem 3.1.2. *(Full Rank and Normal Equations) If the $m \times n$ matrix A has full column rank ($Ax = 0_{m \times 1}$ implies $x = 0_{n \times 1}$), then $A^T A$ is nonsingular and the normal equations have a unique solution.*

Proof. $A^T Ax = 0_{n \times 1}$ implies $x^T (A^T Ax) = (Ax)^T (Ax) = 0$. Then $Ax = 0_{m \times 1}$ and the full column rank gives $x = 0_{n \times 1}$. So, the $n \times n$ matrix $A^T A$ is nonsingular. ∎

Example 3.1.4. $A = \begin{bmatrix} 1 & 1 \\ 1 & 2 \\ 1 & 3 \end{bmatrix}$ and $d = \begin{bmatrix} 10 \\ 9 \\ 7 \end{bmatrix}$. Then $A^T A = \begin{bmatrix} 3 & 6 \\ 6 & 14 \end{bmatrix}$ and $A^T d = \begin{bmatrix} 26 \\ 49 \end{bmatrix}$. So,

$$x = \begin{bmatrix} 3 & 6 \\ 6 & 14 \end{bmatrix}^{-1} \begin{bmatrix} 26 \\ 49 \end{bmatrix} = \begin{bmatrix} 35/3 \\ -3/2 \end{bmatrix}.$$

The slope of the desired line is $m = x_2 = -3/2$ and the y-intercept is $c = x_1 = 35/3$.

Example 3.1.5. Identify the parameters in the Newton cooling $(u(t) =$ temperature) or the Price decay $(p(t) =$ price) models:

$$\frac{du}{dt} = c(u_{sur} - u) \text{ or}$$

$$\frac{dp}{dt} = c(p_{\min} - p).$$

The price in a market place is discrete because the market is a large but finite. The heat model is continuous and so the differential equation may be a better choice! Subsection 12.3 returns to this topic.

A discrete approximation of the Price model is

$$\frac{p^{k+1} - p^{k-1}}{2\Delta t} = (cp_{\min}) + (-c)p^k \text{ where } p^k \simeq p(k\Delta t).$$

The unknown parameters are $(cp_{\min})$ and $(-c)$. If there are six data measurements for the past price, then A will be 4×2 with $x_1 = (-c)$ and $x_2 = (cp_{\min})$. The least squares problem is

$$\begin{bmatrix} p^2 & 1 \\ p^3 & 1 \\ p^4 & 1 \\ p^5 & 1 \end{bmatrix} \begin{bmatrix} x_1 \\ x_2 \end{bmatrix} = \begin{bmatrix} (p^{2+1} - p^{2-1})/(2\Delta t) \\ (p^{3+1} - p^{3-1})/(2\Delta t) \\ (p^{4+1} - p^{4-1})/(2\Delta t) \\ (p^{5+1} - p^{5-1})/(2\Delta t) \end{bmatrix}.$$

When this is solved, one can compute c and $p_{\min}$. The solution of the continuous price model has these values and an exponential function of time. Now predicted prices for future times can be done.

3.1.1   Exercises

1. Verify the row operations in Example 3.1.3.

2. Find the parabola that is closet to the data

x	$ax^2 + bx + c$
1	1
2	7
3	13
4	15

Use $x_1 = c$, $x_2 = b$ and $x_3 = a$.

FIGURE 3.2.1
Parameter identification using least squares

3.2 MATLAB® Code price_expdata.m

The code in this section is an implementation of the above example. The price
data is given in line 8, and the matrix and column vectors are computed in
lines 10–14. Solution of the normal equation is given in line 18. This is used in
lines 19–21 to compute the two parameters and a future price. The numerical
and graphical outputs are given in lines 25–31 and in Figure 3.2.1.

```
1    % Predict the price given more additional past prices.
2    % The exponential model is used.
3    % The method used is the normal equations.
4    %
5    % Input data
6    %
7    clear
8    price = [2080 2000 1950 1910 1875 1855];
9    time = 0:1:15;
10     for i = 2:5
11         d(i-1) = (price(i+1) - price(i-1))/2;
12     end
13     A = [price(2:5)' ones(4,1)]
14     d = d'
```

```
15      %
16      %  The normal equations are solved.
17      %
18      x = (A'*A)\(A'*d)                    % x = A\d
19      c = -x(1)
20      pmin = x(2)/(-x(1))
21      future_price = pmin +(2080 - pmin)*exp(-c*time);
22      %
23      %  Output is in both numerical and graphical form.
24      %
25      plot(time(1:6),price,'*',time, future_price)
26      title('price data and predicted price curve')
27      display('Predicted price at time = 8')
28      future_price(9)
29      display('Residual vector')
30      r = price' - future_price(1:6)'
31      rTr = r'*r
```

```
>>    price_expdata

    A =
       2000            1
       1950            1
       1910            1
       1875            1

    d =
      -65.0000
      -45.0000
      -37.5000
      -27.5000

    x =
       -0.2920
      520.8994

    c =
        0.2920

    pmin =
       1.7839e+03

Predicted price at time = 8
    =
       1.8126e+03
```

Residual vector
```
   r =
            0
      -5.0238
       0.9662
       2.7779
      -0.9983
       2.3187

   rTr =
      40.2620
```

3.2.1   Exercises

1. Consider the discrete price model for the following data

time	price
1	10,000
3	9,000
5	8,200
7	7,400
9	6,700
11	6,200
12	6,000

(a). Find the least squares system $Ax = d$.

(b). Find the solution.

(c). Predict the price at time equal to 15.

2. Consider the discrete Newton cooling model for a well-stirred cup of coffee

time	temperature
1	180
6	170
11	165
16	150
21	145
26	140
31	128

(a). Find the least squares system $Ax = d$.

(b). Find the solution.

(c). Predict the temperature at time equal to 40.

3.3   Basis of Subspace

In the least squares problem $Ax = d$ one can view the d to be the closest to

$$R(A) \equiv \{w : w = Ax \text{ and } x \in \mathbb{R}^m\}.$$

It is important to view the matrix vector product as a linear combination of the columns in the matrix. $R(A)$ is called the *range* of A, and it is a subspace of $\mathbb{R}^m$. Recall $S \subset \mathbb{R}^m$ is a *subspace* means $u+v \in S$ and $cu \in S$ when $u, v \in S$ and $c \in \mathbb{R}$. Consequently, S has the same vector space properties as those for $\mathbb{R}^m$. Another important subspace is the *null space* of A

$$N(A) \equiv \{x : Ax = 0 \text{ and } x \in \mathbb{R}^n\}.$$

Definition. *Let $S \subset \mathbb{R}^m$ be any subspace. The set of vectors $\{w_1, \cdots, w_k\}$ is called a basis of S if and only if*
   (i).   *each $w \in S$ is a linear combination of the basis*

$$w = c_1 w_1 + \cdots + c_k w_k \text{ and}$$

   (ii).   *the set is linearly independent*

$$c_1 w_1 + \cdots + c_k w_k = 0_{m \times 1} \text{ implies } c_1 = \cdots = c_k = 0.$$

The k is called the dimension, and it is unique regardless of the basis.

Remark. Let W be the $m \times k$ matrix formed by the columns of a basis. Linear independence means $Wc = 0_{m \times 1}$ implies $c = 0_{k \times 1}$.

Example 3.3.1. Let $S = \mathbb{R}^2$. Two bases are

$$w_1 = \begin{bmatrix} 1 \\ 0 \end{bmatrix} \text{ and } w_2 = \begin{bmatrix} 0 \\ 1 \end{bmatrix}$$

$$w_1 = \begin{bmatrix} 1 \\ 1 \end{bmatrix} \text{ and } w_2 = \begin{bmatrix} 1 \\ -2 \end{bmatrix}.$$

Example 3.3.2. The columns of the matrix A are linearly independent

$$A = \begin{bmatrix} 2000 & 1 \\ 1950 & 1 \\ 1910 & 1 \\ 1875 & 1 \end{bmatrix}.$$

Example 3.3.3. Consider the subset $S \subset \mathbb{R}^3$

$$S = \{[x_1 \ x_2 \ x_3]^T : x_1 - 6x_2 + x_3 = 0\}.$$

This is a subspace because for $u = [x_1 \ x_2 \ x_3]^T$ and $v = [y_1 \ y_2 \ y_3]^T$

$$u + v = [x_1 + y_1 \ x_2 + y_2 \ x_3 + y_3]^T$$
$$(x_1 + y_1) - 6(x_2 + y_2) + (x_3 + y_3) = (x_1 - 6x_2 + x_3) + (y_1 - 6y_2 + y_3)$$
$$= 0 + 0 = 0$$
$$su = [sx_1 \ sx_2 \ sx_3]^T$$
$$(sx_1) - 6(sx_2) + (sx_3) = s(x_1 - 6x_2 + x_3)$$
$$= s0 = 0.$$

The following two vectors are a basis for this space

$$\begin{bmatrix} 1 \\ 1/6 \\ 0 \end{bmatrix} \text{ and } \begin{bmatrix} 1 \\ 0 \\ -1 \end{bmatrix}.$$

Theorem 3.3.1. *(Basis Properties) If $S \subset \mathbb{R}^m$ has a basis $\{w_1, .., w_k\}$, then*

(i). each $w \in S$ has a unique linear combination and

(ii). $k = dim(S)$ is unique.

Proof. Consider the special case in the first item where $k = 2$. Let $w = c_1 w_1 + c_2 w_2$ and $w = \hat{c}_1 w_1 + \hat{c}_2 w_2$. Subtract these two

$$(c_1 - \hat{c}_1)w_1 + (c_2 - \hat{c}_2)w_2 = 0_{m \times 1}.$$

The linear independence implies $c_1 - \hat{c}_1 = 0$ and $c_2 - \hat{c}_2 = 0$. The general case is similar.

Consider the special case for the second item where $k = 2$ and $\hat{k} = 3$ with bases $\{w_1, w_2\}$ and $\{\hat{w}_1, \hat{w}_2, \hat{w}_3\}$. This will lead to a contradiction that $\hat{w}_3$ is a linear combination of $\hat{w}_1$ and $\hat{w}_2$. Since $\{w_1, w_2\}$ is a basis,

$$\hat{w}_1 = c_{11} w_1 + c_{12} w_2$$
$$\hat{w}_2 = c_{21} w_1 + c_{22} w_2 \text{ and}$$
$$\hat{w}_3 = c_{31} w_1 + c_{32} w_2.$$

At least one of c_{11} and c_{12} is not zero, say $c_{11} \neq 0$. Then

$$w_1 = \frac{\hat{w}_1 - c_{12} w_2}{c_{11}} \text{ and}$$
$$\hat{w}_2 = \frac{c_{21}}{c_{11}} \hat{w}_1 + \frac{-c_{21} c_{12} + c_{11} c_{22}}{c_{11}} w_2.$$

If $-c_{21} c_{12} + c_{11} c_{22} = 0$, then $\hat{w}_2 = \frac{c_{21}}{c_{11}} \hat{w}_1$ contradicts the linear independence of $\{\hat{w}_1, \hat{w}_2, \hat{w}_3\}$. If $-c_{21} c_{12} + c_{11} c_{22} \neq 0$, then one can solve w_2 as a linear combination of $\hat{w}_1$ and $\hat{w}_2$. So, both w_1 and w_2 are linear combinations of $\hat{w}_1$ and $\hat{w}_2$. From $\hat{w}_3 = c_{31} w_1 + c_{32} w_2$, $\hat{w}_3$ is also a linear combination of $\hat{w}_1$ and

$\widehat{w}_2$. This contradicts the linear independence of $\{\widehat{w}_1, \widehat{w}_2, \widehat{w}_3\}$. The proof of the general case is similar.

Another variation of the above paragraph is to show there is a nontrivial solution $x \in \mathbb{R}^3$ such that

$$
\begin{aligned}
0_{m \times 1} &= x_1 \widehat{w}_1 + x_2 \widehat{w}_2 + x_3 \widehat{w}_3 \\
&= x_1 \left(c_{11} w_1 + c_{12} w_2 \right) \\
&\quad + x_2 \left(c_{21} w_1 + c_{22} w_2 \right) \\
&\quad + x_3 \left(c_{31} w_1 + c_{32} w_2 \right) \\
&= \left(x_1 c_{11} + x_2 c_{21} + x_3 c_{31} \right) w_1 \\
&\quad + \left(x_1 c_{12} + x_2 c_{22} + x_3 c_{32} \right) w_2.
\end{aligned}
$$

There exists a nontrivial solution to the following algebraic system with two equations and three unknowns

$$
\begin{bmatrix} c_{11} & c_{21} & c_{31} \\ c_{12} & c_{22} & c_{32} \end{bmatrix} \begin{bmatrix} x_1 \\ x_2 \\ x_3 \end{bmatrix} = \begin{bmatrix} 0 \\ 0 \end{bmatrix}.
$$

In the above paragraph, we argued the $(1,1)$ and $(2,2)$ pivots were not zero. ∎

Example 3.3.4. Consider a 3×2 matrix

$$
A = \begin{bmatrix} 1 & 1 \\ 1 & 2 \\ 1 & 3 \end{bmatrix}.
$$

The two columns are linearly independent, and

$$
R(A) = \left\{ c_1 \begin{bmatrix} 1 \\ 1 \\ 1 \end{bmatrix} + c_2 \begin{bmatrix} 1 \\ 2 \\ 3 \end{bmatrix} : c_1, c_2 \in \mathbb{R} \right\}.
$$

Example 3.3.5. Consider a 3×3 matrix

$$
A = \begin{bmatrix} 1 & 1 & 2 \\ 1 & 2 & 3 \\ 1 & 3 & 4 \end{bmatrix}.
$$

The third column is the sum of the first two columns. The first two columns are linearly independent, and they are a basis for $R(A)$.

Example 3.3.6. Consider the least squares problem $Ax = d$ where

$$
A = \begin{bmatrix} 1 & 2 \\ 2 & 4 \\ 3 & 6 \end{bmatrix} \text{ and } d = \begin{bmatrix} 10 \\ 5 \\ 2 \end{bmatrix}.
$$

The normal equations have multiple solutions

$$x = c \begin{bmatrix} 2 \\ -1 \end{bmatrix} + \begin{bmatrix} 26/14 \\ 0 \end{bmatrix}.$$

This follows from $A \begin{bmatrix} 2 \\ -1 \end{bmatrix} = \begin{bmatrix} 0 \\ 0 \\ 0 \end{bmatrix}$.

$$A^T(d - Ax) = \begin{bmatrix} 1 & 2 & 3 \\ 2 & 4 & 6 \end{bmatrix} \left(\begin{bmatrix} 10 \\ 5 \\ 2 \end{bmatrix} - \begin{bmatrix} 0 \\ 0 \\ 0 \end{bmatrix} - 26/14 \begin{bmatrix} 1 \\ 2 \\ 3 \end{bmatrix} \right)$$

$$= \begin{bmatrix} 0 \\ 0 \end{bmatrix}.$$

We will return to this example.

3.3.1 Exercises

1. Consider the subset $S \subset \mathbb{R}^3$

$$S = \{[x_1 \ x_2 \ x_3]^T : x_1 - x_2 + 2x_3 = 0\}.$$

   (a). Show this is a subspace.

   (b). Show the vectors are a basis

$$\begin{bmatrix} 1 \\ 1 \\ 0 \end{bmatrix} \text{ and } \begin{bmatrix} 1 \\ 0 \\ -1/2 \end{bmatrix}.$$

2. Consider the least squares problem

$$\begin{bmatrix} 1 & 2 & 3 \\ 2 & 3 & 5 \\ 3 & 1 & 4 \\ 4 & 1 & 5 \end{bmatrix} \begin{bmatrix} x_1 \\ x_2 \\ x_3 \end{bmatrix} = \begin{bmatrix} 10 \\ 11 \\ 12 \\ 15 \end{bmatrix}.$$

   (a). Show the columns in the matrix are not linearly independent.

   (b). Verify the general solutions of the normal equations are

$$\begin{bmatrix} 48/15 \\ 32/15 \\ 0 \end{bmatrix} + c \begin{bmatrix} 1 \\ 1 \\ -1 \end{bmatrix} \text{ where } c \in \mathbb{R}.$$

3.4 Projection to Subspace

Consider a two dimensional subspace in $\mathbb{R}^3$. If a point $d \in \mathbb{R}^3$ is not in this subspace, then one can find a point in the subspace that is closest to d. The following is a generalization to $\mathbb{R}^m$. As we shall see this is related to the least squares problem.

Definition. *Let $d \in \mathbb{R}^m$ and $S \subset \mathbb{R}^m$ be a subspace with basis $\{w_1, \cdots, w_k\}$. $P_S(d) \in S$ is called a projection of d onto S if and only if*

$$(d - P_S(d))^T (d - P_S(d)) \le (d - w)^T (d - w) \text{ for all } w \in S.$$

Theorem 3.4.1. *(Projection to Subspace) Let $\{w_1, \cdots, w_k\}$ be a basis for $S \subset \mathbb{R}^m$. $P_S(d) \in S$ if and only if $w_j^T(d - P_S(d)) = 0$ for all $j = 1, \cdots, k$. Moreover,*

(i). the projection is unique so that $P_S(d)$ is a function,

(ii). $w^T(d - P_S(d)) = 0$ for all $w \in S$ and

(iii). $d^T d \ge P_S(d)^T P_S(d)$.

Proof. Use $P_S(d) \in S$ and w are a linear combinations of the basis

$$P_S(d) = \widehat{c}_1 w_1 + \cdots + \widehat{c}_k w_k \text{ and}$$
$$w = c_1 w_1 + \cdots + c_k w_k.$$

Let $W \equiv [w_1 \cdots w_k]$ be an $m \times k$ matrix. Then $P_S(d) = W\widehat{c}$, $w = Wc$ and the projection gives a least squares solution of $W\widehat{c} = d$

$$(d - P_S(d))^T (d - P_S(d)) \le (d - w)^T (d - w)$$
$$(d - W\widehat{c})^T (d - W\widehat{c}) \le (d - Wc)^T (d - Wc).$$

Because $\{w_1, \cdots, w_k\}$ is a basis, W has full column rank and $W^T W$ is nonsingular. In this case the normal equations $W^T(d - Wc) = 0_{k \times 1}$ have a unique solution.

The proof of the inequality follows from $P_S(d) \in S$ and $P_S(d)^T(d - P_S(d)) = 0$. Write $d = (d - P_S(d)) + P_S(d)$ to get

$$d^T d = (d - P_S(d))^T (d - P_S(d)) +$$
$$2 P_S(d)^T (d - P_S(d)) + P_S(d)^T P_S(d)$$
$$= (d - P_S(d))^T (d - P_S(d)) + 0 + P_S(d)^T P_S(d)$$
$$\ge P_S(d)^T P_S(d).$$

■

Example 3.4.1. Consider the subspace $S \subset \mathbb{R}^3$

$$S = \{[x_1 \ x_2 \ x_3]^T : x_1 - 6x_2 + x_3 = 0\}.$$

Find the projection of $d = [1 \ 2 \ 3]^T$ onto S. This is the closest point in S to d. One basis is

$$w_1 = \begin{bmatrix} 6 \\ 1 \\ 0 \end{bmatrix} \text{ and } w_2 = \begin{bmatrix} 1 \\ 0 \\ -1 \end{bmatrix}.$$

The above theorem requires $P_S(d) = \widehat{c}_1 w_1 + \widehat{c}_2 w_2$ and

$$w_1^T(d - P_S(d)) = 0 \text{ and } w_2^T(d - P_S(d)) = 0.$$

The algebraic system for the unknown coefficients is

$$\begin{bmatrix} 37 & 6 \\ 6 & 2 \end{bmatrix} \begin{bmatrix} \widehat{c}_1 \\ \widehat{c}_2 \end{bmatrix} = \begin{bmatrix} 8 \\ -2 \end{bmatrix}.$$

The projection is

$$\begin{aligned}
P_S(d) &= \widehat{c}_1 w_1 + \widehat{c}_2 w_2 \\
&= (0.7368)w_1 + (-3.2105)w_2 \\
&= (0.7368) \begin{bmatrix} 6 \\ 1 \\ 0 \end{bmatrix} + (-3.2105) \begin{bmatrix} 1 \\ 0 \\ -1 \end{bmatrix} \\
&= \begin{bmatrix} 1.2105 \\ 0.7368 \\ 3.2105 \end{bmatrix}.
\end{aligned}$$

The projection of d onto $S = R(A)$ can be used to identify all least squares solutions.

Theorem 3.4.2. *(Projection and Least Squares) Let A be $m \times n$ and $S = R(A)$ be the range of A. Then $x = \widehat{x} + x_N$ are least squares solutions to $Ax = d$ where*

$$\begin{aligned}
P_{R(A)}(d) &\in R(A), \\
A\widehat{x} &= P_{R(A)}(d) \text{ and} \\
Ax_N &= 0_{m \times 1}.
\end{aligned}$$

Proof. Since $P_{R(A)}(d) \in R(A)$, there exists $\widehat{x}$ such that $A\widehat{x} = P_{R(A)}(d)$. Because $P_{R(A)}(d)$ is a projection, for all $w \in R(A)$ with $w = Ay$

$$(d - P_{R(A)}(d))^T(d - P_{R(A)}(d)) \leq (d - w)^T(d - w)$$
$$(d - A\widehat{x})^T(d - A\widehat{x}) \leq (d - Ay)^T(d - Ay).$$

So, $\widehat{x}$ is a least squares solution and must satisfy the normal equations. Because $Ax_N = 0_{m \times 1}$, $\widehat{x} + x_N$ must also satisfy the normal equations

$$A^T A(\widehat{x} + x_N) = A^T(A\widehat{x} + 0_{m \times 1}) = A^T d.$$

■

Example 3.4.2. Revisit Example 3.3.6 where

$$A = \begin{bmatrix} 1 & 2 \\ 2 & 4 \\ 3 & 6 \end{bmatrix} \text{ and } R(A) = \{c \begin{bmatrix} 1 \\ 2 \\ 3 \end{bmatrix} : c \in \mathbb{R}\}.$$

Here $m = 3, n = 2$ and $k = 1$ with

$$w_1 = \begin{bmatrix} 1 \\ 2 \\ 3 \end{bmatrix} \text{ and } d = \begin{bmatrix} 10 \\ 5 \\ 2 \end{bmatrix}.$$

Require $w_1^T(d - P_{R(A)}(d)) = 0$

$$\begin{bmatrix} 1 \\ 2 \\ 3 \end{bmatrix}^T (\begin{bmatrix} 10 \\ 5 \\ 2 \end{bmatrix} - c \begin{bmatrix} 1 \\ 2 \\ 3 \end{bmatrix}) = 0.$$

Then $c = 26/14$ and

$$A\widehat{x} = P_{R(A)}(d)$$

$$\begin{bmatrix} 1 & 2 \\ 2 & 4 \\ 3 & 6 \end{bmatrix} \begin{bmatrix} 26/14 \\ 0 \end{bmatrix} = 26/14 \begin{bmatrix} 1 \\ 2 \\ 3 \end{bmatrix}.$$

Choose $\widehat{x} = \begin{bmatrix} 26/14 \\ 0 \end{bmatrix}$ and $x_N = \widehat{c} \begin{bmatrix} 2 \\ -1 \end{bmatrix}$.

3.4.1   Exercises

1. In Theorem 3.4.1 show the second item is true.

2. Consider the subspace $S \subset \mathbb{R}^3$

$$S = \{[x_1 \ x_2 \ x_3]^T : x_1 - x_2 + 2x_3 = 0\}.$$

   (a). Find a basis for S.

   (b). Use Theorem 3.4.1 to find the point in S that is closest to $d = [4 \ 6 \ 8]^T$.

3. Consider the least squares problem

$$\begin{bmatrix} 1 & 2 & 3 \\ 2 & 3 & 5 \\ 3 & 1 & 4 \\ 4 & 1 & 5 \end{bmatrix} \begin{bmatrix} x_1 \\ x_2 \\ x_3 \end{bmatrix} = \begin{bmatrix} 10 \\ 11 \\ 12 \\ 15 \end{bmatrix}.$$

(a). Show the first two columns are a basis for $R(A)$.

(b). Find the projection of d onto $R(A)$.

(c). Find the $\widehat{x}$ and x_N as given in Theorem 3.4.2.

4

$Ax = d$ with $m < n$

Elementary matrices (row operations) will be used to construct the basis for the null space, $N(A)$, and the range space, $R(A)$. This construction will first be illustrated by several examples. The second section will use row echelon form to find the bases for the range and null spaces. The general case is described in the third section. These bases are not orthonormal, but they still are useful in the general least squares problem. The last section will illustrate how null spaces can be effectively used to solve equilibrium equations that evolve from circuits, structures and fluid flow models.

4.1 Examples in $\mathbb{R}^3$

When there are fewer equations than unknowns, there may be multiple solutions or no solution. The general solution will be a particular solutions plus and homogenous solution.

Example 4.1.1. Consider one equation with three variables

$$x_1 + 2x_2 + 3x_3 = 10.$$

If we set the second and third variable equal to zero, then the first variable must equal 10. In vector notation a particular solution is

$$x_{part} = \begin{bmatrix} 10 \\ 0 \\ 0 \end{bmatrix}.$$

The homogenous or null solution must satisfy

$$x_1 + 2x_2 + 3x_3 = 0.$$

Then the first variable can be solved for in terms of the second and third variables $x_1 = -2x_2 - 3x_3$. In vector notation this is

$$x_{null} = x_2 \begin{bmatrix} -2 \\ 1 \\ 0 \end{bmatrix} + x_3 \begin{bmatrix} -3 \\ 0 \\ 1 \end{bmatrix}.$$

DOI: 10.1201/9781003304128-4

The second and third variables are called *free* and the first variable is called *fixed*. The general solution is $x_{part} + x_{null}$.

Example 4.1.2. Consider two equations with three variables

$$x_1 + 2x_2 + 3x_3 = 10 \ \text{and}$$
$$-x_1 + 2x_2 + 3x_3 = 14.$$

Add the first equation to the second to get $4x_2 + 6x_3 = 24$. A particular solution can be found by setting the third variable equal to zero. This gives $x_2 = 6$ and $x_1 = -2$. The vector form is

$$x_{part} = \begin{bmatrix} -2 \\ 6 \\ 0 \end{bmatrix}.$$

The null solution must satisfy

$$x_1 + 2x_2 + 3x_3 = 0 \ \text{and}$$
$$-x_1 + 2x_2 + 3x_3 = 0.$$

The first and second variables can be solved for in terms of the third variable to get

$$x_2 = (-3/2)x_3 \ \text{and} \ x_1 = (0)x_3$$

The vector form of this has one free variable and is

$$x_{null} = x_3 \begin{bmatrix} 0 \\ -3/2 \\ 1 \end{bmatrix}$$

Example 4.1.3. The system may not have a solution. Think of two planes in $\mathbb{R}^3$ that are parallel such as

$$x_1 + 2x_2 + 3x_3 = 10 \ \text{and}$$
$$-x_1 - 2x_2 - 3x_3 = 1.$$

Add the two equations to get $0 = 11$. Such systems are called *inconsistent*.

4.2 Row Echelon Form

This section presents a systematic way of representing solutions of $Ax = d$ where A is an $m \times n$ matrix with $m < n$. The main step is to use elementary row operations to transform the system to an equivalent system in row echelon form. The examples in the previous section illustrated this for $m = 1$ or 2 equations and $n = 3$ variables. This will allow us to identify the homogeneous (null) and particular solutions. A number of interesting examples will be done by-hand calculations. The bases for the null and range spaces will be constructed from the row echelon form.

4.2.1 Solutions in $\mathbb{R}^4$

The solution in $\mathbb{R}^4$ of $x_1 + 2x_2 + 3x_3 = 10$ is different from the solution of the same equation in $\mathbb{R}^3$. In Example 4.1.1 we derived the solution in $\mathbb{R}^3$ to be

$$x = x_2 \begin{bmatrix} -2 \\ 1 \\ 0 \end{bmatrix} + x_3 \begin{bmatrix} -3 \\ 0 \\ 1 \end{bmatrix} + \begin{bmatrix} 10 \\ 0 \\ 0 \end{bmatrix}.$$

In $\mathbb{R}^4$ the equation is $x_1 + 2x_2 + 3x_3 + 0x_4 = 10$ and there is a third free variable x_4. Hence, the homogeneous solution in $\mathbb{R}^4$ must satisfy $x_1 + 2x_2 + 3x_3 + 0x_4 = 0$. The vector form of the homogeneous (null) solution is

$$x_{null} = x_2 \begin{bmatrix} -2 \\ 1 \\ 0 \\ 0 \end{bmatrix} + x_3 \begin{bmatrix} -3 \\ 0 \\ 1 \\ 0 \end{bmatrix} + x_4 \begin{bmatrix} 0 \\ 0 \\ 0 \\ 1 \end{bmatrix}.$$

The three vectors in $\mathbb{R}^4$ form a basis for the null space of $A = [1 \ 2 \ 3 \ 0]$. A particular solution is now a vector in $\mathbb{R}^4$ and is $x_p = [10 \ 0 \ 0 \ 0]^T$, and so, the general solution in $\mathbb{R}^4$ is

$$\mathbf{x} = x_2 \begin{bmatrix} -2 \\ 1 \\ 0 \\ 0 \end{bmatrix} + x_3 \begin{bmatrix} -3 \\ 0 \\ 1 \\ 0 \end{bmatrix} + x_4 \begin{bmatrix} 0 \\ 0 \\ 0 \\ 1 \end{bmatrix} + \begin{bmatrix} 10 \\ 0 \\ 0 \\ 0 \end{bmatrix}.$$

The next example illustrates that increasing the number of variables can convert an inconsistent smaller variable problem into a problem with multiple solutions! The following system with two equations in $\mathbb{R}^3$ is inconsistent (has no solution)

$$x_1 + 2x_2 + 3x_3 = 1 \text{ and}$$
$$2x_1 + 4x_2 + 6x_3 = 3.$$

Modify these two equations by introducing a fourth variable

$$x_1 + 2x_2 + 3x_3 + 4x_4 = 1 \text{ and} \tag{4.2.1}$$
$$2x_1 + 4x_2 + 6x_3 + 9x_4 = 3$$

Use an elementary matrix transformation on the augmented matrix to obtain the representation of an equivalent system

$$E_{21}(-2)\,[A \ d] = \begin{bmatrix} 1 & 0 \\ -2 & 1 \end{bmatrix} \begin{bmatrix} 1 & 2 & 3 & 4 & 1 \\ 2 & 4 & 6 & 9 & 3 \end{bmatrix}$$
$$= \begin{bmatrix} 1 & 2 & 3 & 4 & 1 \\ 0 & 0 & 0 & 1 & 1 \end{bmatrix}.$$

Choose two "free variables" to be x_2 and x_3. Then in $\mathbb{R}^4$ an equivalent system is

$$\begin{bmatrix} 1 & 4 \\ 0 & 1 \end{bmatrix} \begin{bmatrix} x_1 \\ x_4 \end{bmatrix} = \begin{bmatrix} 1 \\ 1 \end{bmatrix} - x_2 \begin{bmatrix} 2 \\ 0 \end{bmatrix} - x_3 \begin{bmatrix} 3 \\ 0 \end{bmatrix}.$$

Set x_2 and x_3 equal to zero to find a particular solution $x_{part} = [-3 \ 0 \ 0 \ 1]^T$. In order to find the homogeneous solution, consider

$$\begin{bmatrix} 1 & 4 \\ 0 & 1 \end{bmatrix} \begin{bmatrix} x_1 \\ x_4 \end{bmatrix} = \begin{bmatrix} 0 \\ 0 \end{bmatrix} - x_2 \begin{bmatrix} 2 \\ 0 \end{bmatrix} - x_3 \begin{bmatrix} 3 \\ 0 \end{bmatrix}.$$

Then $x_4 = 0$ and $x_1 = -2x_2 - 3x_3$ or in vector notation the homogeneous (null) solution is

$$x_{null} = x_2 \begin{bmatrix} -2 \\ 1 \\ 0 \\ 0 \end{bmatrix} + x_3 \begin{bmatrix} -3 \\ 0 \\ 1 \\ 0 \end{bmatrix}.$$

The general solution in $\mathbb{R}^4$

$$x = x_2 \begin{bmatrix} -2 \\ 1 \\ 0 \\ 0 \end{bmatrix} + x_3 \begin{bmatrix} -3 \\ 0 \\ 1 \\ 0 \end{bmatrix} + \begin{bmatrix} -3 \\ 0 \\ 0 \\ 1 \end{bmatrix}.$$

In the previous examples we used row operations to transform the augmented matrix into another matrix with all zeros in the "lower left" portion of the matrix. The "free variables" were defined to be the variables not located at a "corner" (called a pivot) of the matrix. These were illustrated by the transformed matrices in Examples 4.1.1, 4.1.2 and 4.2.3. A more precise definition of the transformed matrix and one that generalizes beyond two equations is row echelon form, which are found by a sequence of row operations.

Definition. *The $m \times (n+1)$ augmented matrix $[A \ d]$ has been transformed into a row echelon form matrix by a product of elementary matrices E if and only if*

$$E \, [A \ d] = B \text{ where}$$

(i). *any nonzero row in B has more leading zero components than the previous row and*

(ii). *any row below a zero row must also be a zero row.*

In terms of the components b_{ij} of B this means $b_{ij} = 0$ for $j < n_i$, $b_{in_i} \neq 0$ and $n_{i-1} < n_i$. The b_{in_i} are called the pivots, the variables x_{n_i} are called fixed and the remaining variables are called free. The number of pivots in B is called the rank of $[A \ d]$, and the number of pivots in the first n columns of B is called the rank of A.

Example 4.2.1. In the above system in line (4.2.1) with two equations and four variables the augmented matrix is

$$[A \ d] = \begin{bmatrix} 1 & 2 & 3 & 4 & 1 \\ 2 & 4 & 6 & 9 & 3 \end{bmatrix}.$$

The row echelon form matrix is

$$E_{21}(-2) \ [A \ d] = \begin{bmatrix} 1 & 2 & 3 & 4 & 1 \\ 0 & 0 & 0 & 1 & 1 \end{bmatrix} \quad \text{where}$$

the two pivots are $b_{1n_1} = 1$ with $n_1 = 1$ and $b_{2n_2} = 1$ with $n_2 = 4$. The fixed variables are x_1 and x_4, and the free variables are x_2 and x_3. The rank of the 2×4 matrix A is two.

Example 4.2.2. Consider the following system with three equations and four unknowns

$$\begin{aligned} x_1 + 2x_2 + x_3 + 3x_4 &= 1 \\ 2x_1 + 4x_2 + x_3 + 7x_4 &= 3 \\ x_3 + x_4 &= 7. \end{aligned}$$

The augmented matrix is

$$[A \ d] = \begin{bmatrix} 1 & 2 & 1 & 3 & 1 \\ 2 & 4 & 1 & 7 & 3 \\ 0 & 0 & 1 & 1 & 7 \end{bmatrix}.$$

Subtract 2 times row 1 from row 2 and then add row 2 to row 3 to get

$$E_{32}(1) \ E_{21}(-2) \ [A \ d] = \begin{bmatrix} 1 & 2 & 1 & 3 & 1 \\ 0 & 0 & -1 & 1 & 1 \\ 0 & 0 & 0 & 2 & 8 \end{bmatrix}.$$

The three pivots are $b_{1n_1} = 1$ with $n_1 = 1$, $b_{2n_2} = -1$ with $n_2 = 3$ and $b_{3n_3} = 2$ with $n_3 = 4$. The fixed variables are x_1, x_3 and x_4, and the free variable is x_2. The rank of the 3×4 matrix A is three.

The row echelon form matrix is a short way of writing the equivalent system

$$\begin{bmatrix} 1 & 1 & 3 \\ 0 & -1 & 1 \\ 0 & 0 & 2 \end{bmatrix} \begin{bmatrix} x_1 \\ x_3 \\ x_4 \end{bmatrix} = \begin{bmatrix} 1 \\ 1 \\ 8 \end{bmatrix} - x_2 \begin{bmatrix} 2 \\ 0 \\ 0 \end{bmatrix}.$$

A particular solution is found by setting the free variable $x_2 = 0$ and solving the upper triangular system for the fixed variables. Since the diagonal components are the pivots, the upper triangular solve can be done to yield a particular solution $\mathbf{x}_p = [-14 \ 0 \ 3 \ 4]^T$. In order to find the homogeneous solution, replace the right side d by the zero vector and solve

$$\begin{bmatrix} 1 & 1 & 3 \\ 0 & -1 & 1 \\ 0 & 0 & 2 \end{bmatrix} \begin{bmatrix} x_1 \\ x_3 \\ x_4 \end{bmatrix} = -x_2 \begin{bmatrix} 2 \\ 0 \\ 0 \end{bmatrix}.$$

This gives $x_4 = 0$, $x_3 = 0$ and $x_1 = -2x_2$ or in vector notation the homogeneous solution is

$$x_{null} = x_2 \begin{bmatrix} -2 \\ 1 \\ 0 \\ 0 \end{bmatrix}.$$

The general solution in $\mathbb{R}^4$ is

$$x = x_2 \begin{bmatrix} -2 \\ 1 \\ 0 \\ 0 \end{bmatrix} + \begin{bmatrix} -14 \\ 0 \\ 3 \\ 4 \end{bmatrix}.$$

4.2.2 General solution of $Ax = d$

Consider the algebraic system $Ax = d$ where A is an $m \times n$ matrix with $m < n$. Let E be a product of elementary matrices such that

$$E \begin{bmatrix} A & d \end{bmatrix} = \begin{bmatrix} EA & Ed \end{bmatrix} = B$$

is in row echelon form. Since E has an inverse, the solution sets of the algebraic problems $Ax = d$ and $EAx = Ed$ are equal. However, there may not be a solution as in

$$x_1 + 2x_2 + 3x_3 = 10 \text{ and}$$
$$-x_1 - 2x_2 - 3x_3 = 1.$$

The augmented matrix is

$$\begin{bmatrix} A & d \end{bmatrix} = \begin{bmatrix} 1 & 2 & 3 & 10 \\ -1 & -2 & -3 & 1 \end{bmatrix}$$

$$E_{21}(1) \begin{bmatrix} A & d \end{bmatrix} = \begin{bmatrix} 1 & 2 & 3 & 10 \\ 0 & 0 & 0 & 11 \end{bmatrix}$$

The last row implies a contradiction

$$0x_1 + 0x_2 + 0x_3 = 11$$

The rank of the 2×3 matrix A is one, and the rank of the 2×4 augmented matrix $\begin{bmatrix} A & d \end{bmatrix}$ is two.

Theorem 4.2.1. *(Inconsistent Row Echelon Form) The algebraic system $Ax = d$ is inconsistent (has no solution) if and only if any row, say row i, in the row echelon form matrix B has all zeros to the left of component $b_{i,n+1}$ and $b_{i,n+1}$ is not zero.*

The existence of *particular solution* requires that there be some vector x_{part} such that $Ax_{part} = d$. This means d must be a linear combination of the column vectors of A. The *homogeneous(null) solutions* of $Ax_{null} = 0$ are linear combinations of column vectors obtained from the free variables given by the row echelon matrix. The *multiple solutions* are all homogeneous solutions plus a particular solution.

Theorem 4.2.2. *(General Solution of $Ax = d$)* *If $d \in R(A)$, A is $m \times n$ with $m < n$, and $N(A)$ has a basis $\{w_1, \cdots, w_K\}$ where $K = n - rank(A)$ is the number of free variables, the general solution of $Ax = d$ has the form*

$$x = x_{part} + x_{null} = x_{part} + c_1 w_1 + \cdots + c_K w_K.$$

Outline for Solving $Ax = d$.

1. Find the augmented matrix $[A \ d]$.

2. Use elementary matrices to find the row echelon matrix $E[A \ d]$.

3. There is no solution if the row echelon form matrix is inconsistent.

4. Use the pivots to find the fixed and free variables.

5. Find a particular solution of $EAx_{part} = Ed$.

   Set the free variables equal to zero and solve for the fixed variables.

6. Find the homogeneous solutions of $EAx_{null} = E0 = 0$.

   Replace Ed by the zero vector and solve for the fixed variables in terms of the free variables.

7. The general solution is the linear combination the homogeneous solution plus a particular solution.

Example 4.2.3. This example is more complicated since it has four equations and six unknowns

$$\begin{aligned}
x_1 + x_2 + x_3 + x_4 + x_5 + x_6 &= 1 \\
x_1 + 2x_2 + x_4 + x_6 &= 1 \\
x_2 + x_3 + 2x_4 + x_5 + x_6 &= 2 \\
2x_5 + x_6 &= 4.
\end{aligned}$$

The following elementary row operations and the augmented matrix generate the row echelon form matrix

$$[A \ d] = \begin{bmatrix} 1 & 1 & 1 & 1 & 1 & 1 & 1 \\ 1 & 2 & 0 & 1 & 0 & 1 & 1 \\ 0 & 1 & 1 & 2 & 1 & 1 & 2 \\ 0 & 0 & 0 & 0 & 2 & 1 & 4 \end{bmatrix}$$

$$E_{32}(-1) \ E_{21}(-1) \ [A \ d] = \begin{bmatrix} 1 & 1 & 1 & 1 & 1 & 1 & 1 \\ 0 & 1 & -1 & 0 & -1 & 0 & 0 \\ 0 & 0 & 2 & 2 & 2 & 1 & 2 \\ 0 & 0 & 0 & 0 & 2 & 1 & 4 \end{bmatrix}.$$

The four pivots are $b_{1n_1} = 1$ with $n_1 = 1$, $b_{2n_2} = 1$ with $n_2 = 2$, $b_{3n_3} = 2$ with $n_3 = 3$ and $b_{4n_4} = 2$ with $n_4 = 5$. The fixed variables are x_1, x_2, x_3 and x_5, and the free variables are is x_4 and x_6. The rank of the 4×6 matrix A is four. The row echelon matrix gives the following system

$$
\begin{bmatrix} 1 & 1 & 1 & 1 \\ 0 & 1 & -1 & -1 \\ 0 & 0 & 2 & 2 \\ 0 & 0 & 0 & 2 \end{bmatrix}
\begin{bmatrix} x_1 \\ x_2 \\ x_3 \\ x_5 \end{bmatrix}
=
\begin{bmatrix} 1 \\ 0 \\ 2 \\ 4 \end{bmatrix}
- x_4 \begin{bmatrix} 1 \\ 0 \\ 2 \\ 0 \end{bmatrix}
- x_6 \begin{bmatrix} 1 \\ 0 \\ 1 \\ 1 \end{bmatrix}.
$$

By setting the two free variables to zero and solving the upper triangular system, we find a particular solution $x_{part} = [-1 \ \ 1 \ -1 \ \ 0 \ \ 2 \ \ 0]^T$. Set the vector d equal to the zero vector and solve the following for the homogeneous solutions

$$
\begin{bmatrix} 1 & 1 & 1 & 1 \\ 0 & 1 & -1 & -1 \\ 0 & 0 & 2 & 2 \\ 0 & 0 & 0 & 2 \end{bmatrix}
\begin{bmatrix} x_1 \\ x_2 \\ x_3 \\ x_5 \end{bmatrix}
=
- x_4 \begin{bmatrix} 1 \\ 0 \\ 2 \\ 0 \end{bmatrix}
- x_6 \begin{bmatrix} 1 \\ 0 \\ 1 \\ 1 \end{bmatrix}.
$$

The general solution in $\mathbb{R}^6$ is

$$
x = x_{null} + x_{part} = x_4 \begin{bmatrix} 1 \\ -1 \\ -1 \\ 1 \\ 0 \\ 0 \end{bmatrix} + x_6 \begin{bmatrix} 0 \\ -1/2 \\ 0 \\ 0 \\ -1/2 \\ 1 \end{bmatrix} + \begin{bmatrix} -1 \\ 1 \\ -1 \\ 0 \\ 2 \\ 0 \end{bmatrix}.
$$

The null space of A has dimension equal to two, and the dimension of the range space is four.

4.2.3   Exercises

1. Consider the following system with three equations and four unknowns

$$
2x_1 + x_2 + x_3 + 3x_4 = 4
$$
$$
2x_1 + 4x_2 + 2x_3 + 11x_4 = 10
$$
$$
3x_2 + x_3 + 8x_4 = 6.
$$

(a). Find the augmented matrix $[A \ \ d]$.

(b). Use row operations to find the row echelon form, the free variables and fixed variables.

(c). Find the homogeneous (null) solution.

(d). Find a particular solution.

(e). Find the general solution.

4.3 Relationship of $R(A), N(A^T)$ and $R(A^T), N(A)$

The matrix can be viewed as linear map where A is $m \times n$ and A^T is $n \times m$

$$A : \mathbb{R}^n \longrightarrow \mathbb{R}^m \text{ and } A^T : \mathbb{R}^m \longrightarrow \mathbb{R}^n.$$

This means $R(A) \subset \mathbb{R}^m$ and $N(A^T) \subset \mathbb{R}^m$. Furthermore, if $y \in N(A^T)$ and $z \in R(A)$, then $A^T y = 0$ and $z = Ax$. This leads of the following important observation

$$\begin{aligned} z^T y &= (Ax)^T y \\ &= (x^T A^T) y \\ &= x^T (A^T y) = 0. \end{aligned} \tag{4.3.1}$$

This means any element in $N(A^T)$ is perpendicular (orthogonal) to any element in $R(A)$. The following examples illustrate this, show it is related to the consistent condition, and the formation of the bases.

Example 4.3.1. $A = \begin{bmatrix} 1 & 1 & 2 & 3 \\ 2 & 2 & 8 & 10 \\ 3 & 3 & 10 & 13 \end{bmatrix}$ with $m = 3$ and $n = 4$.

$Ax = d$ may be rewritten in compact form as an augmented matrix

$$[A \ d] = \begin{bmatrix} 1 & 1 & 2 & 3 & d_1 \\ 2 & 2 & 8 & 10 & d_2 \\ 3 & 3 & 10 & 13 & d_3 \end{bmatrix}.$$

Use elementary matrices (row operations) to transform it to row echelon form

$$E_{31}(-3)E_{21}(-2)[A \ d] = \begin{bmatrix} 1 & 1 & 2 & 3 & d_1 \\ 0 & 0 & 4 & 4 & d_2 - 2d_1 \\ 0 & 0 & 4 & 4 & d_3 - 3d_1 \end{bmatrix}$$

$$E_{32}(-1)E_{31}(-3)E_{21}(-2)[A \ d] = \begin{bmatrix} 1 & 1 & 2 & 3 & d_1 \\ 0 & 0 & 4 & 4 & d_2 - 2d_1 \\ 0 & 0 & 0 & 0 & d_3 - d_1 - d_2 \end{bmatrix}.$$

The pivots are $(1,1)$ and $(2,3)$, the fixed variables are x_1, x_3 and the free variables are x_2, x_4. Let E be the product of the elementary matrices

$$E = \begin{bmatrix} 1 & 0 & 0 \\ -2 & 1 & 0 \\ -1 & -1 & 1 \end{bmatrix}.$$

$E[A \ d] = [EA \ Ed]$ or $EAx = Ux = Ed$. If $[Ed]_3 = 0$, then $Ux = Ed$ has a solution. Because E has an inverse, $Ux = Ed$ and $Ax = d$ have the same solutions.

In the above example $rank(A) = 2$. Use the notation $ipiv(1) = 1, jpiv(1) = 1$ for the first pivot and $ipiv(2) = 2, jpiv(2) = 3$ for the second pivot.

Find $R(A)$ by setting the free variables to zero and solving $Ux = Ed$

$$\begin{bmatrix} 1 & 2 \\ 0 & 4 \end{bmatrix} \begin{bmatrix} x_1 \\ x_3 \end{bmatrix} = \begin{bmatrix} d_1 \\ d_2 - 2d_1 \end{bmatrix}.$$

Use the pivot columns of A in $E^{-1}Ux = Ax = E^{-1}Ed = d$.

$$R(A) = \left\{ y : y = \begin{bmatrix} 1 \\ 2 \\ 3 \end{bmatrix} x_1 + \begin{bmatrix} 2 \\ 8 \\ 10 \end{bmatrix} x_3 \right\}. \tag{4.3.2}$$

Since the first and third columns are linearly independent, they are a basis and $\dim(R(A)) = 2$.

Find $N(A)$ by using the free variables. Solve the equivalent $Ax = 0_{3 \times 1}$ and $EAx = Ux = 0_{3 \times 1}$. Solve the pivot variables in terms of the free variables

$$\begin{bmatrix} 1 & 2 \\ 0 & 4 \end{bmatrix} \begin{bmatrix} x_1 \\ x_3 \end{bmatrix} = \begin{bmatrix} -x_2 - 3x_4 \\ -4x_4 \end{bmatrix} \text{ and}$$

$$\begin{bmatrix} x_1 \\ x_3 \end{bmatrix} = \begin{bmatrix} -1 \\ 0 \end{bmatrix} x_2 + \begin{bmatrix} -1 \\ -1 \end{bmatrix} x_4.$$

$$N(A) = \left\{ x : x = \begin{bmatrix} -1 \\ 1 \\ 0 \\ 0 \end{bmatrix} x_2 + \begin{bmatrix} -1 \\ 0 \\ -1 \\ 1 \end{bmatrix} x_4 \right\}. \tag{4.3.3}$$

Since the columns are linearly independent, they form a basis and $\dim(N(A)) = 4 - 2 = 2$.

Example 4.3.2. $A^T = \begin{bmatrix} 1 & 2 & 3 \\ 1 & 2 & 3 \\ 2 & 8 & 10 \\ 3 & 10 & 13 \end{bmatrix}$. Solve $A^T y = g$ by using elementary row operations to form $\widehat{E}A^T = \widehat{U}$ and then solve $\widehat{U}y = \widehat{E}g$.

$$\widehat{E}[A^T \ g] = \begin{bmatrix} 1 & 0 & 0 & 0 \\ -2 & 0 & 1 & 0 \\ -1 & 1 & 0 & 0 \\ -1 & 0 & -1 & 1 \end{bmatrix} \begin{bmatrix} 1 & 2 & 3 & g_1 \\ 1 & 2 & 3 & g_2 \\ 2 & 8 & 10 & g_3 \\ 3 & 10 & 13 & g_4 \end{bmatrix}$$

$$= \begin{bmatrix} 1 & 2 & 3 & g_1 \\ 0 & 4 & 4 & -2g_2 + g_3 \\ 0 & 0 & 0 & -g_1 + g_2 \\ 0 & 0 & 0 & -g_1 - g_3 + g_4 \end{bmatrix}.$$

The pivots are $(1,1)$ and $(2,2)$, the fixed variables are y_1, y_2 and the free variable is y_3. In order to solve $\widehat{U}y = \widehat{E}g$, the third and fourth components of $\widehat{E}g$ must be zero. The rank$(A^T) = \widehat{r} = 2$, and use the notation $ipiv(1) = 1, jpiv(1) = 1$ for the first pivot and $ipiv(2) = 2, jpiv(2) = 2$ for the second pivot.

Find $R(A^T)$ by the pivot columns of A^T. Set the free variable $y_3 = 0$ and solve

$$\begin{bmatrix} 1 & 2 \\ 0 & 4 \end{bmatrix} \begin{bmatrix} y_1 \\ y_2 \end{bmatrix} = \begin{bmatrix} g_1 \\ -2g_2 + g_3 \end{bmatrix}.$$

This solves

$$\widehat{E}A^T \begin{bmatrix} y_1 \\ y_2 \\ 0 \end{bmatrix} = \widehat{E}g = \begin{bmatrix} g_1 \\ -2g_2 + g_3 \\ 0 \\ 0 \end{bmatrix},$$

and because $\widehat{E}$ is nonsingular

$$A^T \begin{bmatrix} y_1 \\ y_2 \\ 0 \end{bmatrix} = g$$

$$\begin{bmatrix} 1 \\ 1 \\ 2 \\ 3 \end{bmatrix} y_1 + \begin{bmatrix} 2 \\ 2 \\ 8 \\ 10 \end{bmatrix} y_2 = g.$$

The first two columns of A^T are linearly independent and form a basis for

$$R(A^T) = \left\{ w : w = \begin{bmatrix} 1 \\ 1 \\ 2 \\ 3 \end{bmatrix} y_1 + \begin{bmatrix} 2 \\ 2 \\ 8 \\ 10 \end{bmatrix} y_2 \right\}. \tag{4.3.4}$$

Find $N(A^T)$ by solving $\widehat{U}y = 0_{4 \times 1}$. Find the fixed variables y_1, y_2 in terms of the free variable y_3

$$\begin{bmatrix} 1 & 2 \\ 0 & 4 \end{bmatrix} \begin{bmatrix} y_1 \\ y_2 \end{bmatrix} + \begin{bmatrix} 3 \\ 4 \end{bmatrix} y_3 = \begin{bmatrix} 0 \\ 0 \end{bmatrix}.$$

This gives $y_2 = -y_3$ and $y_1 = -y_3$. The null space has dimension equal to one and

$$N(A^T) = \left\{ y : y = \begin{bmatrix} -1 \\ -1 \\ 1 \end{bmatrix} y_3 \right\}. \tag{4.3.5}$$

Remark. Use Example 4.3.1. $EA = U$ implies

$$A^T E^T = U^T = \begin{bmatrix} 1 & 0 & 0 \\ 1 & 0 & 0 \\ 2 & 4 & 0 \\ 3 & 4 & 0 \end{bmatrix}.$$

So, the third column of E^T must be in the null space of A^T, which agrees with Example 4.3.2.

Remark. Use Example 4.3.2. $\widehat{E}A^T = \widehat{U}$ implies

$$A\widehat{E}^T = \widehat{U}^T = \begin{bmatrix} 1 & 0 & 0 & 0 \\ 2 & 4 & 0 & 0 \\ 3 & 4 & 0 & 0 \end{bmatrix}.$$

So, the third and fourth columns of $\widehat{E}^T$ must be in the null space of A, and this agrees with Example 4.3.1.

Remark. Example 4.3.1 required $d_3 - d_1 - d_2 = 0$. This can be written as

$$\begin{bmatrix} d_1 & d_2 & d_1 \end{bmatrix} \begin{bmatrix} -1 \\ -1 \\ 1 \end{bmatrix} = 0.$$

So, d is perpendicular to any element in $N(A^T)$.

Remark. Example 4.3.2 required $-g_1 + g_2 = 0$ and $-g_1 - g_3 + g_4 = 0$, that is,

$$g^T \begin{bmatrix} -1 \\ 1 \\ 0 \\ 0 \end{bmatrix} = 0 \text{ and } g^T \begin{bmatrix} -1 \\ 0 \\ -1 \\ 1 \end{bmatrix} = 0.$$

This means g must be perpendicular to any element in $N(A)$.

In line (4.3.1) we noted that the elements of $R(A)$ and $N(A^T)$ must be perpendicular. For the special cases in the above examples this is true for the bases given in lines (4.3.2) and (4.3.5) and (4.3.4), (4.3.3) for the spaces $R(A), N(A^T)$ and $R(A^T), N(A)$, respectively. The bases for $R(A)$ and $N(A^T)$ can be combined give a basis for all $\mathbb{R}^3$ where $A : \mathbb{R}^4 \longrightarrow \mathbb{R}^3$. Moreover, the only vector in common to $R(A)$ and $N(A^T)$ is the zero vector. This is called a *direct sum* and written as $R(A) \oplus N(A^T) = \mathbb{R}^3$.

4.3.1 Construction of bases

Consider the general $m \times n$ matrix A and use elementary matrices to find the row echelon form $E[A \ \ d] = [EA \ \ Ed] = [U \ \ Ed]$. Since E is nonsingular, $Ax = d$ and $Ux = Ed$ are equivalent. If the nonfixed variable rows of $[U \ \ Ed]$ are all zeros, then $Ux = Ed$ has a solution.

Let $(ipiv, jpiv)$ be the row and column numbers of the pivots where $1 \le k \le r = rank(A)$ and

$$ipiv(k) = \text{row of pivot } k \text{ and}$$
$$jpiv(k) = \text{column of pivot } k.$$

The fixed variables are $x_{jpiv(k)}$, and the free variables are x_j where $j \neq jpiv(k)$ with $1 \le k \le r$.

Theorem 4.3.1. *(Bases for N(A) and R(A))* Let A be $m \times n$ with $\text{rank}(A)$ $= r$, and E be a product of elementary matrices that give the row echelon form. If d is $m \times 1$ and $[Ed]_i = 0$ for all $i \neq ipiv(k)$ where $1 \leq k \leq r$, then

> *(i).* $d \in R(A)$.

> *(ii).* $R(A)$ has a basis from the pivot columns of A and $\dim(R(A)) = \text{rank}(A) = r$.

> *(iii).* $N(A)$ has a basis by solving the fixed variables in terms of the free variables and $\dim(N(A)) = n - r$.

Proof. Since the nonfixed variable rows of $[U \quad Ed]$ are all zeros, $Ux = Ed$ has a solution and this solution also solves $Ax = d$. This means $d \in R(A)$. $Ux = Ed$ is solved by setting the free variables to zero. The d in $Ax = d$ is a linear combination of the r pivot columns of A. These pivot columns are linearly independent. Thus, $\dim(R(A)) = r$.

Find $N(A)$ by solving the $r \times r$ upper triangular system for the fixed variables, $x_{fixed} = x_{jpiv}$, in terms of the free variables, $x_{free} = x_j$ with $j \neq jpiv(k)$.

$$U(ipiv, jpiv)x_{fixed} + U(ipiv, free)x_{free} = 0_{r \times 1}.$$

$$x_{fixed} = -U(ipiv, jpiv)^{-1}U(ipiv, free)x_{free}$$
$$= -Cx_{free} \text{ and}$$
$$x = \begin{bmatrix} -C \\ I_{n-r} \end{bmatrix} x_{free}.$$

So, the columns of $\begin{bmatrix} -C \\ I_{n-r} \end{bmatrix}$ are linearly independent and $\dim(N(A)) = n - r$. ∎

Theorem 4.3.2. *(Equal Ranks)* Let A be $m \times n$ with $\text{rank}(A) = r$. Then $\text{rank}(A^T) = r$, that is, $\text{rank}(A^T) = \text{rank}(A)$.

Proof. Let $\text{rank}(A^T) = \hat{r}$. Since A^T is $n \times m$ and by Theorem 4.3.1, $\dim(R(A^T)) = \hat{r}$ and $\dim(N(A^T)) = m - \hat{r}$. Let $EA = U$ have $m - r$ zero rows with $\text{rank}(A) = r$. Then $A^T E^T = U^T$ has $m - r$ zero columns and $\dim(N(A^T)) = m - r$. By Theorem 3.3.1 $\dim(N(A^T))$ is unique so that $m - r = m - \hat{r}$ and $r = \hat{r}$. ∎

4.3.2 Exercises

1. Prove the analogue of (4.3.1): any element in $N(A)$ is perpendicular to any element in $R(A^T)$.

2. Explain why the bases in lines (4.3.3) and (4.3.4) combine to form a basis for all of $\mathbb{R}^4$.

3. Prove for $A : \mathbb{R}^m \longrightarrow \mathbb{R}^m$ the vector in common to both $R(A^T)$ and $N(A)$ is the zero vector.

4. Consider the 4×6 matrix in Example 4.2.3. Find the bases for the four fundamental subspaces. Note their relationships to each other.

4.4 Null Space Method for Equilibrium Equations

In Subsection 1.1.5 an application was given to a two-loop circuit. The Kirchhoff voltage law requires the voltage drops in the three vertical portions of the loops to be equal, see Figure 1.1.4. The resulting equation has the form

$$
\begin{bmatrix}
R_1 & 0 & 0 & -1 \\
0 & R_2 & 0 & 1 \\
0 & 0 & R_3 & -1 \\
-1 & 1 & -1 & 0
\end{bmatrix}
\begin{bmatrix}
x_1 \\ x_2 \\ x_3 \\ y_1
\end{bmatrix}
=
\begin{bmatrix}
f_1 \\ f_2 \\ f_3 \\ g_1
\end{bmatrix}
$$

where the first three unknowns are the currents and the fourth unknown is the potential at the top node relative to the ground. This system is a special case of the equilibrium equations. In this section the general equilibrium equations will be defined and solved by both block Gauss elimination and null space methods. A variety of applications to more complicated circuits, structures and fluid flows models will be outlined; select one and study it carefully. Additional material on equilibrium equations can be found in chapter two of [23].

Definition. *Let A be an $n \times n$ matrix and E be an $m \times n$ matrix with $m < n$. The equilibrium equations can be written as*

$$Ax + E^T y = f \text{ and}$$
$$Ex = g \ .$$

Or, use $0 = 0_{m \times m}$ as an $m \times m$ matrix of zeros to write

$$
\begin{bmatrix}
A & E^T \\
E & 0
\end{bmatrix}
\begin{bmatrix}
x \\ y
\end{bmatrix}
=
\begin{bmatrix}
f \\ g
\end{bmatrix} ..
$$

In the above two-loop circuit algebraic system A is the 3×3 diagonal matrix with the three resistor values, and E is the 1×3 row matrix with coefficients given by the Kirchhoff current law. Typically, $Ex = g$ represents a balance or conservation of quantities in x such as currents, mass or forces. The top equation $Ax + E^T y = f$ may be an implementation of an empirical law such as voltage drops, momentum or deformation from forces.

4.4.1 Block Gauss elimination method

In this section assume A is *symmetric* $(A = A^T)$ *positive definite* $(x \neq 0$ implies $x^T A x > 0)$, and E has full row rank (the row echelon form has exactly m nonzero pivots). In other words, the columns of E^T are linearly independent and, hence, $E^T y = 0$ implies $y = 0$. By combining these two assumptions we can conclude that both A and $EA^{-1}E^T$ have inverse matrices.

Solve the first equation for $x = A^{-1}(f - E^T y)$ and then put this into the second equation

$$Ex = g$$
$$EA^{-1}(f - E^T y) = g.$$

This gives an equation for y where $EA^{-1}E^T$ is an $m \times m$ matrix

$$EA^{-1}E^T y = EA^{-1}f - g.$$

Once y is known, x is computed from the first equation

$$x = A^{-1}(f - E^T y).$$

The block Gauss elimination method may also be described by a block row operation

$$\begin{bmatrix} I_n & 0 \\ -EA^{-1} & I_m \end{bmatrix} \begin{bmatrix} A & E^T \\ E & 0 \end{bmatrix} \begin{bmatrix} x \\ y \end{bmatrix} = \begin{bmatrix} I_n & 0 \\ -EA^{-1} & I_m \end{bmatrix} \begin{bmatrix} f \\ g \end{bmatrix}$$

$$\begin{bmatrix} A & E^T \\ 0 & -EA^{-1}E^T \end{bmatrix} \begin{bmatrix} x \\ y \end{bmatrix} = \begin{bmatrix} f \\ -EA^{-1}f + g \end{bmatrix}.$$

Block Gauss Elimination Method for Equilibrium Equations.

1. Solve $Az = f$ and $AZ = E^T$ so that $z = A^{-1}f$ and $Z = A^{-1}E^T$.

2. Compute $Ez - g$ and EZ.

3. Solve $EZy = Ez - g$ so that $y = (EA^{-1}E^T)^{-1}(EA^{-1}f - g)$.

4. Let $x = z - Zy = A^{-1}f - A^{-1}E^T y$.

Example 4.4.1. Consider a particular case of the two-loop circuit

$$\begin{bmatrix} 2 & 0 & 0 & -1 \\ 0 & 3 & 0 & 1 \\ 0 & 0 & 4 & -1 \\ -1 & 1 & -1 & 0 \end{bmatrix} \begin{bmatrix} x_1 \\ x_2 \\ x_3 \\ y_1 \end{bmatrix} = \begin{bmatrix} 9 \\ -9 \\ 0 \\ 0 \end{bmatrix}.$$

Since A is a diagonal matrix, the solves for $Az = f$ and $AZ = E^T$ are easy

$$z = \begin{bmatrix} 9/2 \\ -9/3 \\ 0/4 \end{bmatrix} \text{ and } Z = \begin{bmatrix} -1/2 \\ 1/3 \\ -1/4 \end{bmatrix}.$$

The second step is

$$Ez - g = -1(9/2) + 1(-9/3) - 1(0/4) - 0 = -15/2 \text{ and}$$
$$EZ = -1(-1/2) + 1(1/3) - 1(-1/4) = 13/12.$$

The solution of $EZy = Ez - g$ is also easy to compute $y = (-15/2)/(13/12) = -90/13$. The last step is

$$x = z - Zy$$
$$= \begin{bmatrix} 9/2 \\ -9/3 \\ 0/4 \end{bmatrix} - \begin{bmatrix} -1/2 \\ 1/3 \\ -1/4 \end{bmatrix} (-90/13) = \begin{bmatrix} 27/26 \\ -18/26 \\ -45/26 \end{bmatrix}.$$

Example 4.4.2. If there are more rows in E, then the solves become more complicated as in

$$\begin{bmatrix} 2 & -1 & 0 & 1 & 0 \\ -1 & 2 & -1 & -1 & 1 \\ 0 & -1 & 2 & 0 & 1 \\ 1 & -1 & 0 & 0 & 0 \\ 0 & 1 & 1 & 0 & 0 \end{bmatrix} \begin{bmatrix} x_1 \\ x_2 \\ x_3 \\ y_1 \\ y_2 \end{bmatrix} = \begin{bmatrix} 1 \\ 2 \\ 3 \\ 4 \\ 5 \end{bmatrix}.$$

Here A is tridiagonal and E has two rows

$$A = \begin{bmatrix} 2 & -1 & 0 \\ -1 & 2 & -1 \\ 0 & -1 & 2 \end{bmatrix} \text{ and } E = \begin{bmatrix} 1 & -1 & 0 \\ 0 & 1 & 1 \end{bmatrix}.$$

The inverse of A is

$$A^{-1} = (1/4) \begin{bmatrix} 3 & 2 & 1 \\ 2 & 4 & 2 \\ 1 & 2 & 3 \end{bmatrix}$$

and, hence, $z = A^{-1}f$ and $Z = A^{-1}E^T$ are

$$z = (1/4) \begin{bmatrix} 10 \\ 16 \\ 14 \end{bmatrix} \text{ and } Z = (1/4) \begin{bmatrix} 1 & 3 \\ -2 & 6 \\ -1 & 5 \end{bmatrix}.$$

Now $EZy = Ez - g$ becomes

$$\begin{bmatrix} .75 & -.75 \\ -.75 & 2.75 \end{bmatrix} \begin{bmatrix} y_1 \\ y_2 \end{bmatrix} = \begin{bmatrix} -5.5 \\ 2.5 \end{bmatrix},$$

whose solution is $y_1 = -8.8333$ and $y_2 = -1.500$. The final step requires the computation of $x = z - Zy$

$$\begin{bmatrix} x_1 \\ x_2 \\ x_3 \end{bmatrix} = \begin{bmatrix} 2.5 \\ 4.0 \\ 3.5 \end{bmatrix} + \begin{bmatrix} .25 & .75 \\ -.5 & 1.5 \\ -.25 & 1.25 \end{bmatrix} \begin{bmatrix} -8.3333 \\ -1.5000 \end{bmatrix} = \begin{bmatrix} 5.8333 \\ 1.8333 \\ 3.1667 \end{bmatrix}.$$

4.4.2 Null space method for equilibrium equations

Again, assume A is symmetric positive definite and E has full row rank. The full row rank assumption gives the existence of a particular solution x_{part} so that $Ex_{part} = g$. Moreover, there is a basis for the null space with $n - m$ vectors. Both these can be constructed as illustrated in the previous sections from the row echelon form of the augmented matrix $[E \ g]$.

Let N be the $n \times (n - m)$ matrix formed by the basis vectors for the null space of E. This implies E times each column in N is a zero vector and, consequently, $EN = 0$. The homogeneous solutions are linear combinations of the columns in N, which may be written as a matrix-vector product Nx_0 where the coefficients in the linear combination are the components of x_0. The general solution of $Ex_{part} = g$ is

$$x = x_{null} + x_{part} = Nx_0 + x_{part}.$$

The next step is to put this representation for x into the first equation

$$A(Nx_0 + x_{part}) + E^T y = f.$$

Now multiply both sides by N^T and use $EN = 0$ giving $(EN)^T = N^T E^T = 0^T$ and

$$N^T(A(Nx_0 + x_{part}) + E^T y) = N^T f$$
$$N^T A N x_0 + N^T A x_{part} + N^T E^T y = N^T f$$
$$N^T A N x_0 + N^T A x_{part} + 0 = N^T f.$$

Thus, one must solve

$$N^T A N x_0 = N^T f - N^T A x_{part}.$$

The above two assumptions imply the $(n - m) \times (n - m)$ matrix $N^T A N$ has an inverse.

Finally one can solve for y by multiplying the first equation by E

$$E(A(Nx_0 + x_{part}) + E^T y) = Ef.$$

This requires the solution of

$$EE^T y = Ef - E(A(Nx_0 + x_{part})).$$

The $m \times m$ matrix EE^T has an inverse because the columns in E^T are linearly independent.

Example 4.4.3. Reconsider the two-loop circuit problem in Example 4.4.1. The 1×3 matrix $E = [-1 \ 1 \ -1]$ already is in row echelon form. Since g is zero, the particular solution is zero. The null space basis vectors are the columns in

$$N = \begin{bmatrix} 1 & -1 \\ 1 & 0 \\ 0 & 1 \end{bmatrix}.$$

Next compute w and W

$$w = N^T(f - Ax_{part}) = \begin{bmatrix} 1 & -1 \\ 1 & 0 \\ 0 & 1 \end{bmatrix}^T \left(\begin{bmatrix} 9 \\ -9 \\ 0 \end{bmatrix} - A0 \right) = \begin{bmatrix} 0 \\ -9 \end{bmatrix}$$

$$W = N^T A = \begin{bmatrix} 1 & -1 \\ 1 & 0 \\ 0 & 1 \end{bmatrix}^T \begin{bmatrix} 2 & 0 & 0 \\ 0 & 3 & 0 \\ 0 & 0 & 4 \end{bmatrix} = \begin{bmatrix} 2 & 3 & 0 \\ -2 & 0 & 4 \end{bmatrix}.$$

The solution of

$$WNx_0 = w$$

$$\begin{bmatrix} 2 & 3 & 0 \\ -2 & 0 & 4 \end{bmatrix} \begin{bmatrix} 1 & -1 \\ 1 & 0 \\ 0 & 1 \end{bmatrix} \begin{bmatrix} x_1^0 \\ x_2^0 \end{bmatrix} = \begin{bmatrix} 0 \\ -9 \end{bmatrix}$$

$$\begin{bmatrix} 5 & -2 \\ -2 & 6 \end{bmatrix} \begin{bmatrix} x_1^0 \\ x_2^0 \end{bmatrix} =$$

is $x_1^0 = -18/26$ and $x_2^0 = -45/26$. The x vector is

$$x = Nx_0 + x_{part}$$
$$= \begin{bmatrix} 1 & -1 \\ 1 & 0 \\ 0 & 1 \end{bmatrix} \begin{bmatrix} -18/26 \\ -45/26 \end{bmatrix} + \begin{bmatrix} 0 \\ 0 \\ 0 \end{bmatrix} = \begin{bmatrix} 27/26 \\ -18/26 \\ -45/26 \end{bmatrix}.$$

Example 4.4.4. Reconsider Example 4.4.2 where the augmented matrix is in row echelon form

$$[E \ g] = \begin{bmatrix} 1 & -1 & 0 & 4 \\ 0 & 1 & 1 & 5 \end{bmatrix}.$$

The free variable is x_3 and this is equivalent to

$$\begin{bmatrix} 1 & -1 \\ 0 & 1 \end{bmatrix} \begin{bmatrix} x_1 \\ x_2 \end{bmatrix} = \begin{bmatrix} 4 \\ 5 \end{bmatrix} - x_3 \begin{bmatrix} 0 \\ 1 \end{bmatrix}.$$

Set $x_3 = 0$ and solve for the particular solution $x_p = [9 \ 5 \ 0]^T$. Find the homogeneous solution by solving

$$\begin{bmatrix} 1 & -1 \\ 0 & 1 \end{bmatrix} \begin{bmatrix} x_1 \\ x_2 \end{bmatrix} = \begin{bmatrix} 0 \\ 0 \end{bmatrix} - x_3 \begin{bmatrix} 0 \\ 1 \end{bmatrix}.$$

This gives $x_2 = -x_3$, $x_1 = -x_3$ and $x_3 = x_3$ or in vector form

$$x_{null} = \begin{bmatrix} -1 \\ -1 \\ 1 \end{bmatrix} x_3 \text{ and } N = \begin{bmatrix} -1 \\ -1 \\ 1 \end{bmatrix}.$$

Next compute $w = N^T(f - Ax_{part})$ and $W = N^T A$.

$$w = N^T(f - Ax_{part})$$

$$= \begin{bmatrix} -1 \\ -1 \\ 1 \end{bmatrix}^T \left(\begin{bmatrix} 1 \\ 2 \\ 3 \end{bmatrix} - \begin{bmatrix} 2 & -1 & 0 \\ -1 & 2 & -1 \\ 0 & -1 & 2 \end{bmatrix} \begin{bmatrix} 9 \\ 5 \\ 0 \end{bmatrix} \right)$$

$$= -1(-12) - 1(1) + 1(8) = 19 \text{ and}$$

$$W = N^T A$$

$$= \begin{bmatrix} -1 \\ -1 \\ 1 \end{bmatrix}^T \begin{bmatrix} 2 & -1 & 0 \\ -1 & 2 & -1 \\ 0 & -1 & 2 \end{bmatrix} = \begin{bmatrix} -1 & -2 & 3 \end{bmatrix}.$$

Solve $WNx_0 = w$

$$\begin{bmatrix} -1 & -2 & 3 \end{bmatrix} \begin{bmatrix} -1 \\ -1 \\ 1 \end{bmatrix} x_0 = 19 \text{ so that } x_0 = 19/6.$$

Then $x = Nx_0 + x_{part}$ is as follows

$$x = \begin{bmatrix} -1 \\ -1 \\ 1 \end{bmatrix} (19/6) + \begin{bmatrix} 9 \\ 5 \\ 0 \end{bmatrix} = \begin{bmatrix} 35/6 \\ 11/6 \\ 19/6 \end{bmatrix}.$$

Find y by solving $EE^T y = E^T(f - A(Nx_0 + x_{part}))$

$$\begin{bmatrix} 1 & -1 & 0 \\ 0 & 1 & 1 \end{bmatrix} \begin{bmatrix} 1 & 0 \\ -1 & 1 \\ 0 & 1 \end{bmatrix} \begin{bmatrix} y_1 \\ y_2 \end{bmatrix} = \begin{bmatrix} 1 & -1 & 0 \\ 0 & 1 & 1 \end{bmatrix} \begin{bmatrix} -53/6 \\ 44/6 \\ -9/6 \end{bmatrix}$$

$$\begin{bmatrix} 2 & -1 \\ -1 & 2 \end{bmatrix} \begin{bmatrix} y_1 \\ y_2 \end{bmatrix} = \begin{bmatrix} -97/6 \\ 35/6 \end{bmatrix}$$

whose solution is $y_1 = -159/18$ and $y_2 = -27/18$.

4.4.3 Application to three-loop circuit

Consider the three-loop circuit in Section 2.1, see Figure 2.1.1. The matrix equation for the five currents x and two potentials at the top nodes y is

$$\begin{bmatrix} R_1 & 0 & 0 & 0 & 0 & -1 & 0 \\ 0 & R_2 & 0 & 0 & 0 & 1 & -1 \\ 0 & 0 & R_3 & 0 & 0 & -1 & 0 \\ 0 & 0 & 0 & R_4 & 0 & 0 & -1 \\ 0 & 0 & 0 & 0 & R_5 & 0 & 1 \\ -1 & 1 & -1 & 0 & 0 & 0 & 0 \\ 0 & -1 & 0 & -1 & 1 & 0 & 0 \end{bmatrix} \begin{bmatrix} x_1 \\ x_2 \\ x_3 \\ x_4 \\ x_5 \\ y_1 \\ y_2 \end{bmatrix} = \begin{bmatrix} E_1 \\ -E_2 \\ 0 \\ 0 \\ -E_3 \\ 0 \\ 0 \end{bmatrix}.$$

This is an equilibrium equation with A being a 5×5 diagonal matrix and the augmented matrix being

$$[E \ \ g] = \begin{bmatrix} -1 & 1 & -1 & 0 & 0 & 0 \\ 0 & -1 & 0 & -1 & 1 & 0 \end{bmatrix}.$$

Here we outline the null space method, which can be viewed as a dimension reduction scheme where the size of the matrix solves gets smaller. The following row operations transform the augmented matrix to reduced row echelon form

$$E_{12}(1) \ E_2(-1) \ E_1(-1) \ [E \ \ g] = \begin{bmatrix} 1 & 0 & 1 & 1 & -1 & 0 \\ 0 & 1 & 0 & 1 & -1 & 0 \end{bmatrix}.$$

There are three free variables x_3, x_4 and x_5, and this implies

$$\begin{bmatrix} 1 & 0 \\ 0 & 1 \end{bmatrix} \begin{bmatrix} x_1 \\ x_2 \end{bmatrix} = \begin{bmatrix} 0 \\ 0 \end{bmatrix} - x_3 \begin{bmatrix} 1 \\ 0 \end{bmatrix} - x_4 \begin{bmatrix} 1 \\ 1 \end{bmatrix} - x_5 \begin{bmatrix} -1 \\ -1 \end{bmatrix}.$$

Then the homogeneous solutions are

$$x_{null} = x_3 \begin{bmatrix} -1 \\ 0 \\ 1 \\ 0 \\ 0 \end{bmatrix} + x_4 \begin{bmatrix} -1 \\ -1 \\ 0 \\ 1 \\ 0 \end{bmatrix} + x_5 \begin{bmatrix} 1 \\ 1 \\ 0 \\ 0 \\ 1 \end{bmatrix}.$$

The null space matrix is

$$N = \begin{bmatrix} -1 & -1 & 1 \\ 0 & -1 & 1 \\ 1 & 0 & 0 \\ 0 & 1 & 0 \\ 0 & 0 & 1 \end{bmatrix}.$$

The matrix for the solve in step five is only 3×3, which is smaller than the 7×7 matrix in the equilibrium equation,

$$N^T A N = \begin{bmatrix} R_1 + R_3 & R_1 & -R_1 \\ R_1 & R_1 + R_2 + R_4 & -R_1 - R_2 \\ -R_1 & -R_1 - R_2 & R_1 + R_2 + R_5 \end{bmatrix}.$$

4.4.4 Application to six-bar truss

The six-bar truss in Subsection 1.4.4, see Figure 1.4.1, assumed there was no deformation of the bars when an external force is applied to one or more nodes. If the force is large, then there may be some significant movement of the nodes resulting in extension or compression of the connecting bars. The

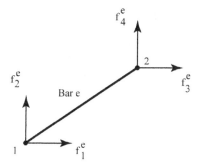

FIGURE 4.4.1
Bar e with four forces.

model of this has the form of the equilibrium equations. The first equation is an enhanced version of Hooke's law, which we will not go into. The second equation manages the assembly of the local forces and requires the sum of the forces at each node to be a zero force vector.

Assume the structure deforms only in one plane so that at each end of the bar (often called an element) there is a two-dimensional force vector, see Figure 4.4.1.

Let the superscript e denote the bar or element number, list the force components in the horizontal direction by odd number subscripts and the vertical forces by the even subscripts so that for each bar there is a four-dimensional force vector

$$\mathbf{f}^e = \begin{bmatrix} f_1^e & f_2^e & f_3^e & f_4^e \end{bmatrix}^T.$$

Using the same labels as in Figure 1.4.1, the sum of the forces at each node gives

Node 1: $[f_3^1 \ f_4^1] + [f_3^2 \ f_2^2] + [0 \ -w] = [0 \ 0],$

Node 2: $[f_1^2 \ f_2^2] + [f_1^3 \ f_2^3] + [f_3^4 \ f_4^4] = [0 \ 0]$ and

Node 3: $[f_1^1 \ f_2^1] + [f_3^3 \ f_4^3] + [f_3^5 \ f_4^5] + [f_3^6 \ f_4^6] = [0 \ 0].$

Now there are still three two-dimensional vector equations, but there are 18 local force components. This can be written as $Ex = g$ where E is an 6×18 matrix and g is a external force vector with six components. In $Ex = g$ the two equations from the two-dimensional vector equation at node one are in rows one and two, and the remaining four rows are obtained from the vector equations at nodes two and three where

$$E = \begin{bmatrix} 0 & 0 & 1 & 0 & 0 & 0 & 1 & 0 & 0 & 0 & 0 & 0 & 0 & 0 & 0 & 0 & 0 & 0 \\ 0 & 0 & 0 & 1 & 0 & 0 & 0 & 1 & 0 & 0 & 0 & 0 & 0 & 0 & 0 & 0 & 0 & 0 \\ 0 & 0 & 0 & 0 & 1 & 0 & 0 & 0 & 1 & 0 & 0 & 0 & 1 & 0 & 0 & 0 & 0 & 0 \\ 0 & 0 & 0 & 0 & 0 & 1 & 0 & 0 & 0 & 1 & 0 & 0 & 0 & 1 & 0 & 0 & 0 & 0 \\ 1 & 0 & 0 & 0 & 0 & 0 & 0 & 0 & 0 & 0 & 1 & 0 & 0 & 0 & 1 & 0 & 1 & 0 \\ 0 & 1 & 0 & 0 & 0 & 0 & 0 & 0 & 0 & 0 & 0 & 1 & 0 & 0 & 0 & 1 & 0 & 1 \end{bmatrix},$$

$g = [0 \; w \; 0 \; 0 \; 0 \; 0]^T$ and

$x = [f_1^1 \; f_2^1 \; f_3^1 \; f_4^1 \; f_1^2 \; f_2^2 \; f_3^2 \; f_4^2 \; f_1^3 \; f_2^3 \; f_3^3 \; f_4^3 \; f_3^4 \; f_4^4 \; f_3^5 \; f_4^5 \; f_3^6 \; f_4^6]^T.$

The corresponding augmented matrix may be transformed to a row echelon form matrix by simply moving rows five and six to rows one and two. The development of the first equation in the equilibrium equations is beyond the scope of this course, but this illustration should give the reader an appreciation for the role of row echelon forms and null spaces.

4.4.5   Application to fluid flow

The fully implicit method gives a sequence of nonlinear algebraic problems that show up in models for fluid flow. Let u, v, and P be unknowns for (u, v) velocity and P pressure in the Navier–Stokes fluid flow model

$$u_t + (u^2)_x + (uv)_y = -P_x + \frac{1}{\text{Re}}(u_{xx} + u_{yy}),$$

$$v_t + (uv)_x + (v^2)_y = -P_y + \frac{1}{\text{Re}}(v_{xx} + v_{yy}) \text{ and}$$

$$u_x + v_y = 0.$$

Use an implicit time step with finite differences in space to get a sequence of nonlinear problems at a fixed time step

$$A_1(u, v)u + E^T P = f_1,$$

$$A_2(u, v)v + E^T P = f_2 \text{ and}$$

$$E \begin{bmatrix} u \\ v \end{bmatrix} = g.$$

The first two equations are for the x-momentum and y-momentum and the third equation is the conservation of mass. The matrices $A_1(u, v)$ and $A_2(u, v)$ have the nonlinear terms. These problems can be solved by a sequence of linear solves where these two matrices are evaluated in a previous step, say, in Newton's method. Consider the grid in Figure 4.4.2. A_1 is 12×12, A_2 is 10×10 and E is 15×22. This often is a dimension reduction method, and in the above example A is 22×22 and $N^T A N$ is 7×7!

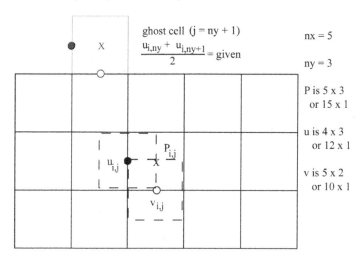

FIGURE 4.4.2
Navier–Stokes grid.

This can be made into a linear problem by evaluating the matrices $A_1(u, v)$ and $A_2(u, v)$ at the previous time step. In either case the linear problems have the form

$$Ax + E^T y = f,$$
$$Ex = g \text{ where}$$

$$A = \begin{bmatrix} A_1 \\ & A_2 \end{bmatrix}, \quad x = \begin{bmatrix} u \\ v \end{bmatrix} \text{ and } y = P.$$

In order to keep the discussion as simple as possible, consider fluid flow in a plane, and the rate of change of mass for four adjacent squares. Each cell has a horizontal and vertical component that are proportional to the velocity components in the fluid. Let cell i have horizontal and vertical components given by u_i and v_i where $i = 1, 2, 3$ and 4. For each cell the mass in minus the mass out yields four equations for the eight unknown components with incoming mass for cells 1-3, we have the following four equations

$$\text{Cell 1:} \quad s_1 - u_1 + v_2 - v_1 = 0,$$
$$\text{Cell 2:} \quad s_2 - u_2 + \bar{s}_2 - v_2 = 0,$$
$$\text{Cell 3:} \quad u_2 - u_3 + s_3 - v_3 = 0 \text{ and}$$
$$\text{Cell 4:} \quad u_1 - u_4 + v_3 - v_4 = 0.$$

This can be written as $Ex = g$ where E is an 4×8 matrix and x has components that are proportional to the velocity components

$$
\begin{bmatrix}
1 & 0 & 0 & 0 & 1 & -1 & 0 & 0 \\
0 & 1 & 0 & 0 & 0 & 1 & 0 & 0 \\
0 & -1 & 1 & 0 & 0 & 0 & 1 & 0 \\
-1 & 0 & 0 & 1 & 0 & 0 & -1 & 1
\end{bmatrix}
\begin{bmatrix}
u_1 \\ u_2 \\ u_3 \\ u_4 \\ v_1 \\ v_2 \\ v_3 \\ v_4
\end{bmatrix}
=
\begin{bmatrix}
s_1 \\ s_2 + \bar{s}_2 \\ s_3 \\ 0
\end{bmatrix}.
$$

The augmented matrix $[E \quad g]$ can be transformed into a reduced row echelon form by adding row 1 to row 4 and then row 2 to row 3. The vertical components will be the free variables, and the resulting representation of the unknowns can be used in the momentum equations for the fluid flow.

4.4.6   Exercises

1. Solve the following equilibrium system

$$
\begin{bmatrix}
2 & -1 & 0 & 0 & 1 & 0 & 1 \\
-1 & 2 & 0 & 0 & 0 & 3 & 0 \\
0 & 0 & 2 & -1 & -1 & 1 & -1 \\
0 & 0 & -1 & 2 & 1 & 1 & 2 \\
1 & 0 & -1 & 1 & 0 & 0 & 0 \\
0 & 3 & 1 & 1 & 0 & 0 & 0 \\
1 & 0 & -1 & 2 & 0 & 0 & 0
\end{bmatrix}
\begin{bmatrix}
x_1 \\ x_2 \\ x_3 \\ x_4 \\ y_1 \\ y_2 \\ y_3
\end{bmatrix}
=
\begin{bmatrix}
1 \\ 2 \\ 3 \\ 4 \\ 0 \\ 0 \\ 7
\end{bmatrix}.
$$

   (a). Show the matrices in equilibrium system satisfy

        (i). A is SPD and
        (ii). E^T has full column rank.

   (b). Use Gauss elimination.

   (c). Use block Gauss elimination.

   (d). Use null space method.

   (e). Compare these methods.

2. Assume A is SPD and E^T has full column rank.

   (a). Show $EA^{-1}E^T$ has an inverse.

   (b). Show $N^T A N$ has an inverse.

5

Orthogonal Subspaces and Bases

If A is $m \times n$, then $A : \mathbb{R}^n \to \mathbb{R}^m$. Any vector in $\mathbb{R}^n$ may be written as a sum of vectors in $N(A)$ and in $R(A^T)$. Moreover, these two vectors are orthogonal (perpendicular), which is part of the fundamental theorem. The bases for these two subspaces may have orthonormal vectors. This is established by using the factors $A = QR$, where the columns of Q are orthonormal and R is upper triangular.

5.1 Orthogonal Subspace

Consider $\mathbb{R}^3$ and let W be the xy-plane. All the vectors in $\mathbb{R}^3$ that are perpendicular to the xy-plane are multiples of the unit vector in the z-direction. This is generalized by the following.

Definition. *Let W be a vector space in $\mathbb{R}^n$. The perpendicular or orthogonal complement is*

$$W^\perp \equiv \left\{ y \in \mathbb{R}^n : y^T x = 0 \text{ for all } x \in W \right\}$$

Example 5.1.1. Consider $\mathbb{R}^3$ and $W = \left\{ c_1 \begin{bmatrix} 1 \\ 1 \\ 0 \end{bmatrix} + c_2 \begin{bmatrix} 1 \\ 0 \\ 1 \end{bmatrix} : c_1, c_2 \in \mathbb{R} \right\}$.

The cross product of the two column vectors is perpendicular to both vectors. Then

$$W^\perp = \left\{ c \begin{bmatrix} 1 \\ -1 \\ -1 \end{bmatrix} : c \in \mathbb{R} \right\}.$$

If $n = 2$ or 3, then the cosine law gives $x^T y = \|x\|_2 \|y\|_2 \cos(\theta)$, where θ is the angle between the two vectors. A generalization of the "angle" between two vectors $x, y \in \mathbb{R}^n$ is

$$\cos(\theta) \equiv \frac{x^T y}{\|x\|_2 \|y\|_2} \text{ where } \|x\|_2 = (x^T x)^{1/2}.$$

This is based on the Cauchy inequality.

DOI: 10.1201/9781003304128-5

Definition. *The Cauchy inequality is*

$$\left| x^T y \right| \leq \|x\|_2 \|y\|_2 .$$

The inequality follows from the quadratic function of the real parameter

$$f(t) \equiv (x + ty)^T (x + ty).$$

Let $t_0 = \frac{-x^T y}{y^T y}$, which gives the minimum of $f(t)$,

$$0 \leq f(t_0) = x^T x - \frac{(x^T y)^2}{y^T y}.$$

Then $(x^T y)^2 \leq x^T x \ \ y^T y = \|x\|_2^2 \|y\|_2^2$ and

$$-1 \leq \frac{x^T y}{\|x\|_2 \|y\|_2} \leq 1.$$

Hence, the expression x is perpendicular to y means $x^T y = 0$.

Theorem 5.1.1. *(Orthogonal Subspace Properties) Let W be a vector space in $\mathbb{R}^n$. The following hold:*

(i). $W^\perp$ *is a subspace of* $\mathbb{R}^n$,

(ii). $W^{\perp\perp} = W$,

(iii). $W^\perp \cap W = \{0_{n \times 1}\}$ *and*

(iv). $W^\perp + W = \mathbb{R}^n$.

Proof. The proof of the first item requires the $W^\perp$ to be closed under addition and scalar multiplication. If $y, \widehat{y} \in W^\perp$, then

$$(y + \widehat{y})^T x = y^T x + \widehat{y}^T x = 0 + 0 = 0.$$

If $y \in W^\perp$ and $c \in \mathbb{R}$, then

$$(cy)^T x = c(y^T x) = c0 = 0.$$

The proofs of the second and third items are easy.

The proof of the last item uses the projection of any $v \in \mathbb{R}^n$ into W, $P_W(v)$. By Theorem 3.4.1 for all $w \in W$

$$w^T (v - P_W(v)) = 0.$$

This means $v - P_W(v) \in W^\perp$ and $v = (v - P_W(v)) + P_W(v) \in W^\perp + W$. ∎

The third and fourth properties are combined to form the *direct sum* denoted by $W^\perp \oplus W$.

5.1.1 Exercises

1. Show $W^{\perp\perp} = W$ and $W^\perp \cap W = \{0_{n \times 1}\}$.

2. Let $W = \{x \in \mathbb{R}^3 : x_1 + x_2 - x_3 = 0\}$.

   (a). Show W is a subspace.

   (b). Find $W^\perp$.

5.2 Fundamental Theorem: $\mathbb{R}^n = N(A) \oplus R(A^T)$

An illustration of the next theorem is given in Examples 4.3.1 and 4.3.2 and Remarks 3 and 4. In the general case, A is $m \times n$ and $W = N(A) \subset \mathbb{R}^n$. The fourth part of Theorem 5.1.1 gives $\mathbb{R}^n = N(A) \oplus N(A)^\perp$.

Theorem 5.2.1. *(Fundamental Decomposition) If* $A : \mathbb{R}^n \to \mathbb{R}^m$ *with* $rank(A) = r$, *then* $A^T : \mathbb{R}^m \to \mathbb{R}^n$ *and*

$$N(A) = R(A^T)^\perp \qquad\qquad N(A^T) = R(A)^\perp,$$
$$\mathbb{R}^n = N(A) \oplus R(A^T) \qquad \mathbb{R}^m = N(A^T) \oplus R(A),$$
$$dim(R(A^T)) = r \qquad\qquad dim(R(A)) = r \text{ and}$$
$$dim(N(A)) = n - r \qquad dim(N(A^T)) = m - r.$$

Proof. The main step is to show $N(A) = (R(A^T))^\perp$. Suppose $x \in N(A)$ and show $x \in (R(A^T))^\perp$.

$$Ax = 0_{n \times 1}$$
$$y^T(Ax) = y^T 0_{n \times 1} \text{ for all } y \in \mathbb{R}^m$$
$$(A^T y)^T x = 0.$$

Suppose $x \in (R(A^T))^\perp$ and show $x \in N(A)$. Then for all $y \in \mathbb{R}^m$ $(A^T y)^T x = y^T(Ax) = 0$. Choose $y = e_i \in \mathbb{R}^m$ to be any unit vector to get

$$e_i^T(Ax) = 0,$$
$$[Ax]_i = 0 \text{ and consequently}$$
$$Ax = 0_{m \times 1}.$$

Apply the above argument to A^T to get $N(A^T) = R(A)^\perp$. The remaining conclusions for both A and A^T follow from Theorem 4.3.1. ∎

The fundamental theorem can be used to analyze least squares problems with multiple solutions. Theorem 3.4.2 showed multiple least squares solution have the form $x = \widehat{x} + x_N$, where $A\widehat{x} = P_{R(A)}(d)$ and any $x_N \in N(A)$. The above Theorem 5.2.1 gives $\widehat{x} = \widehat{x}_R + \widehat{x}_N$ with $\widehat{x}_R = P_{R(A^T)}(\widehat{x}) \in R(A^T)$ and $\widehat{x}_N \in N(A)$.

Theorem 5.2.2. *(Projections and Least Squares)* If $A\widehat{x} = P_{R(A)}(d)$, $\widehat{x} = \widehat{x}_R + \widehat{x}_N$ with $\widehat{x}_R = P_{R(A^T)}(\widehat{x}) \in R(A^T)$ and $\widehat{x}_N \in N(A)$, then $\widehat{x}_R = P_{R(A^T)}(\widehat{x})$ is a least squares solution of $Ax = d$ and $(\widehat{x}_R)^T \widehat{x}_R$ is the smallest of all least squares solutions.*

Proof. $\widehat{x} + x_N = \widehat{x}_R + \widehat{x}_N + x_N$ is a least squares solution. Choose $x_N = -\widehat{x}_N$ to conclude $\widehat{x}_R$ is a least squares solution. For any least squares solution use $x = \widehat{x}_R + x_N$ with $x_N \in N(A)$. By Theorem 5.2.1 $\widehat{x}_R^T x_N = 0$,

$$
\begin{aligned}
x^T x &= (\widehat{x}_R + x_N)^T (\widehat{x}_R + x_N) \\
&= \widehat{x}_R^T \widehat{x}_R + 2\widehat{x}_R^T x_N + x_N^T x_N \\
&= \widehat{x}_R^T \widehat{x}_R + 0 + x_N^T x_N \\
&\geq \widehat{x}_R^T \widehat{x}_R.
\end{aligned}
$$

∎

Example 5.2.1. Return to Example 3.4.2

$$
A = \begin{bmatrix} 1 & 2 \\ 2 & 4 \\ 3 & 6 \end{bmatrix}, \quad A^T = \begin{bmatrix} 1 & 2 & 3 \\ 2 & 4 & 6 \end{bmatrix} \text{ and } \widehat{x} = \begin{bmatrix} 26/14 \\ 0 \end{bmatrix}.
$$

The basis for $R(A^T)$ is $w_1 = \begin{bmatrix} 1 \\ 2 \end{bmatrix}$.

$$
\widehat{x}_R = c \begin{bmatrix} 1 \\ 2 \end{bmatrix} = P_{R(A^T)}(\widehat{x})
$$

$$
\begin{bmatrix} 1 & 2 \end{bmatrix} c \begin{bmatrix} 1 \\ 2 \end{bmatrix} = \begin{bmatrix} 1 & 2 \end{bmatrix} \begin{bmatrix} 26/14 \\ 0 \end{bmatrix}
$$

This gives $c = 13/35$ and $\widehat{x}_R$

$$
\widehat{x}_R = \begin{bmatrix} 13/35 \\ 26/35 \end{bmatrix}.
$$

The general solution is

$$
x = \begin{bmatrix} 26/14 \\ 0 \end{bmatrix} + \widehat{c} \begin{bmatrix} 2 \\ -1 \end{bmatrix}.
$$

Choose $\widehat{c}$ so that $x^T x$ is a minimum to get $\widehat{x}_R$. The set of all least squares solutions is depicted in Figure 5.2.1.

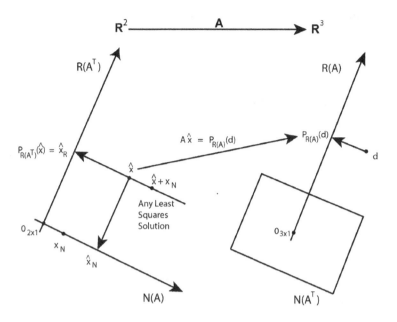

FIGURE 5.2.1
All least squares solutions.

5.2.1   Exercises

1. Return to Exercise 3 in Section 3.4.1.

   (a). Find all least squares solutions to

   $$\begin{bmatrix} 1 & 2 & 3 \\ 2 & 3 & 5 \\ 3 & 1 & 4 \\ 4 & 1 & 5 \end{bmatrix} \begin{bmatrix} x_1 \\ x_2 \\ x_3 \end{bmatrix} = \begin{bmatrix} 10 \\ 11 \\ 12 \\ 15 \end{bmatrix}.$$

   (b). Directly find the smallest least squares solution.

   (c). Use the above theorem to find the least squares solution.

5.3 $A = QR$ Factorization

Any basis can be transformed into an orthonormal basis, whose vectors are perpendicular and unit length. This can be extended to general inner product spaces via the Gram–Schmidt method. Here we will use the QR factors of A. The QR factors are also useful in efficiently solving the normal equations.

Definition. *Let A be $m \times n$ and assume $m \geq n$. $A = QR$ is a QR factorization if and only if*

(i). *Q is $m \times n$ with $Q^T Q = I_n$ and*

(ii). *R is $n \times n$ with R being upper triangular.*

The columns of Q are orthonormal. If the diagonal components of R are not zero, then R is nonsingular and $AR^{-1} = Q$. In the next section, this will be useful in construction of an orthonormal basis from a given basis.

Theorem 5.3.1. *(Full Rank and QR) Let A be $m \times n$. If A has full column rank, $rank(A) = n$, then for $A = QR$*

(i). *R^{-1} exists and*

(ii). *the normal equation reduces to $Rx = Q^T d$.*

Proof. If $Rx = 0_{n \times 1}$, then

$$Q(Rx) = Q0_{n \times 1} \text{ and}$$
$$Ax = 0_{m \times 1}.$$

Because A has full column rank, $x = 0_{n \times 1}$ and R is nonsingular.
Consider the normal equations.

$$A^T A x = A^T d$$
$$(QR)^T (QR)x = (QR)^T x$$
$$R^T (Q^T Q)Rx = R^T Q^T d$$
$$R^T Rx = R^T Q^T d.$$

Because R is nonsingular, R^T is also nonsingular and $Rx = Q^T d$. ∎

In order to find the QR factors, write $A = QR$ as a sequence of equal column vectors

$$\begin{bmatrix} a_1 & a_2 & \cdots & a_n \end{bmatrix} = \begin{bmatrix} q_1 & q_2 & \cdots & q_n \end{bmatrix} \begin{bmatrix} r_{11} & r_{12} & \cdots & r_{1n} \\ & r_{22} & \cdots & r_{2n} \\ & & \ddots & \vdots \\ & & & r_{nn} \end{bmatrix}.$$

$$a_1 = q_1 r_{11}$$
$$a_2 = q_1 r_{12} + q_2 r_{22}$$
$$a_3 = q_1 r_{13} + q_2 r_{23} + q_3 r_{33}$$

We can find q_1 and r_{11} and then r_{12} and r_{13} as follows

$$a_1^T a_1 = q_1^T q_1 r_{11}^2 = 1 r_{11}^2$$
$$a_2^T q_1 = (q_1 r_{12} + q_2 r_{22})^T q_1$$
$$\qquad = 1 r_{12} + 0 r_{22} \text{ and}$$
$$a_3^T q_1 = (q_1 r_{13} + q_2 r_{23} + q_3 r_{33})^T q_1$$
$$\qquad = 1 r_{13} + 0 r_{23} + 0 r_{33}.$$

Now reduce the dimension from m to $m-1$ by moving the first column to the left side

$$a_1 - q_1 r_{11} = 0$$
$$a_2 - q_1 r_{12} = q_2 r_{22}$$
$$a_3 - q_1 r_{13} = q_2 r_{23} + q_3 r_{33}.$$

Repeat the above on the smaller problem. This procedure is called the modified Gram–Schmidt method, and it can be programmed as is illustrated in the next section. There are four methods for finding QR factors: modified Gram–Schmidt, classical Gram–Schmidt, Givens transform and Housholder transforms.

Example 5.3.1. Return to Example 3.1.4 where

$$A = \begin{bmatrix} 1 & 1 \\ 1 & 2 \\ 1 & 3 \end{bmatrix} \text{ and } d = \begin{bmatrix} 10 \\ 9 \\ 7 \end{bmatrix}.$$

$$a_1 = \begin{bmatrix} 1 \\ 1 \\ 1 \end{bmatrix} \text{ gives } q_1 = \begin{bmatrix} 1/\sqrt{3} \\ 1/\sqrt{3} \\ 1/\sqrt{3} \end{bmatrix} \text{ and } r_{11} = \sqrt{3}$$

$$a_2 = \begin{bmatrix} 1 \\ 2 \\ 3 \end{bmatrix} \text{ gives } r_{12} = a_2^T q_1 = 6/\sqrt{3}$$

$$\widehat{q}_2 = a_2 - q_1 r_{12} = \begin{bmatrix} 1 \\ 2 \\ 3 \end{bmatrix} - \begin{bmatrix} 1/\sqrt{3} \\ 1/\sqrt{3} \\ 1/\sqrt{3} \end{bmatrix} 6/\sqrt{3} = \begin{bmatrix} -1 \\ 0 \\ 1 \end{bmatrix}$$

$$q_2 = \begin{bmatrix} -1/\sqrt{2} \\ 0 \\ 1/\sqrt{2} \end{bmatrix} \text{ and } r_{22} = \sqrt{2}.$$

The QR factors and the normal equations are

$$A = QR$$
$$= \begin{bmatrix} 1/\sqrt{3} & -1/\sqrt{2} \\ 1/\sqrt{3} & 0 \\ 1/\sqrt{3} & 1/\sqrt{2} \end{bmatrix} \begin{bmatrix} \sqrt{3} & 6/\sqrt{3} \\ 0 & \sqrt{2} \end{bmatrix} \text{ and}$$
$$Rx = Q^T f$$
$$\begin{bmatrix} \sqrt{3} & 6/\sqrt{3} \\ 0 & \sqrt{2} \end{bmatrix} \begin{bmatrix} x_1 \\ x_2 \end{bmatrix} = \begin{bmatrix} 1/\sqrt{3} & 1/\sqrt{3} & 1/\sqrt{3} \\ -1/\sqrt{2} & 0 & 1/\sqrt{2} \end{bmatrix} \begin{bmatrix} 10 \\ 9 \\ 7 \end{bmatrix} = \begin{bmatrix} 26/\sqrt{3} \\ -3/\sqrt{2} \end{bmatrix}.$$

The solution of this is $x_2 = -3/2$ and $x_1 = 35/3$, which agrees with the calculation in Example 3.1.4.

5.3.1   MATLAB® code qr_col.m

This is the modified Gram–Schmidt and the code is qr_col.m. A related method is the classical Gram–Schmidt which will be described in the last section, see qr_row.m. The sample inputs are given in lines 18–24. Note the matrix in line 20 will give an error because the columns are not linearly independent. The algorithm is executed in lines 25–35. This is the column version where the columns are moved to the left side in the loop starting at line 31. Lines 38–43 give the output, verify the factorization and compare the calculation with the two MATLAB® intrinsic commands.

Notice the difference in the "fat" and "skinny" versions of the QR factors. The QR factors as defined above is the "skinny" and the "fat" is derived from the "skinny":

$$A = QR \text{ where } Q \text{ is } m \times n \text{ and } R \text{ is } n \times n$$

$$A = \begin{bmatrix} Q & \widehat{Q} \end{bmatrix} \begin{bmatrix} R \\ 0_{(m-n) \times n} \end{bmatrix} \text{ where } \widehat{Q} \text{ is } m \times (m-n).$$

```
1     %  This code finds the qr factorization of a matrix.
2     %  The column version of the Gram-Schmidt method is
3     %  used to generate the small qr factors.
4     %  a is mxn, q is mxn and r is nxn where m>n and
5     %          q'*q = eye(n)
6     %
7     %  [a1 a2 a3 ...] = [q1 q2 q2 ...][r11 r12 r13 ...]
8     %                                     r22 r23 ...]
9     %                                         r33 ...]
10    %  OR
11    %
12    %  a1 = q1 r11
13    %  a2 = q1 r12 + q2 r22
14    %  a3 = q1 r13 + q2 r23 + q3 r33
15    %
16    clear
17    %
18    %  Input data
19    %
20    % a = [1 1 3; 1 2 4 ; 1 3 5; 1 4 6]
21    % Above fails because a3 = 2 a1 + a2!
22    a = [1 1;1 2; 1 3]
23    q = a;
24    [m n] = size(a)
25    %
26    %  Execute the column or modified Gram-Schmidt
27    %
```

```
28     for k = 1:n
29         r(k,k) = (q(:,k)'*q(:,k))^.5;
30         q(:,k) = q(:,k)/r(k,k);
31         for j = k+1:n      %  update the next column
32             r(k,j) = a(:,j)'*q(:,k);
33             q(:,j) = q(:,j) - q(:,k)*r(k,j);
34         end
35     end
36     %
37     %  Output Q R factors and compare with qr()
38     %
38     q
40     r
41     q*r
42     [Q R] = qr(a)       % fat version
43     [Q1 R1] = qr(a,0)  % skinny version

>> qr_col

   a =
      1    1
      1    2
      1    3

   m =
      3

   n =
      2

   q =
      0.5774   -0.7071
      0.5774   -0.0000
      0.5774    0.7071

   r =
      1.7321    3.4641
           0    1.4142

   A =
      1    1
      1    2
      1    3

   Q =
```

```
    -0.5774      0.7071      0.4082
    -0.5774     -0.0000     -0.8165
    -0.5774     -0.7071      0.4082


R =
    -1.7321     -3.4641
          0     -1.4142
          0           0


Q1 =
    -0.5774      0.7071
    -0.5774     -0.0000
    -0.5774     -0.7071


R1 =
    -1.7321     -3.4641
          0     -1.4142
```

5.3.2 Exercises

1. Consider the least squares problem

$$\begin{bmatrix} 1 & 2 \\ 2 & 3 \\ 3 & 1 \\ 4 & 1 \end{bmatrix} \begin{bmatrix} x_1 \\ x_2 \\ 2 \end{bmatrix} = \begin{bmatrix} 10 \\ 11 \\ 12 \\ 15 \end{bmatrix}.$$

(a). Show the columns in the matrix are linearly independent.

(b). Find the QR factors for the matrix.

(c). Solve the least squares problem using the QR factors.

5.4 Orthonormal Basis

The QR factorization can be used to generate an orthonormal basis.

Definition. *A basis $\{w_1, .., w_k\}$ in $\mathbb{R}^m$ is orthonormal if and only if*

(i). $w_i^T w_i = 1$ *for all* $1 \leq i \leq k$ *and*

(ii). $w_i^T w_j = 0$ *when* $i \neq j$.

Remark. Let $W = \begin{bmatrix} w_1 & \cdots & w_k \end{bmatrix}$ be an $m \times k$ matrix. $W^T W = I_k$ means the columns are orthonormal.

Theorem 5.4.1. *(Orthonormal Basis) Any basis of an k dimensional subspace $V \subset \mathbb{R}^m$ can be used to generate an orthonormal basis of V. Moreover, this basis can be extended to an orthonormal basis of $\mathbb{R}^m$.*

Proof. Let A be an $m \times k$ matrix formed by the column vectors of the basis for $V \subset \mathbb{R}^m$. Since the basis vectors are linearly independent, A has full column rank and $A = QR$ where R is nonsingular. We claim the columns of Q are an orthonormal basis. First, note the columns of Q are linearly independent. Let

$$Qx = (AR^{-1})x = A(R^{-1}x) = 0_{k \times 1}.$$

Since A has full column rank, $R^{-1}x = 0_{k \times 1}$. Then $x = 0_{k \times 1}$ and the columns are linearly independent. Second, the columns span the subspace $V \subset \mathbb{R}^m$. Let $d \in V = R(A)$ and

$$Ax = Q(Rx) = d \text{ and choose } \widehat{x} = R(x).$$

$Q(\widehat{x}) = d$ means d is a linear combination of the columns of Q.

In order to extend a basis from V to all of $\mathbb{R}^m$, choose a vector not in V and augment the matrix A by this new vector. Apply the QR factorization to the augmented matrix. Continue this process until there is an orthonormal basis $\mathbb{R}^m$. ∎

Example 5.4.1. Consider the 3×4 matrix considered in Examples 4.3.1 and 4.3.2

$$A = \begin{bmatrix} 1 & 1 & 2 & 3 \\ 2 & 2 & 8 & 10 \\ 3 & 3 & 10 & 13 \end{bmatrix}.$$

$$R(A) = \left\{ c_1 \begin{bmatrix} 1 \\ 2 \\ 3 \end{bmatrix} + c_2 \begin{bmatrix} 2 \\ 8 \\ 10 \end{bmatrix} : c_1, c_2 \in \mathbb{R} \right\}$$

$$N(A^T) = \left\{ c \begin{bmatrix} -1 \\ -1 \\ 1 \end{bmatrix} : c \in \mathbb{R} \right\}$$

Find the QR factorization of the 3×2 matrix from the basis of $R(A)$

$$\begin{bmatrix} a_1 & a_2 \end{bmatrix} = \begin{bmatrix} 1 & 2 \\ 2 & 8 \\ 3 & 10 \end{bmatrix}$$

$$= \begin{bmatrix} q_1 & q_2 \end{bmatrix} \begin{bmatrix} r_{11} & r_{12} \\ 0 & r_{22} \end{bmatrix}$$

$$= \begin{bmatrix} 1/\sqrt{14} & -10/\sqrt{168} \\ 2/\sqrt{14} & 8/\sqrt{168} \\ 3/\sqrt{14} & -2/\sqrt{168} \end{bmatrix} \begin{bmatrix} \sqrt{14} & 48/\sqrt{14} \\ 0 & \sqrt{168}/7 \end{bmatrix}.$$

The orthonormal basis is q_1, q_2. In order to extend this to all of $\mathbb{R}^3 = R(A) \oplus N(A^T)$, choose $a_3 = \begin{bmatrix} -1 & -1 & 1 \end{bmatrix}^T$. Since this is already perpendicular to both q_1, q_2, and we only need to normalize it to get a third basis vector

$$\begin{bmatrix} -1/\sqrt{3} \\ -1/\sqrt{3} \\ 1/\sqrt{3} \end{bmatrix}.$$

If a_3 is not perpendicular to both q_1, q_2, then continue with the QR procedure

$$a_3 = q_1 r_{13} + q_2 r_{23} + q_3 r_{33}.$$

For example, suppose $a_3^T = \begin{bmatrix} 0 & 0 & 1 \end{bmatrix}$ and compute

$$a_3^T q_1 = r_{13} = 3/\sqrt{14}$$
$$a_3^T q_2 = r_{23} = -2/\sqrt{168}$$
$$\widehat{q}_3 = a_3 - q_1 r_{13} - q_2 r_{23}$$
$$= \begin{bmatrix} -1/3 \\ -1/3 \\ 1/3 \end{bmatrix} \text{ with } r_{33} = 1/\sqrt{3} \text{ and}$$
$$q_3 = \begin{bmatrix} -1/\sqrt{3} \\ -1/\sqrt{3} \\ 1/\sqrt{3} \end{bmatrix}.$$

The "fat" QR factorization is

$$A = \begin{bmatrix} q_1 & q_2 & q_3 \end{bmatrix} \begin{bmatrix} r_{11} & r_{12} \\ 0 & r_{22} \\ 0 & 0 \end{bmatrix}.$$

5.4.1   Exercises

1. Consider the 3×4 matrix

$$A = \begin{bmatrix} 1 & 2 & 3 & 5 \\ 2 & 4 & 8 & 12 \\ 3 & 6 & 10 & 16 \end{bmatrix}.$$

   (a). Find $R(A)$ and $N(A^T)$.

   (b). Find a basis for the $R(A)$.

   (c). Use the QR factors to find an orthonormal basis.

5.5 Four Methods for QR Factors

The classical Gram–Schmidt starts with set of vectors and by using projections converts them into orthonormal vectors. It can be applied to more general vectors such as functions. However, the modified Gram–Schmidt tends to be less prone to the accumulation of roundoff errors, see Meyer [17, section 5.5] and Kelley [11, page 40]. The Givens transforms are used for sparse matrices. The Householder transforms are typically used for general matrix applications.

5.5.1 Classical Gram–Schmidt

Start with a set of vectors $a^1, a^2, a^3, \cdots$. Goal is to find orthonormal vector $q^1, q^2, q^3, \cdots$

The first one is easy $q^1 = a^1 / \|a^1\|$. The next step uses a projection

$$\hat{q}^2 = a^2 + c_{21}q^1 \text{ and } c_{21} \text{ so that } (q^1)^T \hat{q}^2 = 0.$$
$$q^2 = \hat{q}^2 / \|\hat{q}^2\| \text{ and } 0 = (q^1)^T a^2 + c_{21}1.$$

Continue this with

$$\hat{q}^{m+1} = a^{m+1} + c_{m+1,1}q^1 + \cdots + c_{m+1,m}q^m \text{ and } c_{m+1,i} \text{ so that } (q^i)^T \hat{q}^{m+1} = 0.$$
$$q^{m+1} = \hat{q}^{m+1} / \|\hat{q}^{m+1}\| \text{ and } 0 = (q^i)^T a^{m+1} + c_{m+1,i}1.$$

This is related to the QR factors because

$$a^{m+1} = \hat{q}^{m+1} - (c_{m+1,1}q^1 + \cdots + c_{m+1,m}q^m)$$
$$= \hat{q}^{m+1} + ((q^1)^T a^{m+1}q^1 + \cdots + (q^m)^T a^{m+1}q^m).$$

Example 5.5.1. Consider Example 5.3.1 and compare the classical and modified methods

$$A = \begin{bmatrix} 1 & 1 \\ 1 & 2 \\ 1 & 3 \end{bmatrix} \text{ and } d = \begin{bmatrix} 10 \\ 9 \\ 7 \end{bmatrix}$$

Classical.

$$q^1 = a^1 / \|a^1\| = \begin{bmatrix} 1 \\ 1 \\ 1 \end{bmatrix} / \sqrt{3}.$$

$$\hat{q}^2 = a^2 + c_1 q^1$$

$$= \begin{bmatrix} 1 \\ 2 \\ 3 \end{bmatrix} + c_1 \begin{bmatrix} 1 \\ 1 \\ 1 \end{bmatrix} / \sqrt{3} \text{ where } c_1 = -(q^1)^T a^2 = -6/\sqrt{3}$$

$$= \begin{bmatrix} -1 \\ 0 \\ 1 \end{bmatrix}.$$

$$q^2 = \begin{bmatrix} -1 \\ 0 \\ 1 \end{bmatrix} / \sqrt{2}.$$

Modified.

$$A = QR$$

$$\begin{bmatrix} 1 & 1 \\ 1 & 2 \\ 1 & 3 \end{bmatrix} = \begin{bmatrix} q^1 & q^2 \end{bmatrix} \begin{bmatrix} r_{11} & r_{12} \\ 0 & r_{22} \end{bmatrix}$$

$$= \begin{bmatrix} q^1 r_{11} & q^1 r_{12} + q^2 r_{22} \end{bmatrix}.$$

$$\begin{bmatrix} 1 \\ 1 \\ 1 \end{bmatrix} = q^1 r_{11}, \text{ and } r_{11} = \sqrt{3}.$$

$$\begin{bmatrix} 1 \\ 2 \\ 3 \end{bmatrix} - q^1 r_{12} = q^2 r_{22} \text{ and use } (q^1)^T q^2 = 0. \text{ Then } r_{12} = 6/\sqrt{3}.$$

Solve $[1 \ 2 \ 3]^T - q^1 r_{12} = q^2 r_{22}$

$$\begin{bmatrix} 1 \\ 2 \\ 3 \end{bmatrix} - \begin{bmatrix} 1/\sqrt{3} \\ 1/\sqrt{3} \\ 1/\sqrt{3} \end{bmatrix} 6/\sqrt{3} = \begin{bmatrix} -1 \\ 0 \\ 1 \end{bmatrix} = q^2 r_{22}.$$

This gives $q^2 = [-1 \ 0 \ 1]^T / \sqrt{2}$ and $r_{22} = \sqrt{2}$.

$$\begin{bmatrix} 1 & 1 \\ 1 & 2 \\ 1 & 3 \end{bmatrix} = \begin{bmatrix} 1/\sqrt{3} & -1/\sqrt{2} \\ 1/\sqrt{3} & 0 \\ 1/\sqrt{3} & 1/\sqrt{2} \end{bmatrix} \begin{bmatrix} \sqrt{3} & 2\sqrt{3} \\ 0 & \sqrt{2} \end{bmatrix}$$

5.5.2 Givens transform

This transform is used to obtain the upper triangular matrix by sequence of "2×2" transformations

$$G^T \equiv \begin{bmatrix} c & -s \\ s & c \end{bmatrix}$$

$$G^T \begin{bmatrix} a \\ b \end{bmatrix} \equiv \begin{bmatrix} \widehat{a} \\ 0 \end{bmatrix} \text{ where } s = \frac{-b}{\sqrt{a^2 + b^2}}, c = \frac{a}{\sqrt{a^2 + b^2}}.$$

The s and c are defined so that

$$ca - sb = \widehat{a} \text{ and}$$
$$sa + cb = 0.$$

In this case $\widehat{a} = \sqrt{a^2 + b^2}$. Also, they satisfy $G^T G = I$ and so any product will be unitary.

Example 5.5.2. Apply this transformation to the above matrix three times to obtain the upper triangular matrix.

$$\widehat{G}_{12}^T = \begin{bmatrix} 1/\sqrt{2} & 1/\sqrt{2} \\ -1/\sqrt{2} & 1/\sqrt{2} \end{bmatrix} \text{ with } a = 1, b = 1$$

$$G_{12}^T = \begin{bmatrix} 1/\sqrt{2} & 1/\sqrt{2} & 0 \\ -1/\sqrt{2} & 1/\sqrt{2} & 0 \\ 0 & 0 & 1 \end{bmatrix}$$

This will gives a zero in the 21-component of the matrix

$$G_{12}^T A = \begin{bmatrix} \sqrt{2} & 3/\sqrt{2} \\ 0 & 1/\sqrt{2} \\ 1 & 3 \end{bmatrix}.$$

Next get a zero in the 31-component with $a = \sqrt{2}, b = 1$

$$G_{13}^T G_{12}^T A = \begin{bmatrix} \sqrt{2}/\sqrt{3} & 0 & 1/\sqrt{3} \\ 0 & 1 & 0 \\ -1/\sqrt{3} & 0 & \sqrt{2}/\sqrt{3} \end{bmatrix} \begin{bmatrix} \sqrt{2} & 3/\sqrt{2} \\ 0 & 1/\sqrt{2} \\ 1 & 3 \end{bmatrix}$$

$$= \begin{bmatrix} \sqrt{3} & 2\sqrt{3} \\ 0 & 1/\sqrt{2} \\ 0 & \sqrt{3}/\sqrt{2} \end{bmatrix}.$$

Finally get a zero in the 32-component with $a = 1/\sqrt{2}, b = \sqrt{3}/\sqrt{2}$

$$G_{23}^T G_{13}^T G_{12}^T A = \begin{bmatrix} 1 & 0 & 0 \\ 0 & 1/2 & \sqrt{3}/2 \\ 0 & -\sqrt{3}/2 & 1/2 \end{bmatrix} \begin{bmatrix} \sqrt{3} & 2\sqrt{3} \\ 0 & 1/\sqrt{2} \\ 0 & \sqrt{3}/\sqrt{2} \end{bmatrix}$$

$$= \begin{bmatrix} \sqrt{3} & 2\sqrt{3} \\ 0 & \sqrt{2} \\ 0 & 0 \end{bmatrix} = R.$$

The matrix $Q = G_{12}G_{13}G_{23}$.

5.5.3 Householder transform

The Householder transforms are used in sequence of n products to give the orthonormal matrix

$$Q^T = H_n \cdots H_2 H_1 \text{ where}$$

$$H_1 x = \begin{bmatrix} \widehat{x}_1 \\ 0 \\ 0 \\ \vdots \end{bmatrix} = \widehat{x}_1 e_1 \ , \ H_2 y = \begin{bmatrix} y_1 \\ \widehat{y}_2 \\ 0 \\ 0 \end{bmatrix}, \cdots$$

The Householder transform has the form

$$H \equiv I - 2uu^T \text{ where } u^T u = 1.$$

The particular choice of u depends on the x and the requirement $Hx = \hat{x}_1 e_1$.

Theorem 5.5.1. *(Householder Transform)* *If $u^T u = 1$ and $Hx = \hat{x}_1 e_1$, then $H^T H = I$ and $\hat{x}_1 = \pm (x^T x)^{1/2}$.*

Proof.

$$\begin{aligned}
H^T H &= \left(I - 2uu^T \right)^T \left(I - 2uu^T \right) \\
&= \left(I - 2uu^T \right) - 2 \left(I - 2uu^T \right) uu^T \\
&= \left(I - 2uu^T \right) - 2uu^T + 4u(u^T u)u^T \\
&= I.
\end{aligned}$$

$(Hx)^T (Hx) = x^T (H^T H)x = x^T x$ and $(Hx)^T (Hx) = (\hat{x}_1 e_1)^T (\hat{x}_1 e_1) = (\hat{x}_1)^2$. ∎

The choice of u follows from

$$\begin{aligned}
x &= H^T H x \\
&= H^T \hat{x}_1 e_1 \\
&= \hat{x}_1 \left(I - 2uu^T \right) e_1 \\
&= \hat{x}_1 (e_1 - 2uu_1).
\end{aligned}$$

If $i = 1$, this is $x_1 = \hat{x}_1 (1 - 2(u_1)^2)$.
If $i > 1$, this is $x_i = \hat{x}_1 (0 - 2u_i u_1)$.
The choice for u is assuming $x_1 \neq 0$

$$\hat{x}_1 \equiv -(x^T x)^{1/2} \text{ for } x_1 > 0, \text{ and } \equiv (x^T x)^{1/2} \text{ for } x_1 < 0,$$

$$u_1 \equiv \left(\frac{\hat{x}_1 - x_1}{2\hat{x}_1} \right)^{1/2} \text{ and}$$

$$u_i \equiv \frac{x_i}{-2\hat{x}_i u_1} \text{ for } i > 1.$$

Example 5.5.3. Apply the Householder transform to the above matrix two times to obtain the upper triangular matrix.

$$H_1 \begin{bmatrix} 1 \\ 1 \\ 1 \end{bmatrix} = \begin{bmatrix} -\sqrt{3} \\ 0 \\ 0 \end{bmatrix} \text{ with } u = \begin{bmatrix} 0.8881 \\ 0.3250 \\ 0.3250 \end{bmatrix}$$

$$H_1 A = \begin{bmatrix} -\sqrt{3} & -3.4641 \\ 0 & 0.3661 \\ 0 & 1.3661 \end{bmatrix}.$$

Apply the second Householder transform to the lower two components in the second column.

$$H_2 = \begin{bmatrix} 1 & \\ & \widehat{H}_2 \end{bmatrix} \text{ with } \widehat{H}_2 \begin{bmatrix} 0.3661 \\ 1.3661 \end{bmatrix} = \begin{bmatrix} \widehat{x}_2 \\ 0 \end{bmatrix}.$$

Combine the two transforms to get

$$Q^T A = H_2 H_1 A = \begin{bmatrix} -1.7321 & -3.44641 \\ 0 & -1.4143 \\ 0 & 0 \end{bmatrix} = R.$$

MATLAB® code planerot is an implementation of the Givens transform:

```
>> help planerot
 planerot Givens plane rotation.
    [G,Y] = planerot(X), where X is a 2-component column vector,
    returns a 2-by-2 orthogonal matrix G so that Y = G*X has
    Y(2) = 0.
```

The MATLAB® code qr is an implementation of the sequence of Householder transforms:

```
>> help qr
    qr Orthogonal-triangular decomposition.
    [Q,R] = qr(A) performs a qr decomposition on m-by-n matrix A
    such that A = Q*R. The factor R is an m-by-n upper triangular
    matrix and Q is an m-by-m unitary matrix.
```

If the matrix A is a single column, then $[Q,R] = qr(A)$ returns the Householder transform:

```
>> [Q R] = qr([ 1 1 1]')

Q =
    -0.5774    -0.5774    -0.5774
    -0.5774     0.7887    -0.2113
    -0.5774    -0.2113     0.7887

R =
    -1.7321
          0
          0.
```

Note the columns of Q are orthonormal, form a basis and are generated from a single column vector. This is true for all column vectors with nonzero first component.

5.5.4 Exercises

1. Consider the matrix and find the QR factors

$$A = \begin{bmatrix} 1 & 1 \\ 1 & 2 \\ 1 & 3 \\ 1 & 4 \end{bmatrix}.$$

   (a). Use modified Gram–Schmidt.

   (b). Use classical Gram–Schmidt.

   (c). Use Givens transforms.

   (d). Use Householder transforms.

2. Consider finding the QR factors of an augmented matrix where r is a row vector

$$\begin{bmatrix} A \\ r \end{bmatrix}.$$

   Assume the matrix $A = QR$ is known. For example, use the 3×2 matrix in Examples 5.5.1–5.5.3, and the row vector $r = \begin{bmatrix} 1 & 4 \end{bmatrix}$. How can you use $A = QR$ or $Q^T A = R$ to factor the augmented matrix?

3. Consider finding the QR factors of an augmented matrix where c is a column vector

$$\begin{bmatrix} A & c \end{bmatrix}.$$

   Assume the matrix $A = QR$ is known. How can you use $A = QR$ or $Q^T A = R$ to factor the augmented matrix? Try the finding QR factors of

$$\begin{bmatrix} A & c \end{bmatrix} = \begin{bmatrix} 1 & 1 & 3 \\ 1 & 2 & 4 \\ 1 & 3 & 0 \\ 1 & 4 & 1 \end{bmatrix}.$$

6

Eigenvectors and Orthonormal Basis

Another way to find orthonormal vectors is to consider eigenvectors of a symmetric matrix. An important case is $A^T A$ in the normal equations, and this will form the core the singular value decomposition of an $m \times n$ matrix $A = U\Sigma V^T$. The spectral theorem for real symmetric $n \times n$ matrices gives $A = UDU^T$ where D is the diagonal of eigenvalues and the columns of Q are the corresponding eigenvectors.

6.1 Eigenvectors of Symmetric Matrix

Let $A = [a_{ij}]$ be an $n \times n$ real symmetric matrix ($a_{ij} = a_{ji}$ for $1 \leq i, j \leq n$, or $A^T = A$). Eigenvectors play an important role in generating orthonormal bases. The matrix in least squares problem generally is not symmetric. However, the $A^T A$ in the normal equations is because $(A^T A)^T = A^T A^{TT} = A^T A$. Another important class of symmetric matrices come from differential equation. For example, consider $-u_{xx} = f$ on the interval $[0 \quad L]$ with boundary conditions $u(0) = 0 = u(L)$. A finite difference approximation for $u_i \cong u(i\Delta x)$ with $n = 4$ and $\Delta x = L/n$ is

$$
\begin{bmatrix} 2 & -1 & 0 \\ -1 & 2 & -1 \\ 0 & -1 & 2 \end{bmatrix} \begin{bmatrix} u_1 \\ u_2 \\ u_3 \end{bmatrix} = \Delta x^2 \begin{bmatrix} f(1\Delta x) \\ f(2\Delta x) \\ f(3\Delta x) \end{bmatrix}.
$$

Definition. *Let A be an $n \times n$ matrix. The eigenvector, $x \in \mathbb{C}^n$, associated with an eigenvalue, $\lambda \in \mathbb{C}$, is a nonzero vector such that $Ax = \lambda x$.*

An eigenvector is an element of the null space of $A - \lambda I_n$

$$Ax - \lambda x = 0_{n \times 1}$$
$$Ax - \lambda I_n x = 0_{n \times 1}$$
$$(A - \lambda I_n)x = 0_{n \times 1}.$$

Since x is a nonzero vector, $\det(A - \lambda I_n) = 0$. This allows us to solve for the eigenvalues and then the corresponding solutions of $(A - \lambda I_n)x = 0_{n \times 1}$.

DOI: 10.1201/9781003304128-6

Example 6.1.1. Let $A = \begin{bmatrix} 1 & -2 \\ 1 & 3 \end{bmatrix}$. This matrix is not symmetric and has complex numbers for eigenvalues.

$$\det(A - \lambda I_2) = \det(\begin{bmatrix} 1 - \lambda & -2 \\ 1 & 3 - \lambda \end{bmatrix})$$

$$= \lambda^2 - 4\lambda + 5 = 0$$

and $\lambda = 2 + i, 2 - i$ with $i \equiv \sqrt{-1}$.

Example 6.1.2. Let $A = \begin{bmatrix} 1 & 2 \\ 2 & 4 \end{bmatrix}$. This matrix is symmetric and does not have an inverse.

$$\det(A - \lambda I_2) = \det(\begin{bmatrix} 1 - \lambda & 2 \\ 2 & 4 - \lambda \end{bmatrix}) = \lambda^2 - 5\lambda = 0$$

and $\lambda = 0, 5$.

Find the eigenvector associated with $\lambda = 0$.

$$(A - 0I_2)x = 0_2$$

$$\begin{bmatrix} 1 & 2 \\ 2 & 4 \end{bmatrix} \begin{bmatrix} x_1 \\ x_2 \end{bmatrix} = \begin{bmatrix} 0 \\ 0 \end{bmatrix} \text{ gives}$$

$$\begin{bmatrix} x_1 \\ x_2 \end{bmatrix} = \begin{bmatrix} 2 \\ -1 \end{bmatrix}.$$

Find the eigenvector associated with $\lambda = 5$.

$$(A - 5I_2)x = 0_2$$

$$\begin{bmatrix} -4 & 2 \\ 2 & -1 \end{bmatrix} \begin{bmatrix} x_1 \\ x_2 \end{bmatrix} = \begin{bmatrix} 0 \\ 0 \end{bmatrix} \text{ gives}$$

$$\begin{bmatrix} x_1 \\ x_2 \end{bmatrix} = \begin{bmatrix} 1 \\ 2 \end{bmatrix}.$$

Note the eigenvalues are real, distinct and the eigenvectors are perpendicular. Since a nonzero multiple of an eigenvector is also an eigenvector, we normalize the eigenvectors to get an orthonormal basis of $\mathbb{R}^2$

$$q_1 = \begin{bmatrix} 2/\sqrt{5} \\ -1/\sqrt{5} \end{bmatrix} \text{ and } q_2 = \begin{bmatrix} 1/\sqrt{5} \\ 2/\sqrt{5} \end{bmatrix}.$$

Theorem 6.1.1. *(Real Eigenvalues) If A is symmetric and has real components, then*

  (i). *all the eigenvalues are real and*

  (ii). *eigenvectors with different eigenvalues are orthogonal.*

Proof. Let $\lambda \in \mathbb{C}$, $x \in \mathbb{C}^n$ and $Ax = \lambda x$. Use the notation (*conjugate transpose*)

$$x^* \equiv \bar{x}^T \text{ and } A^* \equiv \bar{A}^T = [\bar{a}_{ji}].$$

A real and symmetric means $\bar{a}_{ji} = a_{ij}$ so that $A^* = A$.

In order to show λ is real, show $\lambda = \bar{\lambda}$. $Ax = \lambda x$ implies $x^* A x = \lambda x^* x$. Also, $\overline{Ax} = \overline{\lambda x}$, $A\bar{x} = \bar{\lambda}\bar{x}$ and $\bar{x}^T A = \bar{\lambda}\bar{x}^T$. Hence,

$$x^* A x = \bar{\lambda} x^* x = \lambda x^* x.$$

Since x is a nonzero vector, $x^* x \neq 0$ and $\lambda = \bar{\lambda}$.

In order to show eigenvectors with different eigenvalues must be orthogonal, assume $Ax = \lambda x$, $A\hat{x} = \hat{\lambda}\hat{x}$ and $\lambda \neq \hat{\lambda}$.

$$\hat{x}^T A x = \hat{x}^T \lambda x = \lambda \hat{x}^T x$$
$$x^T A \hat{x} = x^T \hat{\lambda}\hat{x} = \hat{\lambda}\hat{x}^T x$$

Since A is symmetric, $x^T A \hat{x} = x^T A^T \hat{x} = (Ax)^T \hat{x} = \hat{x}^T A x$. This means $\lambda \hat{x}^T x = \hat{\lambda}\hat{x}^T x$ and $(\lambda - \hat{\lambda})\hat{x}^T x = 0$. Because $\lambda \neq \hat{\lambda}$, $\hat{x}^T x = 0$. ∎

Definition. *The spectrum of A is the set of all eigenvalues, $\sigma(A)$. The spectral radius of A is the largest modulus of the eigenvalues, $\rho(A)$.*

In Example 6.1.1 $\sigma(A) = \{2+i, 2-i\}$ and $\rho(A) = \sqrt{5}$. Example 6.1.2 has $\sigma(A) = \{0, 5\}$ and $\rho(A) = 5$. If the matrix is symmetric, then $\sigma(A) \subset \mathbb{R}$. The spectral radius is not a norm because there are nonzero matrices with $\rho(A) = 0$, for example,

$$A = \begin{bmatrix} 0 & 1 \\ 0 & 0 \end{bmatrix}.$$

Theorem 6.1.2. *(Positive Eigenvalues) Let S and A be $n \times n$ matrices.*

(i). If S has an inverse, then $\sigma(A) = \sigma(S^{-1}AS)$.

(ii). If A is SPD, then $\sigma(A) \subset (0, \infty)$.

(iii). $\rho(A) \leq \|A\| \equiv \sup_{\|x\| \neq 0} \frac{\|Ax\|}{\|x\|}$.

Proof. Let $x \neq 0$ be an eigenvector of A. Because S has an inverse, $S^{-1}x \neq 0$ and

$$Ax = \lambda x$$
$$S^{-1}Ax = S^{-1}\lambda x$$
$$S^{-1}AS(S^{-1}x) = \lambda(S^{-1}x).$$

If A SPD, then for all $x \neq 0$ $x^T A x > 0$. Choose x to be an eigenvector.

$$x^T A x = x^T \lambda x = \lambda x^T x > 0.$$

Since $x^T x > 0$, this proves item (ii). Let x be the eigenvector such that $Ax = \lambda x$ and $|\lambda| = \rho(A)$. Use $\|Ax\| \leq \|A\| \|x\|$ to conclude

$$|\lambda| \|x\| = \|\lambda x\| = \|Ax\| \leq \|A\| \|x\|$$
$$|\lambda| \leq \|A\|.$$

■

6.1.1   Exercises

1. Consider the matrix $A = \begin{bmatrix} 3 & -1 \\ -1 & 3 \end{bmatrix}$.

   (a). Find all the eigenvalues.

   (b). Compute the spectral radius.

   (c). Find all the eigenvectors.

2. Consider the matrix $A = \begin{bmatrix} 3 & 1 \\ -1 & 2 \end{bmatrix}$.

   (a). Find all the eigenvalues.

   (b). Compute the spectral radius.

   (c). Find all the eigenvectors.

3. Let x be an eigenvector of A with $Ax = \lambda x$.

   (a). Prove: if A has an inverse and $\lambda \neq 0$, then an eigenvalue of A^{-1} is $1/\lambda$.

   (b). Prove: if $I - cA$ has an inverse and $c\lambda \neq 1$, then an eigenvalue of $(I - cA)^{-1}$ is $1/(1 - c\lambda)$.

   (c). Prove: if $P(x)$ is a polynomial, then the eigenvalue of $P(A)$ is $P(\lambda)$.

6.2   Approximation of Eigenvalues

The estimation of eigenvalues and their eigenvectors is well studied. Here three approximation methods will be introduced: Gerschgorin circles, power iterations and iteration of QR factors. These are the introductory topics that are required for the extensive study of eigenvalue/eigenvectors. The following MATLAB® commands are good implementations.

```
>> help eig
  eig    Eigenvalues and eigenvectors.
     E = eig(A) produces a column vector E containing
     the eigenvalues of a square matrix A.

     [V,D] = eig(A) produces a diagonal matrix D of eigenvalues
     and eigenvalues and a full matrix V whose columns are the
     corresponding eigenvectors so that A*V = V*D.\smallskip
```

6.2.1   Gerschgorin circles

Let $A = [a_{ij}]$ where $1 \leq i, j \leq n$ and $Ax = \lambda x$. The component form of $Ax = \lambda x$ is $\sum_j a_{i,j} x_j = \lambda x_i$. Choose i_0 so that $|x_{i_0}| \geq |x_j|$ for all j.

$$|a_{i_0,i_0} - \lambda| \, |x_{i_0}| \leq \sum_{j \neq i_0} |a_{i_0,j} \, x_j|$$

$$|a_{i_0,i_0} - \lambda| \, |x_{i_0}| \leq \sum_{j \neq i_0} |a_{i_0,j}| \, |x_j|$$

$$|a_{i_0,i_0} - \lambda| \leq \sum_{j \neq i_0} |a_{i_0,j}| \, |x_j| / |x_{i_0}|$$

$$\leq \sum_{j \neq i_0} |a_{i_0,j}| \, .$$

Thus each eigenvalue must be inside some circle centered at a_{i_0,i_0} with radius $\sum_{j \neq i_0} |a_{i_0,j}|$.

Example 6.2.1. Consider the following 3×3 matrix with three possible circles

$$A = \begin{bmatrix} 10 & 2 & 3 \\ -1 & 0 & 2 \\ 1 & -2 & 1 \end{bmatrix} \text{ and } \sigma(A) = \{10.2600, 0.3870 \pm i2.2216\}.$$

$$D_1 = \{z : |z - 10| \leq 2 + 3\}$$
$$D_2 = \{z : |z - 0| \leq 1 + 2\} \text{ and }$$
$$D_3 = \{z : |z - 1| \leq 1 + 1\}.$$

6.2.2   Power iterations

This is an iterative method which typically approximates the largest eigenvalue. Note the eigenvalue equation can be divided by any nonzero constant, in particular, by $\|x\|$

$$A(\frac{x}{\|x\|}) = \lambda(\frac{x}{\|x\|}).$$

The iterative method attempts to find a fixed point of

$$A(\frac{x^k}{\|x^k\|}) = \lambda^{k+1}(\frac{x^{k+1}}{\|x^{k+1}\|})$$

Compute the norm of both sides to get

$$\left\|A(\frac{x^k}{\|x^k\|})\right\| = \left|\lambda^{k+1}\right| 1.$$

Next let

$$\frac{x^{k+1}}{\|x^{k+1}\|} = A(\frac{x^k}{\|x^k\|})/\left\|A(\frac{x^k}{\|x^k\|})\right\|.$$

This is implemented in the first part of MATLAB® code power_qrone.m where

$$v \approx \frac{x^k}{\|x^k\|} \text{ and } \lambda \approx \left\|A(\frac{x^k}{\|x^k\|})\right\|.$$

```
1     % Test matrix for power and qr methods
2     A = [-261 209 -49;...
3     -530 422 -98;...
4     -800 631 -144]
5     %A = [20 -1 1;3 2 -5; 1 4 1]
6     % A = [10 2 3;-1 0 2;; 1 -2 1]
7     eig(A)
8     %
9     % Power Method
10    %
11    v = [1 0 0]'
12    for i = 1:10
13        z = A*v
14        %z = A\v
15        lam = norm(z,2)
16        v = z/lam
17        pause
18    end
19    %
20    % QR Algorithm (version one)
21    %
22    for i = 1:10
23        [Q R] = qr(A)
24        A = R*Q
25        pause
26    end
```

6.2.3   QR iteration

This is the first version of the QR algorithm. It makes use of the QR factor $A = QR$ where $Q^T Q = I$. The first few iteration are as follows.

$$A = Q_1 R_1$$
$$A_1 \equiv R_1 Q_1.$$

Note $Q_1 A_1 Q_1^T = A$ and $\sigma(A) = \sigma(A_1)$. The new matrix tends to approach a triangle matrix whose eigenvalues are on the diagonal. Repeat the above

$$A_1 = Q_2 R_2$$
$$A_2 \equiv R_2 Q_2$$

Note $Q_2 A_2 Q_2^T = A_1$ and $\sigma(A) = \sigma(A_1) = \sigma(A_2)$. Continue and stop at some point where little change in the matrix is observed. This is implemented in the second part of the above code.

6.2.4   Exercises

1. Consider Example 6.2.1. Use the power and QR algorithm. Notice the imaginary eigenvalues were not directly computed. However, the QR algorithm did get to smaller 2×2 blocks where the imaginary eigenvalues can be computed.

2. Estimate the eigenvalues $\{19.9292, 1.5304 + i4.5418, 1.5304 - i4.5418\}$ of

$$A = \begin{bmatrix} 20 & -1 & 1 \\ 3 & 2 & -5 \\ 1 & 4 & 1 \end{bmatrix}.$$

   (a). Use Gerschgorin circles.

   (b). Use the power method.

   (c). Use the first version of the QR algorithm.

6.3   Spectral Theorem Factors $AQ = QD$

Example 6.1.2 shows there exists two orthonormal eigenvectors such that

$$Aq_1 = 0q_1 \text{ and } Aq_2 = 5q_2.$$

The matrix form of this is

$$A \begin{bmatrix} q_1 & q_2 \end{bmatrix} = \begin{bmatrix} q_1 & q_2 \end{bmatrix} \begin{bmatrix} 0 & 0 \\ 0 & 5 \end{bmatrix} \text{ with}$$

$$\begin{bmatrix} q_1 & q_2 \end{bmatrix}^T \begin{bmatrix} q_1 & q_2 \end{bmatrix} = \begin{bmatrix} q_1^T q_1 & q_1^T q_2 \\ q_2^T q_1 & q_2^T q_2 \end{bmatrix} = \begin{bmatrix} 1 & 0 \\ 0 & 1 \end{bmatrix}.$$

So, this 2×2 matrix has the form $A = QDQ^T$ where Q is the matrix of the orthonormal eigenvectors and the D is the diagonal matrix of the corresponding eigenvalues. We shall show this is true for any symmetric $n \times n$ matrix.

Definition. *Let A be an $n \times n$ matrix. A is called diagonalizable if only if there exists an orthonormal $n \times n$ matrix Q such that*

$$A = QDQ^T \text{ and } D \text{ is an } n \times n \text{ diagonal matrix.}$$

The following proof does not construct the eigenvectors, but it does show the existence of the desired eigenvectors. Numerical linear algebra is used to find the eigenvectors, and the MATLAB® command eig(A) is a good implementation of these methods. An important generalization of the following is the Schur decomposition for matrices with complex components, see Section 10.3.

Theorem 6.3.1. *(Diagonalizable Matrix) All real symmetric $n \times n$ matrices are diagonalizable.*

Proof. Mathematical induction on n will be used to prove this. Choose one eigenvector x where $Ax = \lambda x$ and $q_1 \equiv x/(x^T x)^{1/2}$. If $n = 2$, choose another vector q_2 with such that $q_1^T q_2 = 0$ and $q_2^T q_2 = 1$. Let $Q \equiv \begin{bmatrix} q_1 & q_2 \end{bmatrix}$ and use $A = A^T$

$$
\begin{aligned}
Q^T A Q &= \begin{bmatrix} q_1^T A q_1 & q_1^T A q_2 \\ q_2^T A q_1 & q_2^T A q_2 \end{bmatrix} \\
&= \begin{bmatrix} q_1^T \lambda q_1 & q_1^T A^T q_2 \\ q_2^T \lambda q_1 & q_2^T A q_2 \end{bmatrix} \\
&= \begin{bmatrix} \lambda & 0 \\ 0 & q_2^T A q_2 \end{bmatrix}.
\end{aligned}
$$

For the inductive step extend the orthonormal basis to all of $\mathbb{R}^n$

$$q_2, \cdots, q_n \text{ with } q_i^T q_i = 1, \ q_i^T q_j = 0 \text{ and } i \neq j.$$

Let $Q \equiv \begin{bmatrix} q_1 & q_2 & \cdots & q_n \end{bmatrix}$ so that

$$
Q^T A Q = \begin{bmatrix}
\lambda & q_1^T A q_2 & \cdots & q_1^T A q_n \\
q_2^T \lambda q_1 & q_2^T A q_2 & \cdots & q_2^T A q_n \\
\vdots & \vdots & \ddots & \vdots \\
q_n^T \lambda q_1 & q_n^T A q_2 & \cdots & q_n^T A q_n
\end{bmatrix}.
$$

Since $A = A^T$, $q_1^T A q_j = (A q_1)^T q_j = (\lambda q_1)^T q_j = 0$ for $2 \leq j \leq n$,

$$Q^T A Q = \begin{bmatrix} \lambda & 0_{1 \times (n-1)} \\ 0_{(n-1) \times 1} & \widehat{A} \end{bmatrix}.$$

Since $A = A^T$, the $(n-1) \times (n-1)$ matrix $\widehat{A}$ is also symmetric.

If the proposition is true for $(n-1) \times (n-1)$ matrices, apply it to $\widehat{A}$ to get $\widehat{Q}^T \widehat{A} \widehat{Q} = \widehat{D}$. This gives

$$
\begin{aligned}
Q^T A Q &= \begin{bmatrix} \lambda & 0_{1 \times (n-1)} \\ 0_{(n-1) \times 1} & \widehat{Q}^T \widehat{D} \widehat{Q} \end{bmatrix} \text{ and for } J \equiv \begin{bmatrix} 1 & 0_{1 \times (n-1)} \\ 0_{(n-1) \times 1} & \widehat{Q}^T \end{bmatrix} \\
&= J \begin{bmatrix} \lambda & 0_{1 \times (n-1)} \\ 0_{(n-1) \times 1} & \widehat{D} \end{bmatrix} J^T.
\end{aligned}
$$

Let $\widetilde{Q} \equiv QJ$ and note it is an orthonormal $n \times n$ matrix and satisfies

$$\widetilde{Q}^T A \widetilde{Q} = \begin{bmatrix} \lambda & 0_{1 \times (n-1)} \\ 0_{(n-1) \times 1} & \widehat{D} \end{bmatrix}.$$

■

6.3.1 Exercises

1. Find the orthonormal eigenvectors of

$$A = \begin{bmatrix} 2 & -1 & 0 \\ -1 & 2 & -1 \\ 0 & -1 & 2 \end{bmatrix}.$$

(a). Find the eigenvalues.

(b). Find the eigenvectors.

(c). Normalize the eigenvectors.

2. The proof of Theorem 6.3.1 contains the statement "If $n = 2$, choose another vector q_2 with such that $q_1^T q_2 = 0$ and $q_2^T q_2 = 1$."

Explain how to do this.

6.4 Applications

Eigenvectors are very useful in a number of problems. The first application is to symmetric matrices that have inverses. This includes to SPD matrices. The second application is a first look at the singular value decomposition, SVD, of the general $m \times n$ matrix, which will be further studied in the next three chapters.

6.4.1 Nonsingular $Ax = d$

Eigenvectors are very useful in a number of problems. In this section they will be used to find solution of algebraic problems where there n unknowns, n equations and the associated matrix is nonsingular. Provided one knows the eigenvectors and eigenvalues, this requires very few computations relative to Gauss elimination. This technique also generalizes to self-adjoint boundary value problems.

Assume the matrix in the algebraic problem $Ax = d$ is nonsingular, real and symmetric. Then the eigenvalues must be real and nonzero. Choose the orthonormal eigenvectors and eigenvalues

$$Aq_i = \lambda_i q_i \text{ with } 1 \le i \le n \text{ and } \lambda_i \ne 0.$$

Since $d \in \mathbb{R}^n$ and the eigenvectors are a basis,

$$d = \sum_{j=1}^{n} c_j q_j.$$

Since the eigenvectors are orthonormal,

$$q_i^T d = q_i^T \sum_{j=1}^{n} c_j q_j = \sum_{j=1}^{n} c_j q_i^T q_j = c_i.$$

The solution must also be a linear combination of the eigenvectors

$$x = \sum_{j=1}^{n} \widehat{c}_j q_j.$$

$$Ax = A \sum_{j=1}^{n} \widehat{c}_j q_j$$

$$= \sum_{j=1}^{n} \widehat{c}_j A q_j$$

$$= \sum_{j=1}^{n} \widehat{c}_j \lambda_j q_j.$$

Since $d = \sum_{j=1}^{n} c_j q_j = \sum_{j=1}^{n} (q_j^T d) q_j$ and by the linear independence, $q_j^T d = \widehat{c}_j \lambda_j$. Since $\lambda_j \neq 0$, $\widehat{c}_j = q_j^T d / \lambda_j$.

Not counting the cost of finding the eigenvalues, this requires n inner products, $q_j^T d$, and n divisions. This is in contrast to Gauss elimination, which requires about $n^3/3$ operations. If one has a large sequence of algebraic problems all with the same matrix and and number of right-hand sides, then the eigenvectors and eigenvalues only need to be found once. Also, for some special matrices, these may be easily found.

6.4.2   Singular value decomposition

Consider an $m \times n$ real matrix A. Then $A^T A$ is real symmetric $n \times n$. Let V be the $n \times n$ matrix formed be the column vectors from the orthonormal eigenvectors of $A^T A$. By the spectral theorem

$$V^T(A^T A)V = D \text{ is a diagonal matrix.}$$

Assume $m \geq n$ and use the "fat" QR factors of AV

$$AV = QR \text{ or } A = QRV^T.$$

The matrix Q is $m \times m$, and the matrix R is $m \times n$. By the spectral theorem

$$V^T A^T AV = V^T(QRV^T)^T(QRV^T)V = R^T R = D.$$

The matrix R is $m \times n$ and has the following form where $k = rank(A)$

$$R = \begin{bmatrix} R_k & 0_{k \times (n-k)} \\ 0_{(m-k) \times k} & 0_{(m-k) \times (n-k)} \end{bmatrix}$$

where the R_k is upper triangular $k \times k$ matrix. The matrix equation $R^T R = D$ is

$$\begin{bmatrix} R_k^T & 0_{k \times (m-k)} \\ 0_{(n-k) \times k} & 0_{(n-k) \times (m-k)} \end{bmatrix} \begin{bmatrix} R_k & 0_{k \times (n-k)} \\ 0_{(m-k) \times k} & 0_{(m-k) \times (n-k)} \end{bmatrix} = D$$

$$\begin{bmatrix} R_k^T R_k & 0_{k \times (n-k)} \\ 0_{(n-k) \times k} & 0_{(n-k) \times (n-k)} \end{bmatrix} = D.$$

Since R_k is upper triangular and D is diagonal, the diagonal components must be positive, and they are called the singular values of matrix A. Since the matrix R is now diagonal, this is a very special QR factorization of the matrix AV

$$A[v_1 \cdots v_k] = [q_1 r_{11} \cdots q_k r_{kk}]$$

where $r_{ii} = (\lambda_i)^{1/2}$. This means $q_i = Av_i / r_{ii}$.

Example 6.4.1. This matrix is 4×3 and has rank equal to two

$$A = \begin{bmatrix} 1 & 2 & 3 \\ 1 & 3 & 4 \\ 1 & 5 & 6 \\ 1 & 8 & 9 \end{bmatrix}.$$

The MATLAB® code svdviaqr.m is used to compute the matrix R

$$R = \begin{bmatrix} 15.7156 & 0 & 0 \\ 0 & 1.0101 & 0 \\ 0 & 0 & 0 \\ 0 & 0 & 0 \end{bmatrix}.$$

The columns in matrices Q and V are orthonormal and give bases for $\mathbb{R}^4$ and $\mathbb{R}^3$, respectively. As noted by the last two lines of the code the third column of V is a basis for $N(A)$, and the third and fourth columns of Q are a basis for $N(A^T)$.

```
1    % Illustrates SVD using QR factors
2    A = [1 2 3; 1 3 4; 1 5 6; 1 8 9]
3    P = [ 0 0 1; 0 1 0; 1 0 0]
4    %A = [1 2 3 2; 1 3 4 2; 1 5 6 2; 1 8 9 2; 1 11 12 2]
5    %P = [0 0 0 1 ;0 0 1 0;0 1 0 0; 1 0 0 0]
6    [m n ] = size(A)
7    K = rank(A)
8    [V D] = eig(A'*A); % choice + or - v
```

```
9      % Permutation is used to reorder eigenvalues
10     % from increasing to decreasing
11     DD = P*D*P
12     VV = V*P
13     for k=1:K
14        QQ(1:m,k)= A*VV(:,k)/sqrt(DD(k,k));
15        RR(k,k) = sqrt(DD(k,k));
16     end
17     QQ
18     RR
19     AA = QQ*RR*VV(:,1:K)'
20     % Compare with svd(A).
21     [u sig v] = svd(A)
22     % The third column of v is basis for N(A).
23     A*v
24     % The columns three and four are a basis for N(A').
25     A'*u
```

```
DD =
   246.9797          0          0
         0     1.0203          0
         0          0    -0.0000
VV =
     0.1162     0.8082     0.5774
     0.6418    -0.5047     0.5774
     0.7580     0.3035    -0.5774
QQ =
     0.2338     0.7021
     0.3228     0.5029
     0.5010     0.1045
     0.7682    -0.4931
RR =
    15.7156          0
         0     1.0101

AA =
     1.0000     2.0000     3.0000
     1.0000     3.0000     4.0000
     1.0000     5.0000     6.0000
     1.0000     8.0000     9.0000
u =
    -0.2338     0.7021     0.3832    -0.5528
    -0.3228     0.5029     0.0369     0.8009
    -0.5010     0.1045    -0.8279    -0.2294
    -0.7682    -0.4931     0.4078    -0.0188
```

sig =
      15.7156   0   0
           0   1.0101   0
           0   0   0.0000
           0   0   0
v =
     -0.1162   0.8082   0.5774
     -0.6418   -0.5047   0.5774
     -0.7580   0.3035   -0.5774

ans =
     -3.6738   0.7092   -0.0000
     -5.0737   0.5080   -0.0000
     -7.8733   0.1055   0
    -12.0728   -0.4981   0

ans =
     -1.8258   0.8164   0.0000   -0.0000
    -10.0866   -0.5098   -0.0000   -0.0000
    -11.9124   0.3066   0.0000   -0.0000

6.4.3   Exercises

1. Use normalized eigenvectors to solve $Ax = d$

$$\begin{bmatrix} 2 & -1 & 0 \\ -1 & 2 & -1 \\ 0 & -1 & 2 \end{bmatrix} \begin{bmatrix} x_1 \\ x_2 \\ x_3 \end{bmatrix} = \begin{bmatrix} 1 \\ 6 \\ 1 \end{bmatrix}.$$

(a). Find the orthonormal eigenvectors, q_j for $j = 1, 2, 3$, and eigenvalues.

(b). Find the coefficients in the linear combination for

$$d = c_1 q_1 + c_2 q_2 + c_3 q_3.$$

(c). Find the solution as a linear combination of the eigenvectors.

2. Consider the 5×4 matrix

$$A = \begin{bmatrix} 1 & 2 & 3 & 2 \\ 1 & 3 & 4 & 2 \\ 1 & 5 & 6 & 2 \\ 1 & 8 & 9 & 2 \\ 1 & 11 & 12 & 2 \end{bmatrix}.$$

(a). Find the rank of the matrix.

(b). Modify the above code. Adjust the permutation matrix and compare with svd(A).

7

Singular Value Decomposition

In this chapter we consider the more general $m \times n$ real matrix A and its factorization $U\Sigma V^T$. The factors satisfy the following: U is $m \times m$, $U^T U = I_m$, V is $n \times n$, $V^T V = I_n$ and Σ is $m \times n$. This is the "full" SVD. First, we establish the "small" version. Then the "full" and "truncated" versions are described.

Definition. *The ("full") SVD factorization of A is $A = U\Sigma V^T$ where*

$$U \text{ is } m \times m, U^T U = I_m,$$
$$V \text{ is } n \times n, V^T V = I_n,$$
$$\Sigma \text{ is } m \times n \text{ where } \Sigma = \begin{bmatrix} \Sigma_r & 0_{r \times (n-r)} \\ 0_{(m-r) \times r} & 0_{(m-r) \times (n-r)} \end{bmatrix} \text{ and}$$
$$\Sigma_r = \begin{bmatrix} \sigma_1 & & \\ & \ddots & \\ & & \sigma_r \end{bmatrix} \text{ with } \sigma_1 \geq \cdots \geq \sigma_r > 0.$$

7.1 "Small" SVD

Assume $\text{rank}(A) = r$ with $U_1 \equiv U(1:m, 1:r)$ $m \times r$ and $V_1 \equiv U(1:n, 1:r)$ $n \times r$ matrices. The "small" or "compact" SVD is

$$A = U_1 \Sigma_r V_1^T$$

where $U_1^T U_1 = I_r$, $V_1^T V_1 = I_r$ and Σ_r is the $r \times r$ diagonal matrix with positive diagonal components. The existence of these factors will be established in Theorem 7.1.1 and uses the existence of orthonormal eigenvectors for symmetric matrices via Theorem 6.3.1.

The columns of U_1 are a basis for $R(A)$. This follows from $AV_1 = U_1 \Sigma_r$ and Σ_r^{-1}
$$AV_1 \Sigma_r^{-1} = U_1.$$

So the columns of U_1 are in $R(A)$. The columns are orthonormal, $\dim(R(A)) = r$ are a basis of $R(A)$. Likewise, $A^T U_1 \Sigma_r^{-1} = V_1$ and the columns of V_1 are an orthonormal basis of $R(A^T)$ and $\dim(R(A^T)) = r$.

DOI: 10.1201/9781003304128-7

Theorem 7.1.1. *(Singular Values Exist)* *If A is $m \times n$ with rank$(A) = r$, then U_1, V_1 and Σ_r exists where*

$$A = U_1 \Sigma_r V_1^T,$$
$$U_1 \text{ is } m \times r, \ U_1^T U_1 = I_r,$$
$$V_1 \text{ is } n \times r, \ V_1^T V_1 = I_r \text{ and}$$
$$\Sigma_r = \ r \times r \text{ nonsingular diagonal matrix.}$$

Remark. Multiply $A = U_1 \Sigma_r V_1^T$ on the right by V_1 to get $AV_1 = U_1 \Sigma_r$. In other words, $Av_i = u_i \sigma_i$ where v_i are $n \times 1$ with $V_1 = \begin{bmatrix} v_1 & \cdots & v_r \end{bmatrix}$ and u_i are $m \times 1$ with $U_1 = \begin{bmatrix} u_1 & \cdots & u_r \end{bmatrix}$.

Proof. $A^T A$ is a real symmetric $n \times n$ matrix. By Theorem 6.3.1 there are orthonormal eigenvectors. Let v be anyone of these $A^T Av = \lambda v$ where $v \neq 0_{n \times 1}$ and $\lambda \neq 0$. Then

$$A(A^T Av) = A\lambda v \text{ and}$$
$$(AA^T)Av = \lambda Av.$$

Note $v^T v = 1$ and $A^T Av = \lambda v$ implies

$$v^T A^T Av = \lambda v^T v \text{ and}$$
$$(Av)^T (Av) = \lambda \neq 0.$$

Thus, $\lambda > 0$, $Av \neq 0_{m \times 1}$ and Av is an eigenvector of AA^T. Write $\lambda_i = \sigma_i^2$ and $(Av_i)^T (Av_i) = \sigma_i^2$. Let V_1 have columns of the orthonormal eigenvectors of $A^T A$.

Define the columns of U_1, u_i, from the columns of Av_i as follows

$$u_i \equiv \frac{Av_i}{\sigma_i}.$$

u_i are eigenvectors of AA^T:

$$A^T Av_i = \sigma_i^2 v_i,$$
$$A(A^T Av_i) = A(\sigma_i^2 v_i),$$
$$AA^T (Av_i) = \sigma_i^2 (Av_i) \text{ and}$$
$$AA^T \left(\frac{Av_i}{\sigma_i}\right) = \sigma_i^2 \left(\frac{Av_i}{\sigma_i}\right).$$

u_i are unit vectors:

$$A^T Av_i = \sigma_i^2 v_i,$$
$$v_i^T (A^T Av_i) = v_i^T (\sigma_i^2 v_i),$$
$$(Av_i)^T (Av_i) = \sigma_i^2 \text{ and}$$
$$\left(\frac{Av_i}{\sigma_i}\right)^T \left(\frac{Av_i}{\sigma_i}\right) = 1.$$

u_i are orthogonal:

$$u_i^T u_j = (\frac{Av_i}{\sigma_i})^T (\frac{Av_j}{\sigma_j}),$$
$$= \frac{v_i^T A^T A v_j}{\sigma_i \sigma_j},$$
$$= \frac{v_i^T (A^T A v_j)}{\sigma_i \sigma_j},$$
$$= \frac{v_i^T (\sigma_j^2 v_j)}{\sigma_i \sigma_j} \text{ and for } i \neq j$$
$$= 0.$$

Define the matrices V_1 and U_1 where $1 \leq i, j \leq r$

$$V_1 \equiv \begin{bmatrix} v_1 & \cdots & v_r \end{bmatrix} \text{ and } U_1 \equiv \begin{bmatrix} u_1 & \cdots & u_r \end{bmatrix}.$$

Since $Av_i = u_i \sigma_i$, $AV_1 = U_1 \Sigma_r$ and $A = U_1 \Sigma_r V_1^T$. ∎

Remark. The order of the columns in V_1 and U_1 is chosen so that the singular values have decreasing order $\sigma_1 \geq \cdots \geq \sigma_r > 0$.

Example 7.1.1. $A = \begin{bmatrix} 2 & 2 \\ -1 & 1 \end{bmatrix}$ where $\text{rank}(A) = 2, m = 2$ and $n = 2$. $N(A) = \{0_{2 \times 1}\}$ and $N(A^T) = \{0_{2 \times 1}\}$.

$$A^T A = \begin{bmatrix} 5 & 3 \\ 3 & 5 \end{bmatrix} \text{ and } \det(\begin{bmatrix} 5 - \lambda & 3 \\ 3 & 5 - \lambda \end{bmatrix}) = 0.$$

This gives $\lambda_1 = 8, \sigma_1 = 2\sqrt{2}$ and $\lambda_2 = 2, \sigma_2 = \sqrt{2}$ so that

$$v_1 = \begin{bmatrix} 1 \\ 1 \end{bmatrix} / \sqrt{2} \text{ and } v_2 = \begin{bmatrix} 1 \\ -1 \end{bmatrix} / \sqrt{2}.$$

$$u_1 = \frac{Av_1}{\sigma_1} = \begin{bmatrix} 1 \\ 0 \end{bmatrix} \text{ and } u_2 = \frac{Av_2}{\sigma_2} = \begin{bmatrix} 0 \\ -1 \end{bmatrix}.$$

$$U_1 \Sigma_2 V_1^T = \begin{bmatrix} 1 & 0 \\ 0 & -1 \end{bmatrix} \begin{bmatrix} 2\sqrt{2} & 0 \\ 0 & \sqrt{2} \end{bmatrix} \begin{bmatrix} 1/\sqrt{2} & 1/\sqrt{2} \\ 1/\sqrt{2} & -1/\sqrt{2} \end{bmatrix}$$
$$= \begin{bmatrix} 2 & 2 \\ -1 & 1 \end{bmatrix} = A.$$

7.1.1 Exercises

1. Consider $A = \begin{bmatrix} 2 & 3 \\ -1 & 1 \end{bmatrix}$.

   (a). Find the eigenvalues and eigenvectors of $A^T A$.

   (b). Find the "small" SVD of A.

2. Consider $A = \begin{bmatrix} 1 & 2 \\ 2 & 4 \\ 1 & 2 \end{bmatrix}$.

   (a). Find the eigenvalues and eigenvectors of $A^T A$.

   (b). Find the "small" SVD of A.

3. Use the "small" SVD to solve the normal equations $A^T A x = A^T d$.

7.2 "Full" SVD

The "full" SVD $A = U\Sigma V^T$ has orthonormal matrices $m \times m$ U, $n \times n$ V and $m \times n$

$$\Sigma = \begin{bmatrix} \Sigma_r & 0_{r \times (n-r)} \\ 0_{(m-r) \times r} & 0_{(m-r) \times (n-r)} \end{bmatrix}.$$

The "full" SVD version requires the augmentation of $m \times r$ U_1, $n \times r$ V_1 in the "small" SVD.

Theorem 7.2.1. *(SVD Exists)* *If A is $m \times n$ with rank$(A) = r$, then the "full" SVD exists.*

Proof. In the "small" SVD the columns in U_1 and V_1 form bases of $R(A)$ and $R(A^T)$, respectively. By Theorem 5.2.1 $\mathbb{R}^n = R(A^T) \oplus N(A)$ and $\mathbb{R}^m = R(A) \oplus N(A^T)$. By Theorem 5.4.1 we can extend the orthonormal bases of $R(A^T)$ and $R(A)$ to all of $\mathbb{R}^n$ and $\mathbb{R}^m$, respectively. Let U be the orthonormal augmentation of U_1 so that the columns of $U(1:m,(r+1):m)$ are an orthonormal basis of $N(A^T)$. Likewise, let V be the orthonormal augmentation of V_1 so that the columns of $V(1:n,(r+1):n)$ are an orthonormal basis of $N(A)$.

Next we show $AV = U\Sigma$.

$$AV = A \begin{bmatrix} V_1 & V(1:n,(r+1):n) \end{bmatrix}$$
$$= \begin{bmatrix} U_1 \Sigma_r & 0_{m \times (n-r)} \end{bmatrix}.$$
$$U\Sigma = \begin{bmatrix} U_1 & U(1:m,(r+1):m) \end{bmatrix} \begin{bmatrix} \Sigma_r & 0_{r \times (n-r)} \\ 0_{(m-r) \times r} & 0_{(m-r) \times (n-r)} \end{bmatrix}$$
$$= \begin{bmatrix} U_1 \Sigma_r & 0_{m \times (n-r)} \end{bmatrix}. \qquad \blacksquare$$

Example 7.2.1. $A = \begin{bmatrix} 1 & 1 \\ 2 & 2 \end{bmatrix}$ where $n = m = 2$ and $\text{rank}(A) = 1$.

$$A^T A = \begin{bmatrix} 5 & 5 \\ 5 & 5 \end{bmatrix} \text{ and } \det\left(\begin{bmatrix} 5 - \lambda & 5 \\ 5 & 5 - \lambda \end{bmatrix}\right) = 0 \text{ implies}$$

$$\lambda_1 = 10, \ \sigma_1 = \sqrt{10} \text{ and } \lambda_2 = 0, \ \sigma_2 = 0.$$

The first eigenvectors are

$$v_1 = \begin{bmatrix} 1/\sqrt{2} \\ 1/\sqrt{2} \end{bmatrix} \text{ and }$$

$$u_1 = \frac{Av_1}{\sigma_1} = \begin{bmatrix} 1/\sqrt{5} \\ 2/\sqrt{5} \end{bmatrix}.$$

This gives the "small" SVD

$$u_1 \sigma_1 v_1^T = \begin{bmatrix} 1/\sqrt{5} \\ 2/\sqrt{5} \end{bmatrix} \sqrt{10} \begin{bmatrix} 1/\sqrt{2} & 1/\sqrt{2} \end{bmatrix} = \begin{bmatrix} 1 & 1 \\ 2 & 2 \end{bmatrix}.$$

The "full" SVD uses the v_2 basis for $N(A)$

$$v_2 = \begin{bmatrix} 1/\sqrt{2} \\ -1/\sqrt{2} \end{bmatrix} \text{ and }$$

the u_2 basis for $N(A^T)$

$$u_2 = \begin{bmatrix} -2/\sqrt{5} \\ 1/\sqrt{5} \end{bmatrix}.$$

$$U \Sigma V^T = \begin{bmatrix} u_1 & u_2 \end{bmatrix} \begin{bmatrix} \sigma_1 & 0 \\ 0 & 0 \end{bmatrix} \begin{bmatrix} v_1^T \\ v_2^T \end{bmatrix}$$

$$= \begin{bmatrix} 1/\sqrt{5} & -2/\sqrt{5} \\ 2/\sqrt{5} & 1/\sqrt{5} \end{bmatrix} \begin{bmatrix} \sqrt{10} & 0 \\ 0 & 0 \end{bmatrix} \begin{bmatrix} 1/\sqrt{2} & 1/\sqrt{2} \\ 1/\sqrt{2} & -1/\sqrt{2} \end{bmatrix}$$

$$= \begin{bmatrix} 1 & 1 \\ 2 & 2 \end{bmatrix}.$$

Example 7.2.2. $A = \begin{bmatrix} 1 & 1 & 2 \\ 1 & 2 & 3 \\ 1 & 3 & 4 \\ 1 & 4 & 5 \end{bmatrix}$ has rank$(A) = 2 = r$, $m = 4$ and $n = 3$.

The "full" SVD has the form

$$A = U\Sigma V^T$$

$$= \begin{bmatrix} u_1 & u_2 & u_3 & u_4 \end{bmatrix} \begin{bmatrix} \sigma_1 & 0 & 0 \\ 0 & \sigma_2 & 0 \\ 0 & 0 & 0 \\ 0 & 0 & 0 \end{bmatrix} \begin{bmatrix} v_1^T \\ v_2^T \\ v_3^T \end{bmatrix}$$

$$= \begin{bmatrix} u_1\sigma_1 & u_2\sigma_2 & 0_{4\times 1} \end{bmatrix} \begin{bmatrix} v_1^T \\ v_2^T \\ v_3^T \end{bmatrix}$$

$$= u_1\sigma_1 v_1^T + u_2\sigma_2 v_2^T.$$

The MATLAB® command svd(A) can be used to do these calculations, and this is illustrated in the next subsection.

7.2.1 MATLAB® code svd_ex.m

Line 9 inputs the matrix in Example 7.2.2. The MATLAB® command [U S V] = svd(A) in line 14 computes the three factors. The eigenvectors of $A^T A$ are computed in line 24 using [VV Lam_v] = eig(A'*A). Lines 14 and 26–31 confirm the properties of the SVD.

```
1    % SVD of a 4x3 Example.
2    % The rank of this matrix is r = 2.
3    % Then the dim(N(A)) = 1 and dim(N(A^T)) = 2.
4    %
5    clear
6    %
7    % Input data
8    %
9    A = [ 1 1 2; 1 2 3; 1 3 4; 1 4 5]
10   %
11   % Compute the SVD of A using Matlab®.
12   %
13   display('Find the svd factors via the Matlab® command svd()'
14   [U S V] = svd(A)
15   %
16   % Check the properties of SVD.
17   %
18   display('Check A - U*S*V^T is a zero matrix.')
```

```
19    A - U*S*V'
20    display('Compute the eigenvectors and
                          eigenvalues of A^T A and A A^T.')
21    display('Lam_u = Lam_v.')
22    display('Note, s(1,1) = sqrt(Lam_u(4,4) and
                          s(2,2) = sqrt(Lam_u(3,3) and')
23    display('the change in order.')
24    [VV Lam_v] = eig(A'*A)
25    [UU Lam_u] = eig(A*A')
26    display('The last column of V is the
                          orthonormal basis of N(A).')
27    display('A V = ')
28    A*V
29    display('The last two columns of U form the
                          orthonormal basis N(A^T).')
30    display('A^T U = ')
31    A'*U

>> svd_ex

  A =
     1     1     2
     1     2     3
     1     3     4
     1     4     5
```

Find the svd factors via the Matlab® command svd()

```
  U =
    -0.2524   -0.7977    0.0985    0.5388
    -0.3990   -0.3752    0.2703   -0.7918
    -0.5457    0.0473   -0.8360   -0.0328
    -0.6923    0.4698    0.4672    0.2858

  S =
     9.3441         0         0
          0    0.8290         0
          0         0    0.0000
          0         0         0

  V =
    -0.2022   -0.7911    0.5774
    -0.5840    0.5706    0.5774
    -0.7862   -0.2204   -0.5774
```

Check A - U*S*V^T is a zero matrix.

```
    =
    1.0e-14 *
         0         0   -0.0888
         0   -0.0444   -0.0888
    0.0111   -0.0888   -0.0888
   -0.0222   -0.1776   -0.0888
```

Compute the eigenvectors and eigenvalues of A^T A and A A^T.
Lam_u = Lam_v.
Note, s(1,1) = sqrt(Lam_u(4,4) and s(2,2) = sqrt(Lam_u(3,3) and
the change in order.

```
    VV =
     0.5774    0.7911    0.2022
     0.5774   -0.5706    0.5840
    -0.5774    0.2204    0.7862

    Lam_v =
    -0.0000         0         0
         0    0.6872         0
         0         0   87.3128

    UU =
    -0.4877    0.2494   -0.7977    0.2524
     0.8361    0.0310   -0.3752    0.3990
    -0.2092   -0.8101    0.0473    0.5457
    -0.1392    0.5297    0.4698    0.6923

    Lam_u =
    -0.0000         0         0         0
         0    0.0000         0         0
         0         0    0.6872         0
         0         0         0   87.3128
```

The last column of V is the orthonormal basis of N(A).
```
    A V =
    -2.3585   -0.6612    0.0000
    -3.7287   -0.3110    0.0000
    -5.0989    0.0392    0.0000
    -6.4690    0.3894    0.0000
```

The last two columns of U form the orthonormal basis N(A^T).
```
    A^T U =
```

```
-1.8894   -0.6558         0    -0.0000
-5.4568    0.4730         0    -0.0000
-7.3462   -0.1827         0    -0.0000
```

7.2.2   Exercises

1. Consider $A = \begin{bmatrix} 2 & 1 \\ 4 & 2 \end{bmatrix}$.

    (a). Find the eigenvalues and eigenvectors of $A^T A$.

    (b). Find the "full" SVD of A.

2. Consider $A = \begin{bmatrix} 1 & 2 \\ 2 & 4 \\ 1 & 2 \end{bmatrix}$.

    (a). Find the eigenvalues and eigenvectors of $A^T A$.

    (b). Find the "full" SVD of A.

3. Consider Example 7.2.2. Use the Householder transformation to extend the basis and find u_3 and u_4.

4. Reconsider the 5×4 matrix

$$A = \begin{bmatrix} 1 & 2 & 3 & 2 \\ 1 & 3 & 4 & 2 \\ 1 & 5 & 6 & 2 \\ 1 & 8 & 9 & 2 \\ 1 & 11 & 12 & 2 \end{bmatrix}.$$

    (a). Find the rank of the matrix.

    (b). Modify the above code and compare with svd(A).

7.3   "Truncated" SVD

The "truncated" SVD is formed from the "small" SVD by dropping the latter columns of $U_1 = U(:, 1 : r)$ and $V_1 = V(:, 1 : r)$

$$A \cong A^{(k)} \equiv U(:, 1 : k)\Sigma_k V(:, 1 : k)^T \text{ with } k < r = \text{rank}(A).$$

Σ_k is the $k \times k$ diagonal matrix with components $\sigma_1 \geq \cdots \geq \sigma_k > 0$. An alternate way of writing this is

$$A^{(k)} = \sum_{j=1}^{k} u_j \sigma_j v_j^T \text{ where}$$

$$U(\ : \ ,1:k) = \begin{bmatrix} u_1 & \cdots & u_k \end{bmatrix} \text{ and}$$
$$V(\ : \ ,1:k) = \begin{bmatrix} v_1 & \cdots & v_k \end{bmatrix}.$$

An important observation is the first k columns only require the first k eigenvectors of $A^T A$.

Both A and $A^{(k)}$ are $m \times n$ matrices. In order the quantify the error in using the "truncated" SVD, introduce the 2-norm of any linear operator $A : \mathbb{R}^n \to \mathbb{R}^m$ represented by an $m \times n$ matrix. The error will then be measured by $\left\| A - A^{(k)} \right\|_2^2$ where $\|*\|_2$ is the 2-norm.

Definition. $\|A\|_2^2 \equiv \max\limits_{x \neq 0_{n \times 1}} \dfrac{(Ax)^T(Ax)}{x^T x}$

$\qquad\qquad = \max\limits_{\hat{x}^T \hat{x}=1} (A\hat{x})^T(A\hat{x})$ where $\hat{x} \equiv \dfrac{x}{(x^T x)^{1/2}}$.

The 2-norm of A gives an upper bound on Ax

$$\|Ax\|_2 \leq \|A\|_2 \|x\|_2.$$

Theorem 7.3.1. *(Norm of A)* *If A is $m \times n$ with* $\mathrm{rank}(A) = r$,

$A = \sum\limits_{j=1}^{r} u_j \sigma_j v_j^T$ *be the "small" SVD and*

$A^{(k)} = \sum\limits_{j=1}^{k} u_j \sigma_j v_j^T$ *be the "truncated" SVD where* $k < r$,

then $\|A\|_2^2 = \sigma_1^2$ *and* $\left\| A - A^{(k)} \right\|_2^2 = \sigma_{k+1}^2$.

Proof. Let $V = \{v_1, \cdots, v_n\}$ be the orthonormal basis of eigenvectors for $A^T A$. Order the eigenvectors so that $(A^T A)v_j = \lambda_j v_j = \sigma_j^2 v_j$.

First, we show $\|A\|_2^2 \geq \sigma_1^2$:

$$\text{Choose } x = v_1.$$
$$\|A\|_2^2 = \max_{x^T x=1} (Ax)^T(Ax)$$
$$\geq v_1^T A^T A v_1$$
$$= v_1^T \lambda_1 v_1$$
$$= 1\lambda_1 = \sigma_1^2.$$

Second, show $\|A\|_2^2 \le \sigma_1^2$:

$$\text{Let } x = \sum_{j=1}^{n} c_j v_j \text{ with } x^T x = 1. \text{ Then}$$

$$1 = x^T x = (\sum_{i=1}^{n} c_i v_i)^T (\sum_{j=1}^{n} c_j v_j) = \sum_{j=1}^{n} c_j^2 \text{ and}$$

$$A^T A x = \sum_{j=1}^{n} c_j \lambda_j v_j.$$

$$x^T A^T A x = (\sum_{i=1}^{n} c_i v_i)^T (\sum_{j=1}^{n} c_j \lambda_j v_j)$$

$$= \sum_{j=1}^{n} c_j^2 \lambda_j 1 \text{ and use } \lambda_j \le \lambda_1$$

$$\le (\sum_{j=1}^{n} c_j^2) \lambda_1$$

$$= 1\lambda_1 = \sigma_1^2.$$

$$\|A\|_2^2 = \max_{x^T x = 1} (Ax)^T (Ax) \le \max_{x^T x = 1} \sigma_1^2 = \sigma_1^2.$$

Consider $A - A^{(k)}$ where $n < r$, and note

$$A - A^{(k)} = \sum_{j=1}^{r} u_j \sigma_j v_j^T - \sum_{j=1}^{k} u_j \sigma_j v_j^T$$

$$= \sum_{j=k+1}^{r} u_j \sigma_j v_j^T.$$

This is the "small" SVD of $A - A^{(k)}$ and therefore $\left\|A - A^{(k)}\right\|_2^2 = \sigma_{k+1}^2$. ∎

Example 7.3.1. Consider Example 7.2.1 and the MATLAB® code svd_ex.m. Here A is 4×3, rank$(A) = 2$ and we let the truncation be $k = 1$. The following calculations verify Theorem 7.3.1:

```
>> norm(A) =
   9.3441
>> S(1,1)
   9.3441

>> A1 = U(:,1)*S(1,1)*V(:,1)'
   0.4769    1.3773    1.8543
   0.7540    2.1775    2.9314
```

```
     1.0310       2.9776       4.0086
     1.3081       3.7778       5.0858

>> norm(A - A1) =
    0.8290
>> S(2,2)
    0.8290
```

7.3.1 Exercises

1. Let A be $m \times n$. Prove the following properties:

   (a). $\|A\|_2 \geq 0$, and $\|A\|_2 = 0$ if and only if $A = 0_{m \times n}$.

   (b). $\|cA\|_2 = |c| \, \|A\|_2$ where $c \in \mathbb{R}$.

   (c). $\|A + B\|_2 \leq \|A\|_2 + \|B\|_2$ where B is $m \times n$.

   (d). If A and B are $n \times n$ matrices, then $\|AB\|_2 \leq \|A\|_2 \, \|B\|_2$.

2. Consider the 5×4 matrix

$$A = \begin{bmatrix} 1 & 2 & 3 & 2 \\ 1 & 3 & 4 & 7 \\ 1 & 5 & 6 & 9 \\ 1 & 8 & 9 & 2 \\ 1 & 11 & 12 & 1 \end{bmatrix}.$$

   (a). Modify the above code for this matrix.

   (b). Compute the norms of the errors in the "truncated" SVD

$$\left\| A - A^{(k)} \right\|_2^2 = \sigma_{k+1}^2.$$

8

Three Applications of SVD

The last theorem of the previous chapter suggests the "truncated" SVD can be used to approximate the original matrix. Three important applications will be illustrated for image compression, search engines and noise filters. Additional applications will be given to the general least squares problem in Chapter 9, to hazard identification in Chapter 9, and to epidemic models in Chapter 12. Also, consult N. Gillis, [7], for similar low rank approximations of matrices.

8.1 Image Compression

Grayscale images are associated with $m \times n$ matrices whose components are integers. For 8-bit images the integers range from 0 to $255 = 2^8 - 1$, and for 16-bit images they range from 0 to $65535 = 2^{16} - 1$. The black image pixel is associated with 0, and the white image pixel is associated with 255 (8-bit) or 65535 (16-bit). In MATLAB® one can "view" the image in several ways. First, just inspect the matrix components. Second, use the MATLAB® command mesh() to generate a surface of the image where the indices of the matrix are on the xy-plane and the intensity of the image is on the z-axis. Third, one can map the matrix into a standard image file such as a *.jpg file. This can be done by the MATLAB® command imwrite(). The inverse of the imwrite() is imread(), which generates a 8-bit integer matrix from an image file.

Additional material on image compression and image processing can be found in the last two chapters in [26].

Example 8.1.1. We will create alphabetical letters from 50×40 matrices. The letter "U" can be created by defining a 50×40 matrix to initially be zero and then nonzero for some components to form the letter. This will produce a letter with black background and with lighter regions to form the letter. The following MATLAB® function defines the letter "U" where the lighter region has an input value equal to g.

```
1    function letu = letteru(g)
2       letu = zeros(50,40);
3       letu(10:40,4:8) = g;
```

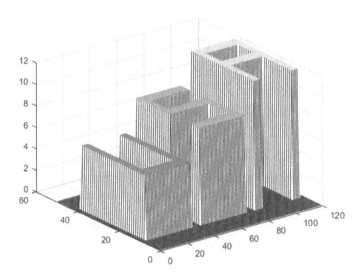

FIGURE 8.1.1
USA matrix via mesh().

```
4       letu(10:40,32:36) = g;
5       letu(10:14,4:36) = g;
```

The general scheme is to convert the image to a matrix, modify the matrix and then to convert the modified matrix back to a new image. The operation of light/dark corresponds to multiplying the matrix by a constant less/larger than one. The operation of cropping/panorama corresponds to deleting/augmenting rows or columns of the matrix. In image compression we use the "truncated" SVD of the image matrix. A "truncated" SVD of and $m \times n$ image matrix requires $km + kn + k$ storage of the k eigenvectors and singular values. If $k << m, n$, then this is a lot less than mn for the full image matrix.

Example 8.1.2. In this example three letters are created by functions similar to the letter function letteru(g) to form the matrix *usa* and then the image USA. The mesh plot of the matrix is given in Figure 8.1.1, the image USA with black background is in Figure 8.1.2, and the "negative" of the image

FIGURE 8.1.2
USA jpg picture.

FIGURE 8.1.3
Negative of USA jpg picture.

USA with white background is in Figure 8.1.3. The "negative" is created by replacing the components in the matrix *usa* by $255 - usa(i,j)$, that is, $negusa(i,j) = 255 - usa(i,j)$. The following MATLAB® code imagusa.m was used to generate these figures.

```
1    usa = [letteru(6) letters(9) lettera(12)];
2    mesh(usa)
3    newusa = 20*usa;
4    newusa = newusa(50:-1:1,:);
5    negusa = 255*ones(size(newusa)) - newusa;
6    newusa1 = uint8(newusa);
7    imwrite(newusa1, 'usa.jpg');
8    negusa1 = uint8(negusa);
9    imwrite(negusa1, 'negusa.jpg');
```

8.1.1 MATLAB® code svdimage.m

Line 16 converts the image to an image matrix. Lines 22, 23 and 27 create the scaled image in the upper left in Figure 8.1.4. The "full" SVD of the

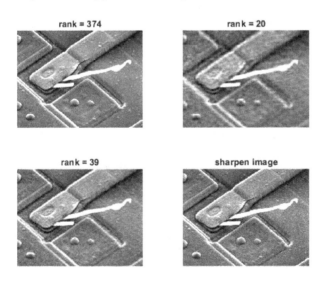

FIGURE 8.1.4
Image compression using SVD.

image matrix is computed in line 35, and the "truncated" SVD with $k = 20$ is computed in line 39. This is displayed in the upper right in Figure 8.1.4. The loop in lines 50–59 varies the $k = 1 : 2 : 40$ and displays the images in the lower left in Figure 8.1.4 (requires one to hit the space bar). The lines 64–68 illustrate how to use portions of the SVD to "sharpen" the resulting image matrix. The reader will find it interesting to experiment with this and compare the images in the lower right in Figure 8.1.4.

```
1    % This code illustrates several images, their associated
2    % matrices,SVD expansions and image enhancements. There are
3    % four modified images, which are obtained by converting the
4    % image (*.jpg) to a matrix of 64-bit double precision
5    % numbers, modifying the matrix by using the SVD expansion,
6    % and then converting the new matrix into a new image (*.jpg).
7    %
8    clear; clf
9    %
10   % Input data
11   %
12   % Select an image. X will be an mxn matrix with 8-bit
13   % components rangingfrom 0 (dark) to 255 = 2^8 - 1 (light).
14   % load detail
15   % X = imread('moon.jpg');
16   X = imread('microchip.jpg');
17   % X = imread('pollen.jpg');
18   % The command double(X) converts the components to double
19   % precision (64-bit) numbers. The scale factor adjusts
20   % the darkness of the image.
21   %
22   scale = 0.3;
23   X = scale*double(X);
24   %
25   subplot(2,2,1)
26   % This is the new image.
27   image(X)
28   colormap(gray(64))
29   axis image, axis off
30   r = rank(X)
31   title(['rank = ' int2str(r)])
32   %
33   % Compute the SVD factors of X = u*s*v'.
34   %
35   [u s v] = svd(X,0);
36   %
37   subplot(2,2,2)
38   % This is the first 20 terms of the SVD expansion.
39   X20 = u(:,1:20)*s(1:20,1:20)*v(:,1:20)';
40   image(X20)
41   colormap(gray(64))
```

```
42      axis image, axis off
43      r20 = rank(X20)
44      title(['rank = ' int2str(r20)])
45      %
46      subplot(2,2,3)
47      %
48      % This loop presents a sequence of partial SVD expansions.
49      %
50      for rnew = 1:2:40
51          Xnew = u(:,1:rnew)*s(1:rnew,1:rnew)*v(:,1:rnew)';
52          image(Xnew)
53          colormap(gray(64))
54          axis image, axis off
55          rnew
56          title(['rank = ' int2str(rnew)])
57          disp('Hit the space bar to move to the next image.')
58          pause
59      end
60      %
61      subplot(2,2,4)
62      % This modifies the last image by adding additional parts
63      % of the SVD expansion.
64      % Xnew = X;
65      for rr = 39-2:39
66          Xnew = Xnew + u(:,rr)*s(rr,rr)*v(:,rr)';
67      end
68      image(Xnew)
69      colormap(gray(64))
70      axis image, axis off
71      title(['sharpen image'])
```

8.2   Search Engines

Consider searching for item i in document j. Let m be the number of items and n be the number of documents.

$$A = m \times n \text{ frequency matrix}$$
$$= [a_{i,j}] \text{ where}$$
$$a_{i,j} = \text{the number of times item } i \text{ appears in document } j.$$

Example 8.2.1. Suppose there are four items and two documents with

	doc. 1	doc. 2
vector	4	1
matrix	0	7
real	3	2
complex	1	4

and define $A = \begin{bmatrix} 4 & 1 \\ 0 & 7 \\ 3 & 2 \\ 1 & 4 \end{bmatrix}$.

The frequency matrix can be sparse and very large. In order to search for one or more items, we use a query vector

$$q = [q_i] \in \mathbb{R}^m \text{ where } q_i = 1 \text{ for a single item } i.$$

Search for item i in document d_j (column j in A) by computing $q^T d_j$, and then normalize it by using the Cauchy inequality

$$-1 \le \frac{q^T d_j}{\|q\|_2 \|d_j\|_2} \le 1.$$

Definition. Let $d_j = A e_j$ be column j in the $m \times n$ frequency matrix A.

$$\cos(\theta_j) \equiv \frac{q^T A e_j}{\|q\|_2 \|A e_j\|_2}.$$

Example 8.2.2. Return to Example 8.2.1 and consider four query vectors.

$q = \begin{bmatrix} 1 & 0 & 0 & 0 \end{bmatrix}^T$ gives $\cos(\theta_1) = 4/\sqrt{26}$ and $\cos(\theta_2) = 1/\sqrt{70}$,

$q = \begin{bmatrix} 0 & 1 & 0 & 0 \end{bmatrix}^T$ gives $\cos(\theta_1) = 0/\sqrt{26}$ and $\cos(\theta_2) = 7/\sqrt{70}$,

$q = \begin{bmatrix} 0 & 0 & 1 & 0 \end{bmatrix}^T$ gives $\cos(\theta_1) = 3/\sqrt{26}$ and $\cos(\theta_2) = 2/\sqrt{70}$ and

$q = \begin{bmatrix} 1 & 0 & 0 & 1 \end{bmatrix}^T$ gives $\cos(\theta_1) = 5/(\sqrt{2}\sqrt{26})$ and $\cos(\theta_2) = 5/(\sqrt{2}\sqrt{70})$.

Because the components of A and q are nonnegative, $0 \le \cos(\theta_j) \le 1$. If $\cos(\theta_j)$ is close to 1, then one can make a judgment that document j has the items in the q. Because the frequency matrix is very very large, we will approximate it by the "truncated" SVD

$$A \approx A^{(k)} = U^{(k)} \Sigma_k (V^{(k)})^T \text{ with } k < \text{rank}(A) = r.$$

This gives a new $\cos(\theta_j)$

$$\frac{q^T U^{(k)} (\Sigma_k (V^{(k)})^T e_j)}{\|q\|_2 \|U^{(k)} (\Sigma_k (V^{(k)})^T e_j)\|_2}.$$

Use the orthonormal property and define $S_j \equiv \Sigma_k (V^{(k)})^T e_j$.

Definition. *First search engine approximation is*

$$\cos(\theta 1_j) \equiv \frac{q^T U^{(k)} S_j}{\|q\|_2 \|S_j\|_2}.$$

This approximation may have negative values, and the choice of k is a judgment to be made. A variation is to project the query q to the range of $A^{(k)}$. In order to find this projection, use the orthonormal basis of $R(A^{(k)})$ given by the columns of $U^{(k)}$. Theorem 3.4.1 yields

$$P_{R(A^{(k)})}(q) = U^{(k)} c \text{ where } (U^{(k)})^T q = c.$$

In the first search engine replace q by $P_{R(A^{(k)})}(q) = U^{(k)} c = U^{(k)} (U^{(k)})^T q$. Use the orthonormal property and Theorem 3.4.1 to conclude

$$P_{R(A^{(k)})}(q)^T P_{R(A^{(k)})}(q) = ((U^{(k)})^T q)^T ((U^{(k)})^T q) \leq q^T q \text{ and}$$
$$P_{R(A^{(k)})}(q)^T A^{(k)} e_j = q^T U^{(k)} S_j.$$

This gives the second version of search engine.

Definition. *Second search engine approximation is*

$$\cos(\theta 2_j) \equiv \frac{q^T U^{(k)} S_j}{\left\|(U^{(k)})^T q\right\|_2 \|S_j\|_2} \geq \cos(\theta 1_j).$$

Both search engines will be illustrated in the next subsection. The first code is a simple example with small dimensions. The second code uses a $11,390 \times 1,265$ frequency matrix, and there can be, for example, a significant difference in $\left\|(U^{(k)})^T q\right\|_2 = 2.3287$ and $\|q^T q\|_2 = 71.5542$.

The search engine literature is extensive, but the reader may find the following to be useful: [5], [2] and [21].

8.2.1 MATLAB® codes sengine.m, senginesparse.m

The first code, sengine.m, illustrates the effect of using variable, k, truncation of the SVD. The sample calculations are for smaller frequency matrices. One case is for a 5×6 matrix whose rank is 5. It is interesting to experiment with different query vectors, q.

Line 13 defines (it is not used) the frequency matrix for Example 8.2.1. In lines 14–18 the larger frequency matrix is defined and uses in the subsequent calculations. A query vector is defined in line 19, and the user of the code may want to experiment with this. The SVD is calculated in line 24. The outer loop in lines 29–43 varies the document number, j. The inner loop varies the truncation number, k. The two search engine values for $cos(\theta 1)$ and $cos(\theta 2)$ are computed inside these nested loops in lines 33 and 34.

```
1    %
2    % Search Engine
3    %
4    % The code uses the SVD to search an mxn frequency matrix
5    % where the j-column contains the frequency of items in
6    % document j. This trival example illustrates two search
7    % methods and variable truncation of the SVD.
8    %
9    clear
10    %
11    %Input data
12    %
13    %  A = [ 4 1;0 7;3 2;1 4];              % frequency matrix
14    A = [ 4 1 0 0 1 0;
15         0 7 0 1 2 0;
16         3 2 0 0 0 1;
17         1 4 1 0 3 2;
18         0 1 0 0 1 0];
19    q = [ 1 0  0 0 1]';                     % query
20    r = rank(A);                           % rank of A
21    %
22    % Compute the SVD
23    %
24    [U S V] = svd(A);
25    [m n]   = size(A);
26    S
27    % For each document (j) compute uss all the truncated SVDs.
28    % Compute both the versions of the cos theta.
29    for j = 1:n
30       for k = 1:r
31          Sv = S(1:k,1:k)*V(:,1:k)';
32          SV = Sv(:,j);
33          costhjone(k) = q'*U(:,1:k)*SV/(norm(q)*norm(SV));
34          costhjtwo(k) = q'*U(:,1:k)*SV/...
                              (norm(U(:,1:k)'*q)*norm(SV));
35       end
36    %
37    % Output all possible cos theta
38    %
39       display('document = ' )
40       j
41       display('variable truncation for cos theta one and two =')
42       [costhjone' costhjtwo']
43    end
```

```
>> sengine

S =
   Columns 1 through 6
      9.4508          0          0          0          0          0
           0     4.7367          0          0          0          0
           0          0     2.3673          0          0          0
           0          0          0     1.2675          0          0
           0          0          0          0     0.1883          0

   document j =
      1

   variable truncation for cos theta one and two =
      0.2609     1.0000
      0.5461     0.9998
      0.5459     0.9989
      0.5556     0.6364
      0.5547     0.5547
```

The second code, senginesparse.m, uses a larger frequency matrix. It is $11,390 \times 1,265$ and has $109,056$ nonzero components. The rank of the frequency matrix is 988, and the SVD is truncated after $k = 20$. The graphs of the two search engines, $\cos(\theta 1)$ and $\cos(\theta 2)$, are given in Figure 8.2.2. For the particular query in the code, four documents are identified has having the query.

The frequency matrix is imported in lines 14–17. Truncated SVD with $k = 20$ is computed in line 24, and the query is defined in line 27. The singular values are graphed by the command in line 25, see Figure 8.2.1. The loop in lines 35–39 is over all documents, and both $\cos(\theta 1)$ and $\cos(\theta 2)$ are computed for each document, j. The output is given graphical form in Figure 8.2.2.

```
1     %
2     % Search Engine Sparse
3     %
4     % The code uses the SVD to search an mxn frequency matrix
5     % where the j-column contains the frequency of items in
6     % document j. This sparse example two search methods
7     % and k = 20 truncation of the SVD.
8     %
9     clear
10    %
11    % Input data
12    %
13    % download webmatrix_uri            % frequency matrix
14    svdmatrix = uiimport('webmatrix_uri');
```

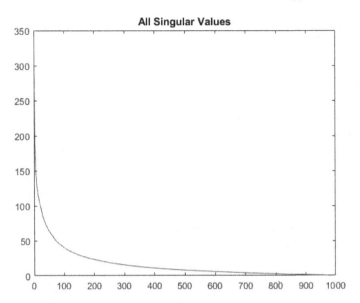

FIGURE 8.2.1
All singular values.

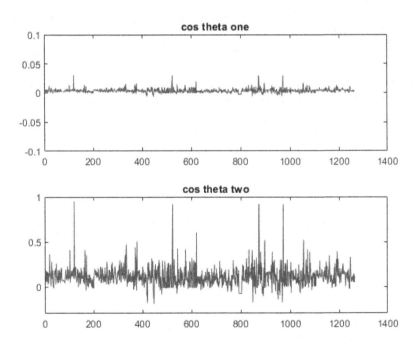

FIGURE 8.2.2
Search using truncated SVD.

```
15      SS = svdmatrix.data;
16      SS(1,:)
17      A = sparse(SS(2:109057,1), SS(2:109057,2), SS(2:109057,3));
18      [m n] = size(A);
19      display('number of search items = ')
20      m
21      display('number of documents = ')
22      n
23      r = rank(full(A))                    % rank of full A
24      sig = svds(A,r);
25      plot(sig); title('all singular values')
26      % choose query q
27      q = zeros(m,1); q(2)= 4; q(55) = 2; q(60:110) = 10;
28      %
29      % For truncated SVD with k = 20 compute the cos theta for
30      % all documents.
31      %
32      for k = 20                           % truncate svd after k
33          [U sig V] = svds(A,k);
34          Sv = sig(1:k,1:k)*V(:,1:k)';
35          for j = 1:n;
36              SV = Sv(:,j);
37              costhjone(j) = q'*U(:,1:k)*SV/(norm(q)*norm(SV));
38              costhjtwo(j) = q'*U(:,1:k)*SV/...
                                    (norm(U(:,1:k)'*q)*norm(SV));
39          end
40      end
41      [costhjone' costhjtwo'];
42      %
43      % Graph the cos theta verse the documents
44      %
45      figure(2)
46      subplot(2,1,1) , plot(1:n, costhjone')
47      axis([ 0 1400 -0.1 0.1])
48      title('cos theta one')
49      subplot(2,1,2) , plot(1:n, costhjtwo')
50      axis([ 0 1400 -0.3 1.0])
51      title('cos theta two')

>> senginesparse

    ans =
        11390          1265          109056

    m =
```

11390

n =
    1265

r =
    988

8.3   Noise Filter

Consider a discrete signal $f \in \mathbb{R}^n$. There is some blurring (K an $n \times n$ matrix) and some signal noise ($\eta \in \mathbb{R}^n$). The data received ($d \in \mathbb{R}^n$) is given by

$$d = Kf + \eta.$$

The objective is to approximate f from the data d. There are two problems: the noise is not known, and the matrix inversion tends to be sensitive to variations in the data.

The matrix K has exponential components

$$K = \left[hCe^{-((i-j)h)^2/2\gamma^2} \right] \text{ with } h = 1/n.$$

It is nonsingular, but it has very small singular values, for example, K in the next section has $\sigma_n = 4.2275 \, 10^{-7}$. The SVD of $K = U\Sigma_n V^T$ has three $n \times n$ matrices and $\sigma_i > 0$. Use the orthonormal properties to get

$$K^{-1} = V\Sigma_n^{-1}U^T \text{ and}$$
$$K^{-1}d = f + K^{-1}\eta$$
$$V(\Sigma_n^{-1}U^T d) = f + V(\Sigma_n^{-1}U^T \eta).$$

Observe the latter column vectors of V have high frequencies (see the next section). Since the noise vector η has higher frequency components, one can drop some of the latter terms in the SVD, that is, use the truncated SVD $K^{(k)} = U^{(k)}\Sigma_k(V^{(k)})^T$ with $k < \text{rank}(K) = n$. The computed signal is

$$\hat{f} = V^{(k)}(\Sigma_k^{-1}(U^{(k)})^T d)$$
$$= \sum_{j=1}^{k} v_j \frac{1}{\sigma_j} u_j^T d$$
$$= \sum_{j=1}^{k} (\frac{1}{\sigma_j} u_j^T d) v_j.$$

Depending in the choice the truncation, the σ_j may still be small so that variations in the data can cause large errors. This is more carefully discussed in Chapter 9 on ill-conditioned matrices. One possible remedy is to replace

$$\frac{1}{\sigma_j} \text{ by } \frac{1}{\sigma_j}\frac{\sigma_j^2}{\sigma_j^2 + \alpha} \text{ where } \alpha > 0.$$

When α is near 0.0, then this approximates $1/\sigma_j$. When α gets large, then this approximates 0.0.

Definition. *Let K be $n \times n$ and choose $\alpha > 0$ and $k < n$. The Tikhonov-Phillips regularization is*

$$(K^T K + \alpha I_n)^{-1} K^T d.$$

The truncated Tikhonov-Phillips regularization is

$$\widetilde{f} = ((K^{(k)})^T K^{(k)} + \alpha I_k)^{-1} (K^{(k)})^T d$$

$$= \sum_{j=1}^{k} (\frac{\sigma_j}{\sigma_j^2 + \alpha} u_j^T d) v_j.$$

Matrix and summation versions of the truncated Tikhonov-Phillips regularization are the same, and this can be established by using the orthonormal properties in the SVD. The choice of $\alpha > 0$ and $k < n$ can be judgmental, and this is illustrated in the next section.

For additional information and analysis the reader should consult [25].

8.3.1 MATLAB® code Image1dsvd.m

The matrix K is 100×100 and is defined by Setup1dsvd.m. The following MATLAB® code Kmatrix.m uses this and computes the SVD of K. The singular values vary from largest to very very small

$$\sigma_1 = 0.9985, \ \sigma_{40} = 0.0944 \text{ and } \sigma_{100} = 4.2267 \ 10^{-7}.$$

The eigenvectors u_j and v_j have more oscillations as j increases. This is illustrated in Figure 8.3.1.

```
1    Setup1dsvd;
2    [U S V] = svd(K);
3    S = diag(S);
4    S(1)
5    S(40)
6    S(100)
7    figure(3)
8    subplot(1,2,1)
```

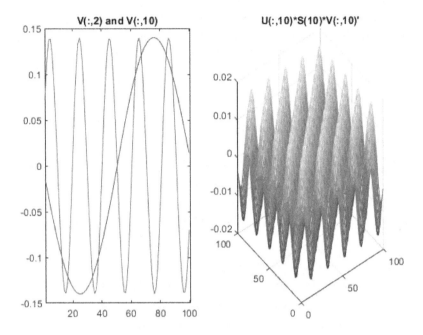

FIGURE 8.3.1
Vectors from SVD of K.

```
 9      plot(V(:,2));
10      hold on
11      plot(V(:,10));
12      subplot(1,2,2)
13      mesh(U(:,10)*S(10)*V(:,10)');
```

The matrix K and the signal with noise is defined in line 8 by
Setup1dsvd.m. This is illustrated in the upper left graph in Figure 8.3.2. The
full SVD of the matrix is computed in line 9. The truncation $k = 40$ is used
in line 11, and the user may want to experiment with the truncation number.
Three values for α are used as indicated by the loop starting at line 16. The
truncated Tikhonov-Phillips approximations are computed in the inner loop
in lines 18–21. The graphs of the approximated signal are given by lines 25–
26 and are displayed in the upper right, lower left and lower right graphs in
Figure 8.3.2.

```
1      % This code illustrates noise filters using the SVD.
2      % The 1D data is given by Setup1dsvd.
3      %
4      clear;
```

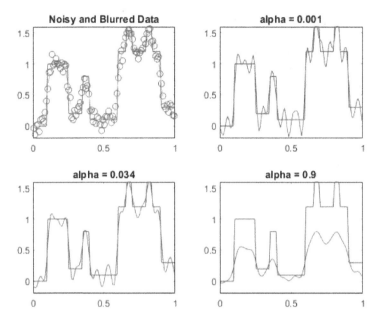

FIGURE 8.3.2
Noise filter with truncation k = 40.

```
5    %
6    % Input data from Setup1dsvd.m
7    %
8    Setup1dsvd
9    [U S V] = svd(K);
10   S = diag(S);
11   k = 40                        % truncation of SVD
12   %
13   % Compute SVD filter using variable parameter alpha.
14   %
15   num = 2;
16   for alpha = [0.001 0.034 0.900]
17      newimage = zeros(n,1);
18      for j = 1:k
19         newimage = newimage + ...
20            (S(j)*(U(:,j)'*d(:))/(S(j)*S(j) + alpha))*V(:,j);
21      end
22   %
23   % Graphaical output is given for each alpha.
24   %
25      subplot(2,2,num)
```

```
26      plot(x,f_true,'-',x,newimage)
27      axis([0 1 -.2 1.6])
28      title([ 'alpha = ' num2str(alpha)])
29      num = num + 1;
30   end
```

9

Pseudoinverse of A

The pseudoinverse of an $m \times n$ matrix is an $n \times m$ matrix $A^\dagger$. It is important because $A^\dagger d$ is a minimal 2-norm solution to the general least squares problem. Moreover, when $A^\dagger$ is restricted to $R(A)$, it is a right inverse, that is, $A(A^\dagger d) = d$ when $d = A\widetilde{x}$.

9.1 $\quad \Sigma^\dagger$ and $A^\dagger = V\Sigma^\dagger U^T$

As motivation for the definition of the pseudoinverse of an $m \times n$ matrix with rank$(A) = r$, consider the least squares problem. Let $A = U_1\Sigma_r V_1^T$ be the "small" SVD where the columns of U_1 and V_1 are orthonormal bases for $R(A)$ and $R(A^T)$, respectively. Theorem 3.4.2 gives the least squares solution based on two projections

$$A\widehat{x} = P_{R(A)}(d) \text{ and } x_R = P_{R(A^T)}(\widehat{x}).$$

Because the bases are orthonormal, the projections are easily computed

$$P_{R(A)}(d) = U_1(U_1^T d) \text{ and } P_{R(A^T)}(\widehat{x}) = V_1(V_1^T\widehat{x}). \tag{9.1.1}$$

Combine these equations to get

$$A\widehat{x} = P_{R(A)}(d)$$
$$U_1\Sigma_r V_1^T\widehat{x} = U_1(U_1^T d) \text{ and }$$
$$V_1^T\widehat{x} = \Sigma_r^{-1}U_1^T d.$$

This gives a minimal 2-norm, see Theorem 9.2.1, least squares solution

$$x_R = P_{R(A^T)}(\widehat{x}) = V_1(V_1^T\widehat{x}) = (V_1\Sigma_r^{-1}U_1^T)d.$$

Definition. *Let $A = U\Sigma V^T$ be the SVD of an $m \times n$ matrix with rank$(A) = r$. The pseudoinverse is an $n \times m$ matrix*

$$A^\dagger \equiv V\Sigma^\dagger U^T \text{ with}$$
$$\Sigma^\dagger \equiv \begin{bmatrix} \Sigma_r^{-1} & 0_{r \times (m-r)} \\ 0_{(n-r) \times r} & 0_{(n-r) \times (m-r)} \end{bmatrix} \text{ is an } n \times m \text{ matrix.}$$

DOI: 10.1201/9781003304128-9

The proofs of the following identities are routine, but they do lead to the fundamental properties of the general pseudoinverse $A^\dagger$.

Theorem 9.1.1. *(Singular Values Identities) Identities with Σ and $\Sigma^\dagger$.*

1. $\Sigma^T \Sigma \Sigma^\dagger = \Sigma^T$.

2. $\Sigma^\dagger \Sigma \Sigma^T = \Sigma^T$.

3. $\Sigma \Sigma^\dagger \Sigma = \Sigma$.

4. *If A has full column rank, $rank(A) = r = n$ with $m > n$, then*

$$\Sigma = \begin{bmatrix} \Sigma_r \\ 0_{(m-r)\times r} \end{bmatrix}, \Sigma^\dagger \Sigma = I_n \text{ and } \Sigma^T \Sigma = \Sigma_n^2.$$

5. *If A has full row rank, $rank(A) = r = m$ with $m < n$, then*

$$\Sigma = \begin{bmatrix} \Sigma_r & 0_{r\times(n-r)} \end{bmatrix}, \Sigma\Sigma^\dagger = I_m \text{ and } \Sigma\Sigma^T = \Sigma_m^2.$$

6. $(\Sigma^\dagger)^T \Sigma^\dagger = \begin{bmatrix} \Sigma_r^{-2} & 0_{r\times(m-r)} \\ 0_{(m-r)\times r} & 0_{(m-r)\times(m-r)} \end{bmatrix}$ *is $m \times m$.*

7. $\Sigma^T \Sigma = \begin{bmatrix} \Sigma_r^2 & 0_{r\times(n-r)} \\ 0_{(n-r)\times r} & 0_{(n-r)\times(n-r)} \end{bmatrix}$ *is $n \times n$.*

Proof. Show the third identity is true.

$$\Sigma^\dagger \Sigma = \begin{bmatrix} \Sigma_r^{-1} & 0_{r\times(m-r)} \\ 0_{(n-r)\times r} & 0_{(n-r)\times(m-r)} \end{bmatrix} \begin{bmatrix} \Sigma_r & 0_{r\times(n-r)} \\ 0_{(m-r)\times r} & 0_{(m-r)\times(n-r)} \end{bmatrix}$$

$$= \begin{bmatrix} I_r & 0_{r\times(n-r)} \\ 0_{(n-r)\times r} & 0_{(n-r)\times(n-r)} \end{bmatrix} \text{ is } n \times n.$$

$$\Sigma(\Sigma^\dagger \Sigma)^T = \begin{bmatrix} \Sigma_r & 0_{r\times(n-r)} \\ 0_{(m-r)\times r} & 0_{(m-r)\times(n-r)} \end{bmatrix} \begin{bmatrix} I_r & 0_{r\times(n-r)} \\ 0_{(n-r)\times r} & 0_{(n-r)\times(n-r)} \end{bmatrix}$$

$$= \begin{bmatrix} \Sigma_r & 0_{r\times(n-r)} \\ 0_{(m-r)\times r} & 0_{(m-r)\times(n-r)} \end{bmatrix}$$

$$= \Sigma.$$

∎

Use the above identities and the orthonormal properties to establish the following theorem.

Theorem 9.1.2. *(Pseudoinverse Identities)* *Let A be $m \times n$ matrix with $rank(A) = r$.*

1. $(A^T A)A^\dagger = A^T$.

2. $A^\dagger(AA^T) = A^T$.

3. $AA^\dagger A = A$.

4. $A(A^\dagger d) = d$ when $d \in R(A)$.

5. If A has full column rank, then $A^T A$ is nonsingular and

$$A^\dagger = (A^T A)^{-1} A^T.$$

6. If A has full row rank, then AA^T is nonsingular and

$$A^\dagger = A^T (AA^T)^{-1}.$$

Proof. The first and second items follow from the first and second identities in Theorem 9.1.1 and the orthonormal properties. In order to show the third item, use $A = U\Sigma V^T$ and $A^\dagger = V\Sigma^\dagger U^T$.

$$\begin{aligned}
AA^\dagger A &= \left(U\Sigma V^T\right)\left(V\Sigma^\dagger U^T\right)\left(U\Sigma V^T\right) \\
&= U\Sigma I_n \Sigma^\dagger I_m \Sigma V^T \\
&= U(\Sigma\Sigma^\dagger\Sigma)V^T \\
&= U\Sigma V^T \text{ by identity 3 in Theorem 9.1.1} \\
&= A.
\end{aligned}$$

The proof of the fourth part follows from $AA^\dagger A = A$. $d \in R(A)$ means there is an $\widehat{x} \in \mathbb{R}^n$ such that $A\widehat{x} = d$.

$$\begin{aligned}
A(A^\dagger d) &= A(A^\dagger (A\widehat{x})) \\
&= (AA^\dagger A)\widehat{x} \\
&= A\widehat{x} \\
&= d.
\end{aligned}$$

$\blacksquare$

Example 9.1.1. Return to Example 3.4.1

$A = \begin{bmatrix} 1 & 2 \\ 2 & 4 \\ 3 & 6 \end{bmatrix}$. Compute $A = U\Sigma V^T$, $A^\dagger = V\Sigma^\dagger U^T$ and use $A^\dagger$ to solve the least squares problem $Ax = d = \begin{bmatrix} 10 & 5 & 2 \end{bmatrix}^T$.

First, compute $A = U\Sigma V^T$ by finding the eigenvalues of

$$A^T A = 14 \begin{bmatrix} 1 & 2 \\ 2 & 4 \end{bmatrix}.$$

$$\det(14 \begin{bmatrix} 1-\lambda & 2 \\ 2 & 4-\lambda \end{bmatrix}) = 0 \text{ implies } \lambda_1 = 70 \text{ and } \lambda_0 = 0.$$

This gives $\sigma_1 = \sqrt{70}$, $\sigma_2 = 0$ and

$$v_1 = \begin{bmatrix} 1 \\ 2 \end{bmatrix} / \sqrt{5} \text{ and } u_1 = \frac{Av_1}{\sigma_1} = \begin{bmatrix} 1 \\ 2 \\ 3 \end{bmatrix} / \sqrt{14}.$$

The "small" SVD is $A = u_1 \sigma_1 v_1^T$. In order to find the "full" SVD, we need to extend the bases to $N(A)$ and $N(A^T)$. $Av_2 = 0_{3\times1}$ and $A^T u_2 = A^T u_3 = 0_{2\times1}$ gives

$$v_2 = \begin{bmatrix} -2 \\ 1 \end{bmatrix} / \sqrt{5} \text{ and } u_2 = \begin{bmatrix} -2 \\ 1 \\ 0 \end{bmatrix} / \sqrt{5}, u_3 = \begin{bmatrix} -3 \\ 0 \\ 1 \end{bmatrix} / \sqrt{10}.$$

The "full" SVD is

$$A = \begin{bmatrix} u_1 & u_2 & u_3 \end{bmatrix} \begin{bmatrix} \sqrt{70} & 0 \\ 0 & 0 \\ 0 & 0 \end{bmatrix} \begin{bmatrix} v_1^T \\ v_2^T \end{bmatrix}.$$

Second, compute $A^\dagger$ and $A^\dagger d$

$$A^\dagger = \begin{bmatrix} v_1 & v_2 \end{bmatrix} \begin{bmatrix} 1/\sqrt{70} & 0 & 0 \\ 0 & 0 & 0 \end{bmatrix} \begin{bmatrix} u_1^T \\ u_2^T \\ u_3^T \end{bmatrix} = v_1(1/\sqrt{70})u_1^T$$

$$A^\dagger d = \begin{bmatrix} 1/\sqrt{5} \\ 2/\sqrt{5} \end{bmatrix} (1/\sqrt{70}) \begin{bmatrix} 1 & 2 & 3/\sqrt{14} \end{bmatrix} \begin{bmatrix} 10 \\ 5 \\ 2 \end{bmatrix}$$

$$= \begin{bmatrix} 13/35 \\ 26/35 \end{bmatrix}.$$

Third, relate this to the general least squares solution

$$x = \begin{bmatrix} 26/14 \\ 0 \end{bmatrix} + c \begin{bmatrix} 2 \\ -1 \end{bmatrix}.$$

Let $f(c) = x^T x = (2c + 26/14)^2 + c^2$. This has a minimum at $c = -26/35$ so that

$$x = \begin{bmatrix} 26/14 \\ 0 \end{bmatrix} + (-26/35) \begin{bmatrix} 2 \\ -1 \end{bmatrix}$$

$$= \begin{bmatrix} 13/35 \\ 26/35 \end{bmatrix} = A^\dagger d.$$

9.1.1   Exercises

1. Justify the two projection equations in line (9.1.1).

2. Prove items 1 and 4 in Theorem 9.1.1.

3. Prove items 1 and 5 in Theorem 9.1.2.

4. Reconsider Exercise 2 in Section 7.2 where

$$A = \begin{bmatrix} 1 & 2 \\ 2 & 4 \\ 1 & 2 \end{bmatrix}.$$

(a). Find $A = U\Sigma V^T$.

(b). Find $A^\dagger = V\Sigma^\dagger U^T$.

(c). Use $A^\dagger$ to solve the least squares problem

$$Ax = d = \begin{bmatrix} 10 & 5 & 2 \end{bmatrix}^T.$$

5. Show $A^\dagger$ satisfies the Moore-Penrose inverse conditions:

$$AA^\dagger A = A, \ (AA^\dagger)^T = AA^\dagger, \ (A^\dagger A)^T = A^\dagger A \text{ and } A^\dagger AA^\dagger = A^\dagger.$$

9.2 $A^\dagger$ and Least Squares

Example 9.1.1 generalizes to any matrix. $A^\dagger d$ can be written in summation form

$$A^\dagger d = \sum_{j=1}^{r} v_j \left(\frac{1}{\sigma_j}\right) u_i^T d \text{ where } \operatorname{rank}(A) = r.$$

Since $1 \leq j \leq r$, this means $A^\dagger d \in R(A^T)$.

Theorem 9.2.1. *(Minimal Least Squares)   Let A be $m \times n$ matrix with $\operatorname{rank}(A) = r$.*

(i). $x^+ = A^\dagger d$ is a solution of the normal equations.

(ii). Moreover, it is a minimal solution of the normal equations, that is,

$$(x^+)^T (x^+) \leq \min_{x \in N(A)} (x^+ + x)^T (x^+ + x).$$

Proof. $A^\dagger d$ is a solution of the normal equations follows from $A^T AA^\dagger = A^T$ in the previous theorem.

$$A^T A(A^\dagger d) = (A^T AA^\dagger)d = A^T d.$$

The proof that $A^\dagger d$ is a minimal least squares solution uses $x^+ = A^\dagger d = V(\Sigma^\dagger U^T d) \in R(A^T)$. If $x \in N(A)$, then $x^+ + x$ is also a solution of the normal equations. By Theorem 5.2.1 $N(A) = R(A^T)^\perp$ and note

$$(x^+ + x)^T(x^+ + x) = (x^+)^T x^+ + 2x^T x^+ + x^T x$$
$$= (x^+)^T x^+ + 0 + x^T x$$
$$\geq (x^+)^T x^+ \text{ for all } x \in N(A).$$

∎

Example 9.2.1. Return to Example 7.2.2.

$A = \begin{bmatrix} 1 & 1 & 2 \\ 1 & 2 & 3 \\ 1 & 3 & 4 \\ 1 & 4 & 5 \end{bmatrix}$ has rank$(A) = 2$. The first two columns of A are a basis

for $R(A)$, and the vector $\begin{bmatrix} -1 & -1 & 1 \end{bmatrix}^T$ is a basis for $N(A)$. The general least squares solution of $Ax = d = \begin{bmatrix} 1 & 2 & 5 & 4 \end{bmatrix}^T$ is

$$x = \begin{bmatrix} 0 \\ 1.2 \\ 0 \end{bmatrix} + c \begin{bmatrix} -1 \\ -1 \\ 1 \end{bmatrix} \quad \text{where}$$

$$A \begin{bmatrix} 0 \\ 1.2 \\ 0 \end{bmatrix} = P_{R(A)}(d) = \begin{bmatrix} 1.2 \\ 2.4 \\ 3.6 \\ 4.8 \end{bmatrix}.$$

$P_{R(A)}(d) = c_1 w_1 + c_2 w_2$ where w_1 and w_2 are the first and second columns of A. The $c_1 = 0.0$ and $c_2 = 1.2$ are found by solving

$$w_1^T(d - c_1 w_1 - c_2 w_2) = 0 \text{ and}$$
$$w_2^T(d - c_1 w_1 - c_2 w_2) = 0.$$

The smallest $x^T x$ is given by $c = 0.4$ and $x = \begin{bmatrix} -.4 & .8 & .4 \end{bmatrix}^T$.
  This agrees with the following MATLAB® computations.

```
A =
      1     1     2
      1     2     3
      1     3     4
      1     4     5

d = [ 1 2 5 4]'
      1
      2
      5
      4
```

```
[U S V] = svd(A);
```

```
>> pinv(A)
   =
       0.7667      0.3667     -0.0333     -0.4333
      -0.5333     -0.2333      0.0667      0.3667
       0.2333      0.1333      0.0333     -0.0667
```

```
>> V*pinv(S)*U'
   =
       0.7667      0.3667     -0.0333     -0.4333
      -0.5333     -0.2333      0.0667      0.3667
       0.2333      0.1333      0.0333     -0.0667
```

```
>> pinv(A)*d
   =
      -0.4000
       0.8000
       0.4000
```

Example 9.2.2. This example exposes some possible difficulties with the pseudoinverse. Let ϵ be a small positive number and consider the 3×2 matrix

$$A = \begin{bmatrix} 1 & 0 \\ 0 & \epsilon \\ 1 & 0 \end{bmatrix}.$$

Since A has full column rank,

$$A^\dagger = (A^T A)^{-1} A^T = \begin{bmatrix} 1/2 & 0 & 1/2 \\ 0 & 1/\epsilon & 0 \end{bmatrix}.$$

First, note A converges as $\epsilon \to 0$, but $A^\dagger$ does not. Second, a small variation in the right side of $Ax = d$ can give a large variation in the least squares solution

$$A^\dagger d = \begin{bmatrix} 1/2 & 0 & 1/2 \\ 0 & 1/\epsilon & 0 \end{bmatrix} \begin{bmatrix} d_1 \\ d_2 \\ d_3 \end{bmatrix} = \begin{bmatrix} (d_1 + d_3)/2 \\ d_2/\epsilon \end{bmatrix}.$$

If $\epsilon = 10^{-6}$ and d_2 varies by 1.0, the second component varies by 10^6.

9.2.1   Exercises

1. Reconsider Exercise 2 in Section 3.4

$$\begin{bmatrix} 1 & 2 & 3 \\ 2 & 3 & 5 \\ 3 & 1 & 4 \\ 4 & 1 & 5 \end{bmatrix} \begin{bmatrix} x_1 \\ x_2 \\ x_3 \end{bmatrix} = \begin{bmatrix} 10 \\ 11 \\ 12 \\ 15 \end{bmatrix}.$$

(a). Find $A = U\Sigma V^T$.

(b). Find $A^\dagger = V\Sigma^\dagger U^T$.

(c). Use $A^\dagger$ to solve the least squares.

9.3   Ill-Conditioned Least Squares

Ill-conditioned matrices have the unpleasant property that small perturbations in the $n \times n$ matrix or right side can give large variations to the solution to the unperturbed system. This will be generalized to certain $m \times n$ matrices and the least squares problem. One application is to hazard identification such as a pollutant in a river with point sources along the river. Can one find the location and intensities of the point sources from observations at other locations along the river? The location of the observations can generate ill-conditioned least squares problems. This will be discussed in more detail in the next section.

Consider a nonsingular $n \times n$ matrix and use any norm. The classical analysis with $Ax = d$ and $A(x + \Delta x) = d + \Delta d$ estimates Δx relative to x.

$$A\Delta x = \Delta d \text{ and } \Delta x = A^{-1}\Delta d.$$
$$\|d\| \leq \|A\| \|x\| \text{ or } \frac{1}{\|x\|} \leq \|A\| \frac{1}{\|d\|}.$$
$$\|\Delta x\| \leq \|A^{-1}\| \|\Delta d\|.$$

Combine these to get

$$\frac{\|\Delta x\|}{\|x\|} \leq \|A\| \|A^{-1}\| \frac{\|\Delta d\|}{\|d\|}.$$

The classical condition number is $\|A\| \|A^{-1}\|$, but it is dependent on the choice of the norm. If the 2-norm is used, then by Theorem 7.3.1 $\|A\|_2 \|A^{-1}\|_2 = \sigma_1/\sigma_r$. For general $m \times n$ use this ratio as condition number.

Definition. *Let A be $m \times n$ matrix with $\text{rank}(A) = r$ with $\sigma_1 \geq \sigma_2 \geq \cdots \geq \sigma_r > 0$.*

$$cond(A) \equiv \sigma_1/\sigma_r.$$

Example 9.3.1. $A = \begin{bmatrix} 1 & 1/2 & 1/3 \\ 1/2 & 1/3 & 1/4 \\ 1/3 & 1/4 & 1/5 \end{bmatrix}$ or, more generally, $A = H(n) = \left[\frac{1}{1+i+j}\right]$ where $i, j = 0, ..., n-1$. The condition numbers increases rapidly with n:

$$cond(H(3)) = 524.0568,$$
$$cond(H(4)) = 1.5514 \ 10^4 \text{ and}$$
$$cond(H(5)) = 4.7661 \ 10^5.$$

If the right sides of $Ax = d$ are close, then the solutions may not be so close. For example,

$$H(4) \begin{bmatrix} 1 \\ 1 \\ 1 \\ 1 \end{bmatrix} = \begin{bmatrix} 2.0837 \\ 1.2833 \\ 0.9500 \\ 0.7595 \end{bmatrix} \text{ and}$$

$$H(4) \begin{bmatrix} -8 \\ 114 \\ -288 \\ 196 \end{bmatrix} = \begin{bmatrix} 2.0000 \\ 1.2000 \\ 0.9000 \\ 0.8000 \end{bmatrix}.$$

This is the Hilbert matrix, and it evolves from approximation of general function by a polynomial. Choose the coefficients of the polynomial to minimize the mean square integral of the difference.

$$F(c) \equiv \int_0^1 (f(x) - \sum_{j=0}^{n-1} c_j x^j)^2 dx$$

$$0 = \frac{\partial F}{\partial c_i} = 2 \int_0^1 (f(x) - \sum_{j=0}^{n-1} c_j x^j)^{2-1} x^i dx$$

$$\int_0^1 f(x) x^i dx = \sum_{j=0}^{n-1} c_j \int_0^1 x^j x^i dx = \sum_{j=0}^{n-1} c_j \frac{1}{1+j+i}.$$

Example 9.3.2. Let $A = K$ be the blurring matrix in the noise filter application in Section 8.3. In the calculations in Section 8.3 the K matrix is 100×100 with $\sigma_1 = 0.9985$, $\sigma_{100} = 4.2267 \ 10^{-7}$ and $cond(K) = 2.3620 \ 10^6$.

Theorem 9.3.1. *(Norm of $A^\dagger$) Let A be an $m \times n$ matrix with rank$(A) = r$ with $\sigma_1 \geq \sigma_2 \geq \cdots \geq \sigma_r > 0$. $A^\dagger \equiv V\Sigma^\dagger U^T$ is almost in the SVD form and*

$$\|A\|_2 = \sigma_1 \text{ and } \|A^\dagger\|_2 = 1/\sigma_r.$$

Proof. The eigenvectors in $A^\dagger$ need to be reordered so that the singular values in $\Sigma^\dagger$ have decreasing order. The proof of $\|A\|_2 = \sigma_1$ was given in Theorem 7.3.1. The proof of $\|A^\dagger\|_2 = 1/\sigma_r$ is similar. Let $x \in \mathbb{R}^m$ be a unit vector and

$$(A^\dagger x)^T A^\dagger x = x^T (V\Sigma^\dagger U^T)^T V\Sigma^\dagger U^T x$$
$$= x^T U(\Sigma^\dagger)^T V^T V\Sigma^\dagger U^T x$$
$$= x^T U(\Sigma^\dagger)^T \Sigma^\dagger U^T x.$$

Use identity 6 in Theorem 9.1.1 to note the largest diagonal component in $(\Sigma^\dagger)^T \Sigma^\dagger$ is $1/\sigma_r^2$. Since the columns of U are an orthonormal basis, $x = Uc$ with $c = U^T x$ and $c^T c = x^T x = 1$. Thus

$$(A^\dagger x)^T A^\dagger x \leq x^T x (1/\sigma_r^2) \text{ and } \left\| A^\dagger \right\|_2 \leq 1/\sigma_r.$$

In order to obtain the inequality in the other direction, choose $x = u_r$ to be column r in U. Then $U^T x = e_r$ and $(A^\dagger x)^T A^\dagger x = 1/\sigma_r^2$. ■

Consider $x^+ = A^\dagger d$ where there is a variation in the right side

$$\begin{aligned}
(x + \Delta x)^+ &= A^\dagger (d + \Delta d) \\
&= A^\dagger d + A^\dagger \Delta d \\
&= x^+ + A^\dagger \Delta d \\
\Delta(x^+) \equiv (x + \Delta x)^+ - x^+ &= A^\dagger \Delta d.
\end{aligned}$$

Theorem 9.3.2. *(Relative Error for LS)* *If A is an $m \times n$ matrix with* rank$(A) = r$ *and $d, \Delta d \in R(A)$, then for $x^+ = A^\dagger d$*

$$\frac{1}{cond(A)} \frac{\left\| \Delta d \right\|_2}{\left\| d \right\|_2} \leq \frac{\left\| \Delta(x^+) \right\|_2}{\left\| x^+ \right\|_2} \leq cond(A) \frac{\left\| \Delta d \right\|_2}{\left\| d \right\|_2}.$$

Proof. Apply the 2-norm to $\Delta(x^+) = A^\dagger \Delta d$

$$\left\| \Delta(x^+) \right\|_2 \leq \left\| A^\dagger \right\|_2 \left\| \Delta d \right\|_2 \leq (1/\sigma_r) \left\| \Delta d \right\|_2.$$

Use $d \in R(A)$ and Theorem 9.1.2 to write $Ax^+ = A(A^\dagger d) = d$. Apply the 2-norm

$$\left\| d \right\|_2 \leq \left\| A \right\|_2 \left\| x^+ \right\|_2 \leq \sigma_1 \left\| x^+ \right\|_2 \text{ or}$$
$$\frac{1}{\left\| x^+ \right\|_2} \leq \frac{\sigma_1}{\left\| d \right\|_2}.$$

Combine these to get the desired upper inequality.

Apply the 2-norm to $x^+ = A^\dagger d$

$$\left\| x^+ \right\|_2 \leq \left\| A^\dagger \right\|_2 \left\| d \right\|_2 \leq (1/\sigma_r) \left\| d \right\|_2 \text{ or}$$
$$\sigma_r / \left\| d \right\|_2 \leq 1/ \left\| x^+ \right\|_2$$

Use $\Delta d \in R(A)$ and Theorem 9.1.2 to write $A\Delta(x^+) = A(A^\dagger \Delta d) = \Delta d$. Apply the 2-norm

$$\left\| \Delta d \right\|_2 \leq \left\| A \right\|_2 \left\| \Delta(x^+) \right\|_2 \leq \sigma_1 \left\| \Delta(x^+) \right\|_2 \text{ or}$$
$$(1/\sigma_1) \left\| \Delta d \right\|_2 \leq \left\| \Delta(x^+) \right\|_2.$$

These two inequalities give the lower inequality. ■

The above result for the least squares solution does not include variations in the least squares solution because of the possible changes in the matrix. The following result can be established for nonsingular $n \times n$ matrices. Consider $Ax = d$ and $(A + \Delta A)(x + \Delta x) = d + \Delta d$ and assume

$$\left\|A^{-1}\right\|_2 \left\|\Delta A\right\|_2 < 1.$$

Then $A + \Delta A$ has an inverse and one can eventually show

$$\frac{\|\Delta x\|_2}{\|x\|_2} \leq \frac{cond(A)}{1 - \|A^{-1}\|_2 \|\Delta A\|_2} \left(\frac{\|\Delta d\|_2}{\|d\|_2} + \frac{\|\Delta A\|_2}{\|A\|_2} \right). \tag{9.3.1}$$

A possible extension of this to the general leastsquares problem is interesting!

9.3.1 Exercises

1. Find $cond(A)$ where

$$A = \begin{bmatrix} 1 & 2 & 3 \\ 2 & 3 & 5 \\ 3 & 1 & 4 \\ 4 & 1 & 5 \end{bmatrix}.$$

2. Prove the inequalities in line (9.3.1) by using the following:

   (a). Prove: if $\|B\|_2 < 1$, then $I - B$ has an inverse.

   (b). Prove: if $\|B\|_2 < 1$, then

   $$\left\|(I - B)^{-1}\right\|_2 \leq \frac{1}{1 - \|B\|_2}.$$

   Use the identity

   $$(I - B)(I + B + \cdots + B^k) = I - B^{k+1}.$$

   (c). Apply the above to $A + \Delta A = A(I - A^{-1}(-\Delta A))$.

9.4 Application to Hazard Identification

An interesting application is to hazard identification such as a pollutant in a river with point sources along the river. Consider a governing differential equation for the hazard, and discretize it to obtain a linear system

$$Au = d.$$

The coefficient matrix A is derived using finite differences with upwind finite differences on the velocity term. The coefficient matrix will be $n \times n$, and the sites or nodes will be partitioned into three ordered sets, whose order represents a reordering of the nodes,

$$osites \quad \text{observe sites}$$
$$ssites \quad \text{source sites and}$$
$$rsites \quad \text{remaining sites.}$$

The *hazard identification* problem is given data *data* at *osites*, find $d(ssites)$ such that

$$Au = d \text{ and } u(osites) = data.$$

In general, the reordered matrix has the following 2×2 block structure with $other = [ssites \quad rsites]$

$$A = \begin{bmatrix} a & e \\ f & b \end{bmatrix} \text{ where}$$

$$a = A(osites, osites),$$
$$e = A(osites, other),$$
$$f = A(other, osites) \text{ and}$$
$$b = A(other, other).$$

Assume the following are true:

(i). A and a have inverses.

(ii). $\#(osites) = k$, $\#(ssites) = l < k$ and $\#(rsites) = r$ so that $n = k + l + r$.

(iii). $d = \begin{bmatrix} 0 & d_1^T \end{bmatrix}^T$ and the only nonzero terms in d_1 are at the nodes in *ssites*, that is,

$$d \equiv \begin{bmatrix} 0 & d_1(ssites)^T & 0 \end{bmatrix}^T.$$

Multiply the above equation by a block elementary matrix

$$\begin{bmatrix} I_k & 0 \\ -fa^{-1} & I_{n-k} \end{bmatrix} \begin{bmatrix} a & e \\ f & b \end{bmatrix} \begin{bmatrix} u_0 \\ u_1 \end{bmatrix} = \begin{bmatrix} I_k & 0 \\ -fa^{-1} & I_{n-k} \end{bmatrix} \begin{bmatrix} 0 \\ d_1 \end{bmatrix}$$

$$\begin{bmatrix} a & e \\ 0 & \widehat{b} \end{bmatrix} \begin{bmatrix} u_0 \\ u_1 \end{bmatrix} = \begin{bmatrix} 0 \\ d_1 \end{bmatrix} \text{ and } \widehat{b} \equiv b - fa^{-1}e. \qquad (9.4.1)$$

Solve for u_1 and then $u_0 = -a^{-1}e(\widehat{b})^{-1}d_1$. Define

$$d_1 \equiv \begin{bmatrix} d_1(ssites)^T & 0 \end{bmatrix}^T \text{ and } C \equiv -a^{-1}e(\widehat{b})^{-1}.$$

The computed approximations of the observed *data* is in u_0 and the source data is in $z = d_1(ssites)$. This gives a *least squares* problem for z

$$C(:, ssites)z = data.$$

Let $d_1(ssites) = z$ and use the notation $d(z) = [0 \quad z^T \quad 0]^T$.

Since a is $k \times k$, e is $k \times (n - k)$ and $\hat{b}$ is $(n - k) \times (n - k)$, the matrix C is $k \times (n - k)$ and the *least squares matrix* $C(:, ssites)$ is $k \times l$ where $k > l$. The least squares problem will have a unique solution if the columns of $C(:, ssites)$ are linearly independent ($C(:, ssites)z = 0$ implies $z = 0$). The condition number will depend on the parameters in the differential equation as well as the position of the observation sites relative to the source sites. Here both the least squares matrix and the right side have variations. If there are insufficient source sites, $k > l$, then the pseudoinverse of $C(:, ssites)$ can be used to find the minimal 2-norm sources.

A more general discussion of the topic can be found in [27].and [3]

9.4.1   MATLAB® code hazidsvd1.m

Lines 1–33 define the 100×100 matrix associated with the differential equation. Line 34 locates the three source sites, and lines 35–38 give three possible observation sites and intensities. Lines 39–53 computes the new source numbers, which are caused by the reordering. The reordered matrix is computed in lines 57–61. The least squares matrix is computed in lines 62–64. The loop in lines 71–76 solves the least square problem 100 times with variable noise in the data. Lines 82 and 83 display the mean and standard deviation of the computed intensities at the source sites.

```
1     clear; clf(figure(1));
2     % hold on
3     % This code illustrates the linear least squares problem to
4     % identify the sources given observations. This is from
5     %    -(K u_x)_x + vel u_x + ru = point sources.
6     % The intensities at the point source sites are to be found
7     % from data at observation sites.
8     % The 1D steady state problem with fixed locations and
9     % unknown intensities is solved.
10    %
11    % Input data and system matrix
12    %
13    n = 100; L = 2.0; dx = L/n;
14    K     = 0.005;     % diffusion
15    vel   = 0.10;      % flow rate
16    r     = 0.0200;    % decay rate
17    noise = 0.10       % percent noise level
18    A = zeros(n);
```

```
19      for i = 1:n
20          if i == 1
21              A(i,i) = K*2/dx^2 + vel/dx + r;
22              A(i,i+1) = -K/dx^2;
23          end
24          if i>1 && i<n
25              A(i,i) = K*2/dx^2 + vel/dx + r;
26              A(i,i+1) = -K/dx^2;
27              A(i,i-1) = -K/dx^2 - vel/dx;
28          end
29          if i == n
30              A(i,i) = K*2/dx^2 + 2*vel/dx + r;
31              A(i,i-1) = -K*2/dx^2 - 2*vel/dx;
32          end
33      end
34      ssites = [20 35 80 ];        % source sites
35      osites = [10 30 60 90];      % observation sites
36      % osites = [10 30 60 70];
37      % osites = [10 30 ];
38      p = [1 5 4]                  % intensities of sources
39      [ms ns] = size(ssites)
40      [mo no] = size(osites)
41      % Find shifted ssites
42      newssites = ssites;
43      for js = 1:ns
44          for jo = 1:no
45              if osites(jo) < ssites(js)
46                  newssites(js) = newssites(js) - 1;
47              end
48          end
49      end
50      newssites
51      other = setdiff(1:n, osites);
52      set1 = union(osites,ssites);
53      rsites = setdiff(1:n, set1);
54      %
55      % Computation of reordered matrix and least squares matrix.
56      %
57      a = A(osites,osites);
58      b = A(other,other);
59      e = A(osites,other);
60      f = A(other,osites);
61      AA   = [ a e; f b];
62      bhat = b - f*inv(a)*e;
63      C    = -inv(a)*e*inv(bhat);
```

```
64      CLS   = C(:,newssites);
65      d     = zeros(n,1);
66      [U S V] = svd(CLS)
67      % Dirac delta appoximations
68      d(ssites)= p/dx;
69      u = A\d;
70      % Simulations with noise 100 executions
71      for kk = 1:100
72          data = u(osites);
73          data = data + data*(noise).*(rand(jo,1) - .5);
74          z = pinv(CLS)*data;
75          zz(1:ns,kk) = z;
76      end
77      display('U^T*data = ')
78      U'*data
79      %
80      % Numeric and graphical output
81      %
82      meanzz = mean(zz')
83      stdzz  = std(zz')
84      display('u at ssites = ')
85      u(ssites)
86      res_error = data - CLS*z
87      rank_num = rank(CLS)
88      cond_num = cond(CLS)
89      x = (1:n)*dx;
90      plot(u)
91      dd = zeros(n,1);
92      dd(ssites) = z;
93      uu = A\dd;
94      hold on
95      plot(uu,'k:')
96      plot(osites,4,'*')
```

In the following numerical experiments, there are four or two observation sites and three source sites. So the leastsquare matrix CLS will be either 4×3 or 2×3. The pseudoinverse times *data* can be written as

$$CLS^\dagger \, data = v_1(\frac{u_1^T d}{\sigma_1}) + v_2(\frac{u_2^T d}{\sigma_2}) + v_3(\frac{u_3^T d}{\sigma_3}).$$

In the second numerical experiment, the fourth observation site is poorly placed. This results in small third singular value, and the third term in the above being prone to computation errors.

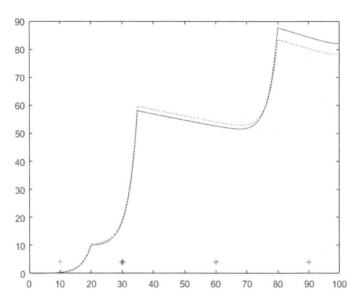

FIGURE 9.4.1
Hazard Id with osites = [10 30 60 90].

The first numerical experiment uses observation sites 10, 30, 60 and 90 and source sites 20, 35 and 80. The graphical output is given in Figure 9.4.1 and some of numerical output is given below. The mean computed intensities are close to the exact values of 50, 250 and 200. Moreover, the standard deviation is small.

```
>> hazidsvd1
noise =
    0.1000
ssites =
    20     35     80
osites =
    10     30     60     90
p =
     1      5      4
newssites =
    19     33     77
U =
   -0.0120     0.0203     0.0339    -0.9991
   -0.4116     0.5815     0.7006     0.0405
   -0.5921     0.4142    -0.6912    -0.0079
   -0.6927    -0.6999     0.1739     0.0000
```

```
S =
     0.3830          0          0
          0     0.1520          0
          0          0     0.0880
          0          0          0
V =
    -0.7272     0.4910     0.4797
    -0.5963    -0.1056    -0.7958
    -0.3401    -0.8647     0.3695
U^T*data =
   -96.8285
   -28.6826
    -6.7047
     0.0194
meanzz =
    49.6993   250.0798   201.6292
stdzz =
     3.8101    10.8717    14.8688
u at ssites =
    10.0538
    58.0373
    87.6532
rank_num =
     3
cond_num =
     4.3520
```

The second numerical experiment uses observation sites 10, 30, 60 and 70 and source sites 20, 35 and 80. The graphical output is given in Figure 9.4.2 and some of numerical output is given below. The mean computed intensities are close to the exact values of 50, 250 and 200. However, the standard deviation is larger, and this is more pronounced in the third intensity where the third singular value is small. The condition number for this least squares matrix is larger, and the rank is still equal to 3.

```
>> hazidsvd1
noise =
     0.1000
ssites =
    20    35    80
osites =
    10    30    60    70
p =
     1     5     4
newssites =
    19    33    76
```

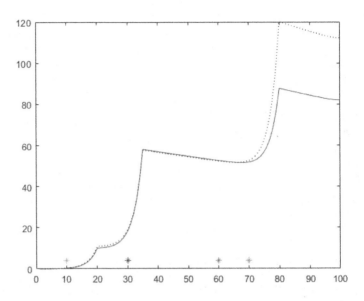

FIGURE 9.4.2
Hazard Id with osites = [10 30 60 70].

```
U =
    -0.0131      0.0388      0.0059     -0.9991
    -0.4497      0.8922      0.0005      0.0405
    -0.6438     -0.3237     -0.6933     -0.0082
    -0.6190     -0.3124      0.7206      0.0003
S =
     0.3741           0           0
          0      0.1012           0
          0           0      0.0045
          0           0           0
V =
    -0.7775      0.6289      0.0044
    -0.6288     -0.7772     -0.0231
    -0.0111     -0.0207      0.9997
U^T*data =
   -73.1552
   -16.6344
     2.7738
     0.0029
meanzz =
    49.6001    249.5465    214.6239
stdzz =
     4.1013     12.0458    380.1449
```

```
u at ssites =
    10.0538
    58.0373
    87.6532
rank_num =
     3
cond_num =
    82.5861
```

The third numerical experiment uses observation sites 10 and 30 and source sites 20, 35 and 80. The graphical output is given in Figure 9.4.3 and some of numerical output is given below. The mean computed intensities are close to the exact values of 50, 250 but not 200. The condition number of the leastsquare matrix has increased and the rank is 2. The general leastsquares solution is

$$CLS^\dagger \, data + c \, v_3 \text{ where } CLS \, v_3 = 0_{2\times 1}.$$

The minimal 2-norm least squares solution follows from

$$z + cv_3 = \begin{bmatrix} z_1 & z_2 & z_3 \end{bmatrix}^T + c\begin{bmatrix} 0 & 0 & 1 \end{bmatrix}^T.$$

Then $c = -z_3$ and $z + cv_3 = \begin{bmatrix} z_1 & z_2 & 0 \end{bmatrix}^T$.

```
>>hazidsvd1
```

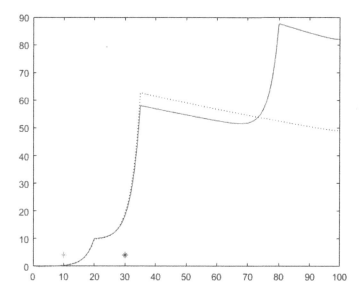

FIGURE 9.4.3
Hazard Id with osites = [10 30].

```
noise =
    0.1000
ssites =
    20      35      80
osites =
    10      30
p =
     1       5       4
newssites =
    19      33      78
U =
   -0.0324    0.9995
   -0.9995   -0.0324
S =
    0.1911        0           0
         0   0.0011           0
V =
   -0.9825    0.1863    0.0000
   -0.1863   -0.9825   -0.0000
   -0.0000   -0.0000    1.0000
U^T*data =
  -17.6769
   -0.2616
meanzz =
   49.6716  251.3730    0.0001
stdzz =
    1.5527   17.3989    0.0000
u at ssites =
   10.0538
   58.0373
   87.6532
rank_num =
     2
cond_num =
  168.4529
```

10

General Inner Product Vector Spaces

The previous chapters were restricted to the real vector space $\mathbb{R}^n$. Now we will consider more general sets where the operations of addition and scaler multiplication are defined. These include complex vectors, real-valued functions and linear maps such as integral and differential operators. Inner products will be generalized so that orthogonal concepts, eigenvectors, eigenvectors, self-adjoint and positive definite operations can be extended.

10.1 Vector Spaces

Vectors will be a more general term than vectors in $\mathbb{R}^n$ or $\mathbb{C}^n$. For example, they may be continuous functions from any set to the real numbers: $f, g : S \to \mathbb{R}$ where

$$(f + g)(x) \equiv f(x) + g(x) \text{ and}$$
$$(cf)(x) \equiv cf(x) \text{ and } c \in \mathbb{R}.$$

Here the $c \in \mathbb{R}$ is the scaler. The scalers may be more general and can be any field, such as $\mathbb{R}$ or $\mathbb{C}$ or the finite field $\mathbb{Z}_p$ where p is a prime number. For a general and more axiomatic presentation consider book by Serge Lang, [12]. Usually we will use the field $\mathbb{F} = \mathbb{C}$.

Definition. *Let V be a set, whose elements are called vectors, and let $\mathbb{F}$ be a field, whose elements are called scalers. Let $+ : V \times V \to V$ be addition and $\cdot : \mathbb{F} \times V \to V$ be scaler multiplication. The quadruple $(V, +, \cdot, \mathbb{F})$ is called a vector space over $\mathbb{F}$ if and only if the following five addition and five scaler properties hold:*

1. *$u, v \in V$ implies $u + v \in V$,*

2. *$u + v = v + u$,*

3. *$u + (v + w) = (u + v) + w$,*

4. *there is a zero vector $z \in V$ where $u + z = u$ for all $u \in V$,*

DOI: 10.1201/9781003304128-10

5. *there is an inverse vector* $(-u) \in V$ *where* $u + (-u) = z$ *for all* $u \in V$,

6. $c \in \mathbf{F}$ *and* $u \in V$ *implies* $c \cdot u \in V$,

7. $1 \cdot u = u$ *for all* $u \in V$ *and* 1 *is the unit in the field,*

8. $(c_1 c_2) \cdot u = c_1 \cdot (c_2 \cdot u)$,

9. $c \cdot (u + v) = c \cdot u + c \cdot v$ *and*

10. $(c_1 + c_2) \cdot u = c_1 \cdot u + c_2 \cdot u$.

Notation. The products in $(c_1 c_2) \cdot u = c_1 \cdot (c_2 \cdot u)$, are two types: one is for products in the field and one is for the scaler product. Often the scaler product symbol is deleted so that property eight would be written as

$$(c_1 c_2) u = c_1 (c_2 u).$$

Example 10.1.1.

1. $V = \mathbb{R}^n$ is a vector space of over $\mathbb{R}$ and $z = 0_{n \times 1}$

2. $V = \mathbb{C}^n$ is a vector space of over $\mathbb{C}$ and $z = 0_{n \times 1}$ is a column with $0 + i0$ components.

3. $V = \{x \in \mathbb{R}^3 : x_1 + x_2 + x_3 = 0\}$. Properties 1 and 6 hold and the other are inherited from $\mathbb{R}^3$. V is called a *subspace* of $\mathbb{R}^3$.

4. This set $S = \{x \in \mathbb{R}^3 : x_1 + x_2 + x_3 = 1\}$ is not a vector space because properties 1 and 6 are not true.

5. $V = $ set of $m \times n$ real or complex matrices.

6. $V = P_n = $ set of n degree polynomials over $\mathbb{R}$ or $\mathbb{C}$. The polynomials are a linear combination of $1, x, x^2, ..., x^n$.

7. $V = \{f : [0 \ \ l] \to \mathbb{R}$ and is a function (single valued)$\}$. Here the field is $\mathbb{R}$, and the zero vector is the zero-valued function $z(x) = 0$. The following are also vector subspaces:

$$P_{n-1} \subset P_n \subset C^1 \subset C^0 \subset V.$$

8. $V = (\mathbb{Z}_p)^n = $ set of column vectors with n components in the field $\mathbb{Z}_p$.

Definition. *Let V be a vector space. $S \subset V$ is called a subspace if and only if it is also a vector space using the same vector addition and scaler product.*

Theorem 10.1.1. *(Vector Subspace) Let V be a vector space over $\mathbf{F}$. $S \subset V$ is a subspace if and only if*

$$u, v \in S \text{ and } c \in \mathbf{F} \text{ imply } u + v \in S \text{ and } cu \in S.$$

Proof. Properties 1 and 6 hold by assumption. The other eight properties hold because $S \subset V$ and V is a vector space over $\mathbf{F}$. ∎

The above examples in 3 and 7 are subspaces as well as the null space for matrices, $N(A)$, and the range space, $R(A)$.

Example 10.1.1.

9. The boundary value problems are sources of subspaces. For example, consider

$$L(u) = -u_{xx} + u.$$

Let $V = C^2$ be the set of functions on $[0 \ l]$ with two continuous derivatives. Then L maps from C^2 into C^0. The null space of L requires $L(u) = 0$, and this is a subspace because $L(u+v) = 0+0 = 0$ and $L(cu) = cL(u) = c0 = 0$. An alternative reason uses the explicit form of the null space functions

$$null(L) = \{c_1 e^x + c_2 e^{-x} : c_1, c_2 \in \mathbb{R}\}.$$

This is a vector subspace of C^2.

10. Let V be a vector space over $\mathbb{R}(\text{or } \mathbb{C})$ and with k vectors $v_1, v_2, ..., v_k$. Let S be the set of all linear combinations

$$S = \{v : v = c_1 v_1 + c_2 v_2 + \cdots + c_k v_k \text{ with } c_i \in \mathbb{R}(\text{or } \mathbb{C})\}.$$

Both conditions of the above theorem hold and so S is a vector subspace.

11. In the above example there were no given boundary condition associated with the differential operator. Consider one or two boundary conditions.

(a).
$$L_{1:} : L(u) = -u_{xx} + u \text{ and } u(0) = 0.$$

The null space of the boundary value problem is

$$null(L_1) = \{c_1(e^x - e^{-x}) : c_1 \in \mathbb{R}\}.$$

(b).
$$L_2 : L(u) = -u_{xx} + u \text{ and } u(0) = 0, \ u(l) = 0.$$

The null space of this boundary value problem is

$$null(L_2) = \{zero \text{ function}\}.$$

The set of linear combinations of vectors as in Example 10 may or may not have a unique representation

$$v = c_1 v_1 + c_2 v_2 + \cdots + c_k v_k$$
$$= d_1 v_1 + d_2 v_2 + \cdots + d_k v_k.$$

In order to ensure $c_i = d_i$, the following expression for the zero vector

$$z = v - v = (c_1 - d_1)v_1 + (c_2 - d_2)v_2 + \cdots + (c_k - d_k)v_k$$

must imply each scaler $c_i - d_i = 0$.

Definition. *Let V be a vector space with k vectors $v_i \in V$. The vectors $\{v_1, \cdots, v_k\}$ are called linearly independent if and only if*

$$z = c_1 v_1 + c_2 v_2 + \cdots + c_k v_k \text{ implies each scaler } c_i = 0.$$

Example 10.1.2.

1. Let $V = \mathbb{C}^3$ and the two complex vectors

$$v_1 = \begin{bmatrix} i \\ 2 \\ 3 \end{bmatrix} \text{ and } v_1 = \begin{bmatrix} 2 \\ 2 \\ 3 \end{bmatrix}.$$

$$\begin{bmatrix} 0 \\ 0 \\ 0 \end{bmatrix} = c_1 \begin{bmatrix} i \\ 2 \\ 3 \end{bmatrix} + c_2 \begin{bmatrix} 2 \\ 2 \\ 3 \end{bmatrix}$$

Then equating the first two components gives $0 = c_1 i + c_2 2$ and $0 = c_1 2 + c_2 2$ so that $0 = c_1 i - c_1 2$ and

$$c_1 = 0/(i - 2) = 0 \frac{i + 2}{(i - 2)(i + 2)} = 0.$$

2. Let $V = \mathbb{C}^n$ and consider n column vectors of a complex $n \times n$ invertible matrix A. This means $Ax = 0_{n \times 1}$ implies $x = 0_{n \times 1}$. But Ax is a linear combination of the columns and $x_i = c_i$.

3. Let $V = C^2$ and consider two functions $v_1 = e^x$ and $v_2 = e^{-x}$

$$0 = z(x) = c_1 e^x + c_2 e^{-x}.$$

In order to find a second equation, compute the derivative with respect to x

$$0 = c_1 e^x - c_2 e^{-x}.$$

Then both coefficients must be zero.

4. Let $V = C^0$ and consider quadratic polynomials

$$a_0 + a_1 x + a_2 x^2.$$

They are a linear combination of the three functions $v_1 = 1, v_2 = x$ and $v_3 = x^2$. Consider

$$0 = z(x) = a_0 1 + a_1 x + a_2 x^2.$$

By computing the first and second derivatives, we obtain the system

$$\begin{bmatrix} 1 & x & x^2 \\ 0 & 1 & 2x \\ 0 & 0 & 2 \end{bmatrix} \begin{bmatrix} a_0 \\ a_1 \\ a_2 \end{bmatrix} = \begin{bmatrix} 0 \\ 0 \\ 0 \end{bmatrix}.$$

Since the matrix has an inverse, the solution is $0_{3 \times 1}$.

Definition. *Let V be a vector space. Consider a subset of vectors $\{v_1, v_2, \cdots, v_k\}$. The subset of vectors is called a basis of V if and only if*

(i). *each vector in V is a linear combination the vectors and*

(ii). *the vectors are linearly independent.*

The coefficients in the linear combination are called the coordinates with respect to the basis, and the number k is called the dimension.

Example 10.1.3.

1. Let V be the null space of the differential operator $Lu = -u_{xx} + 4u$. The basis is $\{e^{2x}, e^{-2x}\}$ and the dimension equals 2.

2. Let V be the cubic polynomials. The basis is $\{1, x, x^2, x^3\}$ and the dimension equals 4.

3. Let V be the null space of the complex 3×1 matrix $A = \begin{bmatrix} 1 & 2i & -4 \end{bmatrix}$. The free variables are x_2 and x_3 where $Ax = 1x_1 + 2ix_2 - 4x_3 = 0$. Elements of the null space have the form

$$x = x_2 \begin{bmatrix} -2i \\ 1 \\ 0 \end{bmatrix} + x_3 \begin{bmatrix} 4 \\ 0 \\ 1 \end{bmatrix}$$

The two column vectors are the basis.

The number of vectors in the basis, the dimension, is unique. The proof of this follows the vectors space $\mathbb{R}^n$. Consider the general vector space, and for notation simplification assume there is a two-dimensional basis and show there can not be a basis with dimension three. Let $\{v_1, v_2\}$ be a basis for V and show the three vectors $\{w_1, w_2, w_3\} \subset V$ cannot be linearly independent. Because $\{v_1, v_2,\}$ is a basis for V, three vectors can be written as

$$w_1 = c_{11}v_1 + c_{12}v_2$$
$$w_2 = c_{21}v_1 + c_{22}v_2 \text{ and}$$
$$w_3 = c_{31}v_1 + c_{32}v_2.$$

Let z be the zero vector and form the linear combination of the three vectors

$$z = c_1 w_1 + c_2 w_2 + c_3 w_3$$
$$= (c_1 c_{11} + c_2 c_{21} + c_3 c_{31})v_1 +$$
$$(c_1 c_{12} + c_2 c_{22} + c_3 c_{32})v_2.$$

The following 3×2 matrix problem has a nonzero solution

$$\begin{bmatrix} c_{11} & c_{21} & c_{31} \\ c_{12} & c_{22} & c_{32} \end{bmatrix} \begin{bmatrix} c_1 \\ c_2 \\ c_3 \end{bmatrix} = \begin{bmatrix} 0 \\ 0 \end{bmatrix}.$$

Therefore, there is at least one $c_i \neq 0$ and $\{w_1, w_2, w_3\}$ are linearly dependent and not a basis.

10.1.1   Exercises

1. Consider Example 1 in Examples 10.1.3.

   (a). Show the proposed basis is linearly independent.
   (b). Show $\{e^{2x}, e^{2x} + Ce^{-2x}\}$ is also a basis provided $C \neq 0$.

2. Consider Example 2 in Examples 10.1.3.

   (a). Show the proposed basis is linearly independent.
   (b). Find another basis and verify both conditions for a basis.

3. Consider Example 3 in Examples 10.1.3.

   (a). Show the proposed basis is linearly independent.
   (b). Find another basis that has the form

   $$v_1 = \begin{bmatrix} -2i \\ 1 \\ 0 \end{bmatrix} \text{ and } v_2 = \begin{bmatrix} 4 \\ 0 \\ 1 \end{bmatrix} + C \begin{bmatrix} -2i \\ 1 \\ 0 \end{bmatrix}.$$

   Choose C so that $\overline{v_1}^T v_2 = 0$, and verify both conditions for a basis.

10.2   Inner Products and Orthogonal Vectors

The inner product for real vectors, $x^T y$, can be extended to general vector spaces. First consider the extension of complex vector space. The modus of a single complex number, $w = a + bi$, is given by using the conjugate of the complex number

$$|w|^2 = (a - bi)(a + bi) = a^2 + b^{\cdot 2}.$$

The "length" squared of a complex vector, $x \in \mathbb{C}^n$, is the sum of moduli squared of the complex components

$$\begin{aligned} \|x\|^2 &= |x_1|^2 + \cdots + |x_n|^2 \\ &= \overline{x_1} x_1 + \cdots + \overline{x_n} x_n \\ &= \overline{x}^T x. \end{aligned}$$

The inner product of two complex vectors is defined using the conjugate transpose

$$< x, y > \equiv \bar{x}^T y = x^* y.$$

Because of the conjugate, the properties are a little different

$$< cx, y > = \bar{c} < x, y > \quad \text{and} \quad \overline{< x, y >} = < y, x > .$$

10.2.1 General inner products

The general inner product of two vector space vectors is satisfied by the inner products of both real and complex vectors.

Definition. *Let V be a vector space over the complex numbers. The inner product is a mapping from*

$$< \quad , \quad >: V \times V \to \mathbb{C}$$

that satisfies the following properties:

(i). *$< x, x > \geq 0$ and is a real number;*
   $x = z$ (the zero vector) if and only if $< x, x > = 0$,

(ii). *$\overline{< x, y >} = < y, x >$,*

(iii). *$< x, y + w > = < x, y > + < x, w >$ and*

(iv). *$< x, cy > = c < x, y > .$*

Example 10.2.1.

1. Let V be the vector space of continuous complex-valued function define on the interval from 0 to l.

$$< f, g > \equiv \int_0^l \overline{f(x)} g(x) dx.$$

2. Let V be the vector space of real functions with continuous derivative on the interval $[0 \quad l]$

$$< u, v > \equiv \int_0^l u(x)v(x) + u_x(x)v_x(x) dx.$$

3. Let V be the vector space of real functions with continuous derivatives on a closed bounded set $\Omega \subset \mathbb{R}^2$

$$< u, v > \equiv \int \int_\Omega uv + u_x v_x + u_y v_y dx dy.$$

4. Let $V = \mathbb{R}^n$ and let A be SPD real matrix

$$< x, y > \equiv x^T A y.$$

Inner products are important because of orthogonal properties that evolve from the Cauchy inequality $|x^T y| \leq \|x\| \, \|y\|$. The proof of this uses $f(t) = (x + ty)^T (x + ty) \geq 0$ for all real numbers t, and $t = t_0$ where $f(t_0) = \min f(t)$. In the general inner product $< x, y >= A + Bi$ and $t = a + bi$ are complex numbers. Use the defining properties of the general product to show

$$F(a, b) = < x + ty, x + ty >$$
$$= < x, x > + \overline{t} < x, y > + t < x, y > + \overline{t}t < y, y > . \quad (10.2.1)$$

Find the a and b that minimize F by setting the two partial derivatives equal to zero. This gives $a = -\operatorname{Re}(< y, x > / < y, y >)$ and $b = -\operatorname{Im}(< y, x > / < y, y >)$, that is,

$$t_0 = - < y, x > / < y, y >= -(A - Bi)/ < y, y > .$$

In order to derive the new Cauchy inequality, insert this into line (10.2.1)

$$0 \leq < x, x > + \frac{\overline{-(A - Bi)}}{< y, y >} \overline{A + Bi} + \frac{-(A - Bi)}{< y, y >}(A + Bi) + \frac{A^2 + B^2}{< y, y >^2} < y, y >$$

$$\leq < x, x > - \frac{|< x, y >|^2}{< y, y >}.$$

This may be written using the norm associated with the inner product $\|x\| \equiv < x, x >^{1/2}$ as

$$|< x, y >| \leq \|x\| \, \|y\| .$$

This proves the first part of the following theorem. The proofs of the triangle inequality and the "Pythagorean on Steroids" identity follows from line (10.2.1) with $t = 1$. In this case line (10.2.1) is

$$< x + y, x + y >=< x, x > + 2A + < y, y > \quad \text{and}$$
$$\|x + y\|^2 \leq \|x\|^2 + 2\sqrt{A^2 + B^2} + \|y\|^2 .$$

Apply the Cauchy inequality to get

$$\|x + y\|^2 \leq \|x\|^2 + 2 \|x\| \, \|y\| + \|y\|^2 = (\|x\| + \|y\|)^2$$

The fourth item is the "parallelogram law."

Theorem 10.2.1. *(Inner Product Properties)* *Let V be a vector space over the complex numbers and with an inner product.*

(i) $|< x, y >| \leq \|x\| \, \|y\|$ *(Cauchy inequality),*

(ii). $\|x + y\| \le \|x\| + \|y\|$ *(triangle inequality)*,

(iii). If $< x, y >= 0$, *then* $\|x + y\|^2 = \|x\|^2 + \|y\|^2$ *and*

(iv). $\|x + y\|^2 + \|x - y\|^2 = 2(\|x\|^2 + \|y\|^2)$.

The Cauchy inequality has a long history: Cauchy (1821), Bunyakovsky (1859), Scharz (1888). It is often called the CBS inequality. Because of the orthogonality properties, it is a very important result.

10.2.2 Orthonormal vectors

In Chapter 5 the inner product on the real vector space, $V = \mathbb{R}^n$, was used to generate orthonormal vectors and the existence condition for $Ax = d \in N(A^T)^\perp$. The classical Gram-Schmidt method can be used with general inner product spaces where orthonormal vectors means

$$< x, y >= 0 \text{ and } \|x\| = \|y\| = 1.$$

Start with a set of vectors $a^1, a^2, a^3, \cdots$ in the general vector space V. The objective is to find orthonormal vector $q^1, q^2, q^3, \cdots$ also in V.

The first one is easy $q^1 = a^1 / \|a^1\|$. The next step uses a projection

$$\hat{q}^2 = a^2 + c_{21}q^1 \text{ and } c_{21} \text{ so that } < q^1, \hat{q}^2 >= 0.$$
$$q^2 = \hat{q}^2 / \|\hat{q}^2\| \text{ and } 0 =< q^1, a^2 > +c_{21}1.$$

Continue this with

$$\hat{q}^{m+1} = a^{m+1} + c_{m+1,1}q^1 + \cdots + c_{m+1,m}q^m$$
$$0 = < q^i, a^{m+1} > +c_{m+1,i}1.$$
$$q^{m+1} = \hat{q}^{m+1} / \|\hat{q}^{m+1}\|$$

Example 10.2.2. Let $V = \mathbb{C}^3$ and use $< x, y >= \bar{x}^T y$. Let the three vectors be the columns in

$$A = \begin{bmatrix} i & 2 & 1+1 \\ 0 & i & 2 \\ 3 & i+1 & 3i \end{bmatrix}$$

The first vector is $q^1 = a^1 / \|a^1\| = [i \ 0 \ 3]^T / \sqrt{10}$. In order to find the second vector, compute

$$c_{21} = - < q^1, a^2 >= -[-i \ 0 \ 3] / \sqrt{10} \begin{bmatrix} 2 \\ i \\ i+1 \end{bmatrix} = \frac{-i - 3}{\sqrt{10}}.$$

Then $\hat{q}^2 = a^2 + c_{21}q^1$ and

$$q^2 = \hat{q}^2 / \|\hat{q}^2\| = \begin{bmatrix} 0.8573 - 0.1225i \\ 0.0000 + 0.4082i \\ 0.0408i + 0.2858i \end{bmatrix}$$

.

The code GScomplex.m reduces the computational effort. In MATLAB® $x' *$
$y = \bar{x}^T y$.

```
 1      A = [i 2 1+i; 0 i 2; 3 i+1 3i]
 2      det(A)
 3      q1 = A(:,1)/norm(A(:,1));
 4      % q2 = A(:,2) + c21 q1
 5      c21 = -q1'*A(:,2);
 6      q2 = A(:,2) + c21*q1;
 7      q2 = q2/norm(q2);
 8      % q3 = A(:,3) + c31 q1 + c32 q2
 9      c31 = -q1'*A(:,3);
10      c32 = -q2'*A(:,3);
11      q3 = A(:,3) + c31*q1 + c32*q2;
12      q3 = q3/norm(q3);
13      Q = [q1 q2 q3]
14      Q'*Q
>> GScomplex
A =
    0.0000 + 1.0000i    2.0000 + 0.0000i    1.0000 + 1.0000i
    0.0000 + 0.0000i    0.0000 + 1.0000i    2.0000 + 0.0000i
    3.0000 + 0.0000i    1.0000 + 1.0000i    0.0000 + 3.0000i

det(A)
    17.0000 - 8.0000i

Q =
    0.0000 + 0.3162i    0.8573 - 0.1225i    0.1649 + 0.3504i
    0.0000 + 0.0000i    0.0000 + 0.4082i    0.8727 - 0.2680i
    0.9487 + 0.0000i    0.0408 + 0.2858i   -0.1168 + 0.0550i

Q'*Q
    1.0000 + 0.0000i    0.0000 - 0.0000i   -0.0000 + 0.0000i
    0.0000 + 0.0000i    1.0000 + 0.0000i    0.0000 + 0.0000i
   -0.0000 - 0.0000i    0.0000 - 0.0000i    1.0000 + 0.0000i
```

Example 10.2.3. Let V = the continuous real-valued functions on the interval $[-1 \ 1]$ and with the inner product

$$< f, g >= \int_{-1}^{1} f(x)g(x)dx.$$

Generate the normalized Legendre polynomials starting with the three vector functions $1, x$ and x^2. The first normalized vector is $q^1 = 1/\|1\| = 1/\sqrt{2}$.

$$c_{21} = - <q^1, x> = - \int_{-1}^{1} \frac{1}{\sqrt{2}} x dx = 0 \text{ and}$$

$$q^2 = x/\|x\| = x/\sqrt{\frac{2}{3}} = \sqrt{\frac{3}{2}} x.$$

The third vector evolves from

$$\widehat{q^3} = x^2 + c_{31} q^1 + c_{32} q^2$$

$$= x^2 - <x^2, 1/\sqrt{2}> \frac{1}{\sqrt{2}} - <x^2, \sqrt{\frac{3}{2}} x> \sqrt{\frac{3}{2}} x$$

$$= x^2 - \frac{\sqrt{2}}{3} \frac{1}{\sqrt{2}} - 0 \sqrt{\frac{3}{2}} = x^2 - \frac{1}{3} = \frac{3x^2 - 1}{3}$$

Finally,

$$q^3 = \widehat{q^3}/\|q^3\| = \sqrt{\frac{5}{2}} \frac{1}{2}(3x^2 - 1).$$

10.2.3   Norms on vector spaces

Given an inner product on a vector space, the norm associated with the inner product is $\|x\| \equiv <x, x>^{1/2}$. There are a number of other "norms" which are not associated with an inner product. The formal definition of a norm is a generalization of the absolute value of a single number, and they are important in the analysis of convergence of iterative methods as illustrated in the next chapter.

Definition. *Let V be a vector space. A norm on V is a mapping from V into the nonnegative real numbers $\|*\| : V \to [0 \ \infty)$ such that*

(i). $\|x\| \geq 0$; $x = z$ *(the zero vector) if and only if $\|x\| = 0$,*

(ii). $\|x + y\| \leq \|x\| + \|y\|$ *and*

(iii). $\|cx\| = |c| \|x\|$.

Example 10.2.4.

1. The 2-norm is the norm from the inner product on $\mathbb{C}^n$

$$\|x\|_2 \equiv (\overline{x}^T x)^{1/2} = (\sum_{i=1}^{i=n} |x_i|^2).$$

2. The one and ∞ norms do not evolve from inner products

$$\|x\|_1 \equiv \sum_{i=1}^{i=n} |x_i| \text{ and}$$

$$\|x\|_\infty \equiv \max |x_i| .$$

3. If E has an inverse, then the infinity norm gives

$$\|x\|_E \equiv \max |(Ex)_i| .$$

4. The set of matrices is a vector space, and it has a norm associated with any given norm on $\mathbb{C}^n$

$$\|A\| \equiv \sup_{x \neq 0} \frac{\|Ax\|}{\|x\|} = \sup_{\|x\|=1} \|Ax\| . \text{ Hence, } \|Ax\| \leq \|A\| \, \|x\| .$$

5. Consider a real matrix and vectors in $\mathbb{R}^n$. Use the 2-norm on $\mathbb{R}^n$.

$\|A\|_2 \equiv \sup_{\|x\|_2=1} \|Ax\|_2$ is the largest singular value of A. This norm was introduced in Section 7.3. The largest singular value is the square root of the largest eigenvalue of $A^T A$.

6. Consider a real matrix and vectors in $\mathbb{R}^n$. Use $\|x\|_\infty \equiv \max |x_i|$.

$\|A\|_\infty \equiv \sup_{\|x\|_\infty=1} \|Ax\|_\infty = \max_i \sum_j |a_{ij}|$ (max of row sums). This equality follows from the definition.

7. Consider a real matrix and vectors in $\mathbb{R}^n$. Use $\|x\|_E \equiv \max |(Ex)_i|$ on $\mathbb{R}^n$.

$$\|A\|_E \equiv \sup_{\|y\|_E=1} \|Ax\|_E .$$

Note $\|Ax\|_E = \max |(EAx)_i| = \max |(EAE^{-1}(Ex)_i|$ and $\|x\|_E \equiv \max |(Ex)_i| .$

Let $y = Ex$ and then $\|A\|_E = \sup_{\|y\|_E=1} \|EAE^{-1}y\|_E = \|EAE^{-1}\|_\infty .$

8. Consider a real matrix and vectors in $\mathbb{R}^n$. Use $\|x\|_1 \equiv \sum |x_i|$.

$\|A\|_1 \equiv \sup_{\|x\|_1=1} \|Ax\|_1 = \max_j \sum_i |a_{ij}|$ (max of column sums). This equality follows from the definition.

The utility of the various norms will be illustrated in the following chapters. The convergence of sequence of real vectors, $x^k \to x$, is defined by $\|x^k - x\| \to 0$. Fortunately, this does not depend on the choice of norm, see Ortega and Rheinboldt, [20, Section 2.2]. The following norm properties will be very useful.

Theorem 10.2.2. *(Norm Properties) Let the $\|x\|$ be a norm on $\mathbb{R}^n$ and $\|A\|$ be the corresponding norm on $n \times n$ matrices.*

(i). $|\, \|x\| - \|y\| \,| \leq \|x - y\|$ *(The norm is a continuous function.).*

(ii). *If B is also an $n \times n$ matrix, then $\|AB\| \leq \|A\| \, \|B\|$.*

(iii). *If $\|A\| < 1$, then $I - A$ has an inverse.*

(iv). *If $\|A\| < 1$, then for all x^0 the sequence $x^{k+1} = Ax^k + d$ converges to*

$$x = Ax + d.$$

Proof. The inequality follows from the triangle inequality applied to $y = x + (y - x)$ and then to $x = y + (x - y)$. The second inequality follows directly from the definition and

$$\|(AB)x\| = \|A(Bx)\| \le \|A\| \|Bx\| \le \|A\| \|B\| \|x\|.$$

The third item follow from $(I - A)x = 0$ and $x = Ax$. This gives a contradiction to any singularity because $\|x\| = \|Ax\| \le \|A\| \|x\|$ implies $1 \le \|A\|$ for nonzero x. The convergence uses the second and third inequalities and

$$
\begin{aligned}
x^{k+1} - x &= (Ax^k + d) - (Ax + d) \\
&= A(x^k - x) \\
&\;\;\vdots \\
&= A^{k+1}(x^0 - x)
\end{aligned}
$$

The error on the left side must go to zero because of the assumption and repeated use of $\|AB\| \le \|A\| \|B\|$

$$
\begin{aligned}
\|x^{k+1} - x\| &= \|A^{k+1}(x^0 - x)\| \\
&\le \|A\|^{k+1} \|x^0 - x\|.
\end{aligned}
$$

∎

10.2.4 Exercises

1. Confirm the conditions for the inner products in Examples 10.2.1.

2. Consider the proof of the Cauchy inequality.

   (a). From the inner product properties establish line (10.2.1)
   (b). Confirm the equality for t_0.

3. Prove the fourth item in Theorem 10.2.1.

4. Confirm the conditions for the norms in Examples 10.2.4.

5. Find the various norms for the matrix

$$A = \begin{bmatrix} 1 & 2 \\ 3 & -1 \end{bmatrix}.$$

10.3   Schur Decomposition

The Schur decomposition of an $n \times n$ complex matrix shows that the matrix is unitarily equivalent to an upper triangular matrix. Thus, the eigenvalues of the matrix are the diagonal components of the upper triangular matrix. This gives three additional results relating the spectrum and a special norm, a characterization of diagonalization of a matrix, and $p(A) = 0_{n \times n}$ where $p(\lambda) = \det(A - \lambda I)$.

The existence of the Schur decomposition will follow by mathematical induction. First, we consider the 2×2 case

$$\begin{bmatrix} u_1^* \\ u_2^* \end{bmatrix} A \begin{bmatrix} u_1 & u_2 \end{bmatrix} = \begin{bmatrix} t_{11} & t_{12} \\ 0 & t_{22} \end{bmatrix}$$

$$\begin{bmatrix} u_1^* A u_1 & u_1^* A u_2 \\ u_2^* A u_1 & u_2^* A u_2 \end{bmatrix} =$$

By equating the components we have

$$u_1^* A u_1 = t_{11} \text{ and } u_2^* A u_1 = 0.$$

If $A u_1 = t_{11} u_1$ and $u_1^* u_1 = 1$, then $u_1^* A u_1 = t_{11}$. Put this into $u_2^* A u_1 = u_2^* t_{11} u_1 = 0$ and so u_2 should be orthogonal to u_1.

Example 10.3.1. Consider the real 2×2 matrix

$$A = \begin{bmatrix} 0 & 1/2 \\ 1 & 0 \end{bmatrix}$$

with eigenvalues $\sigma(A) = \{1/\sqrt{2}, -1/\sqrt{2}\}$ and $\rho(A) = 1/\sqrt{2}$. The normalized eigenvector that corresponds to the first eigenvalue is

$$u_1 = \begin{bmatrix} 1/\sqrt{3} \\ \sqrt{2}/\sqrt{3} \end{bmatrix}.$$

By inspection the normalized second vector is

$$u_2 = \begin{bmatrix} -\sqrt{2}/\sqrt{3} \\ 1/\sqrt{3} \end{bmatrix}.$$

The unitary matrix is $U = \begin{bmatrix} u_1 & u_2 \end{bmatrix}$ and

$$U^* A U = \begin{bmatrix} 1/\sqrt{2} & -1/2 \\ 0 & -1/\sqrt{2} \end{bmatrix}.$$

Example 10.3.2. Consider the complex 2×2 matrix

$$A = \begin{bmatrix} 2 & i \\ i & 3 \end{bmatrix}.$$

The characteristic polynomial is $p(\lambda) = \det(A - \lambda I) = \lambda^2 - 5\lambda + 7 = 0$. The eigenvalues are complex numbers $\sigma(A) = \{(5 + i\sqrt{3})/2, (5 - i\sqrt{3})/2\}$ and the normalized eigenvector corresponding to the eigenvalue $(5 + i\sqrt{3})/2$ is

$$u_1 = \begin{bmatrix} (\sqrt{3} + i)/(2\sqrt{2}) \\ 1/\sqrt{2} \end{bmatrix} = \begin{bmatrix} 0.6124 + 0.3561i \\ 0.7071 + 0.0000i \end{bmatrix}.$$

Find the second column vector by requiring $u_1^* \widehat{u_2} = 0$ where

$$\widehat{u_2} = a_2 + cu_1 = \begin{bmatrix} i \\ 3 \end{bmatrix} + c\begin{bmatrix} (\sqrt{3} + i)/(2\sqrt{2}) \\ 1/\sqrt{2} \end{bmatrix}$$

and solving for $c = -u_1^* a_2 = -(7 + i\sqrt{3})/(2\sqrt{2})$. After some complex arithmetic we have

$$u_2 = \begin{bmatrix} -0.6944 - 0.1336i \\ 0.6682 - 0.2315i \end{bmatrix} \text{ and}$$

$$U^* A U = \begin{bmatrix} 2.5000 + 0.8660i & 0.9449 - 0.3274i \\ 0 & 2.5000 - 0.8660i \end{bmatrix}$$

Theorem 10.3.1. *(Schur Decomposition) Let A be an $n \times n$ complex matrix. There exist an $n \times n$ complex matrix unitary matrix U such that*

$$U^* A U = T \text{ where}$$

$U^* U = I$ *and T is upper triangular.*

Proof. This proof is by mathematical induction and the 2×2 matrix case is the starting point and is established as in Example 10.3.2. Assume the decomposition is true for $n \times n$ matrix and A is an $(n+1) \times (n+1)$. Let w_1 be a normalized eigenvector of A. Extend this to be an orthonormal set of vectors $\{w_1, w_2, \cdots, w_{n+1}\}$. One can do this via Gram-Schmidt or the Householder transform. Let W be the $(n + 1) \times (n + 1)$ matrix formed by these column vectors.

$$W^* A w_1 = W^* \lambda_1 w_1 = \lambda_1 e_1 \text{ and}$$

$$W^* A W = \begin{bmatrix} \lambda_1 & \widetilde{} \\ 0_{n \times 1} & \widehat{A} \end{bmatrix}.$$

Use the inductive assumption on $\widehat{A}$ and $\widehat{V}^* \widehat{A} \widehat{V} = \widehat{T}$ where $\widehat{V}^* \widehat{V} = I_n$ and $\widehat{T}$ is upper triangular. Define the $(n + 1) \times (n + 1)$ matrix

$$V = \begin{bmatrix} 1 & 0_{1 \times n} \\ 0_{n \times 1} & \widehat{V} \end{bmatrix}.$$

Define the unitary $(n+1) \times (n+1)$ matrix $U = WV$ and note

$$\begin{aligned}
U^* A U &= (WV)^* A (WV) \\
&= V^*(W^* A W) V \\
&= \begin{bmatrix} \lambda_1 & \widetilde{} \\ 0_{n \times 1} & \widehat{V}^* \widehat{A} \widehat{V} \end{bmatrix} \\
&= \begin{bmatrix} \lambda_1 & \widetilde{} \\ 0_{n \times 1} & \widehat{T} \end{bmatrix}.
\end{aligned}$$

∎

10.3.1   Norms and spectral radius

The spectral radius is not a norm, but one can find a norm that is close to it. As noted in Theorem 10.2.2 the convergence of $x^{k+1} = Ax^k + d$ depends on $\|A\| < 1$, and hence, on the choice of the norm. In order to gain some insight, reconsider Example 10.3.1 where

$$A = \begin{bmatrix} 0 & 1/2 \\ 1 & 0 \end{bmatrix}, \quad \|A\|_\infty = 1 \text{ and } \rho(A) = 1/\sqrt{2}.$$

Use the Schur decomposition and form the following norm

$$\begin{aligned}
\|A\| &\equiv \left\| \Delta^{-1} U^* A U \Delta \right\|_\infty \\
&= \left\| \begin{bmatrix} 1 & 0 \\ 0 & \delta \end{bmatrix}^{-1} \begin{bmatrix} 1/\sqrt{2} & -1/2 \\ 0 & -1/\sqrt{2} \end{bmatrix} \begin{bmatrix} 1 & 0 \\ 0 & \delta \end{bmatrix} \right\|_\infty \\
&= \left\| \begin{bmatrix} 1/\sqrt{2} & (-1/2)\delta \\ 0 & -1/\sqrt{2} \end{bmatrix} \right\|_\infty \\
&\leq \left\| \begin{bmatrix} 1/\sqrt{2} & 0 \\ 0 & -1/\sqrt{2} \end{bmatrix} \right\|_\infty + \left\| \begin{bmatrix} 0 & (-1/2)\delta \\ 0 & 0 \end{bmatrix} \right\|_\infty \\
&\leq \rho(A) + (1/2)\,\delta.
\end{aligned}$$

Since $\delta > 0$ can be as small as we wish, this norm can be close to $\rho(A)$. This approach can be extended to any $n \times n$ complex matrix.

Theorem 10.3.2. *(Special Matrix Norm) Let A be an $n \times n$ complex matrix. For all $\epsilon > 0$ there exist a matrix norm, depending on ϵ, such that*

$$\rho(A) \leq \|A\| < \rho(A) + \epsilon \text{ where}$$

$\rho(A)$ is the spectral radius of A. Moreover, for some $\epsilon > 0$ this norm $\rho(A) < 1$ if and only if $\|A\| < 1$.

Proof. Use the Schur decomposition and use the following notation

$$\Delta \equiv \begin{bmatrix} 1 & & & \\ & \delta & & \\ & & \ddots & \\ & & & \delta^{n-1} \end{bmatrix}.$$

$$\|A\| \equiv \left\| \Delta^{-1} U^* A U \Delta \right\|_\infty$$
$$= \left\| \Delta^{-1} (\lambda + \widehat{T}) \Delta \right\|_\infty$$
$$\leq \|\lambda\|_\infty + \left\| \Delta^{-1} \widehat{T} \Delta \right\|_\infty \quad \text{where}$$

λ is the diagonal of $U^* A U = T$ and $\widehat{T}$ is the strictly upper part of T.

$$\Delta^{-1} \widehat{T} \Delta = \begin{bmatrix} 0 & t_{12}\delta & t_{13}\delta^2 & \cdots & t_{1n}\delta^{n-1} \\ & 0 & t_{23}\delta & \cdots & t_{2n}\delta^{n-2} \\ & & 0 & \vdots & \vdots \\ & & & \ddots & t_{n-1,n}\delta \\ & & & & 0 \end{bmatrix}.$$

Then factor one δ from $\Delta^{-1}\widehat{T}\Delta$ and use $\Delta^{-1}\widehat{T}\Delta = \delta B$

$$\|A\| \leq \|\lambda\|_\infty + \delta \|B\|_\infty .$$
$$\leq \rho(A) + \delta \|B\|_\infty .$$

Since B bounded for bounded $1 \geq \delta > 0$, and choose δ as small as we wish to get $\delta \|B\|_\infty < \epsilon$.

If $\rho(A) < 1$, then choose $\epsilon = (1 - \rho(A))/2$ so that $\|A\| < 1$. ∎

10.3.2 Normal matrices

In Section 6.3 the real symmetric matrices were shown to have eigenvectors such that $AU = UD$ where the columns of U are the eigenvectors and the diagonal matrix has the eigenvalue. This section will generalize this to complex matrices.

Definition. *Let A be an $n \times n$ complex matrix. A is called diagonalizable if and only if there exist unitary matrix such that*

$$U^* A U = D = diagonal\ matrix.$$

A is called a normal matrix if and only if

$$AA^* = A^* A.$$

Example 10.3.3. This matrix is self-adjoint, $A = A^* \equiv \overline{A}^T = [\overline{a_{ji}}]$, and therefore a normal matrix

$$A = \begin{bmatrix} 2 & i \\ -i & 3 \end{bmatrix}.$$

Example 10.3.4. This matrix is not self-adjoint, but it is normal

$$A = \begin{bmatrix} 1 & i \\ i & 1 \end{bmatrix}.$$

The eigenvalues are $\sigma(A) = \{1 + i, 1 - i\}$ and the corresponding eigenvectors are

$$u_1 = \begin{bmatrix} 1/\sqrt{2} \\ 1/\sqrt{2} \end{bmatrix} \text{ and } u_2 = \begin{bmatrix} 1/\sqrt{2} \\ -1/\sqrt{2} \end{bmatrix}.$$

This gives the special Schur decomposition

$$U^* A U = \begin{bmatrix} 1 + i & 0 \\ 0 & 1 - i \end{bmatrix}.$$

Theorem 10.3.3. *(Normal Matrix Characterization) Let A be an $n \times n$ complex matrix. Diagonalizable and normal matrices are equivalent.*

Proof. First assume the matrix is diagonalizable $U^* A U = D$ or $A = UDU^*$. Then use $U^* U = I$ to get

$$AA^* = UDD^* U^* \text{ and } A^* A = UD^* D U^*.$$

Since $DD^* = D^* D$, the matrix is normal.

Next assume the matrix is normal and use the Schur decomposition $U^* A U = T$ or $A = UTU^*$. Again use $U^* U = I$

$$AA^* = A^* A$$
$$UTT^* U^* = UT^* T U^*.$$

This means $TT^* = T^* T$. In order to shown this implies T must be a diagonal matrix, use mathematical induction. The 2×2 version of $TT^* = T^* T$ is

$$\begin{bmatrix} t_{11} & t_{12} \\ 0 & t_{22} \end{bmatrix} \begin{bmatrix} \overline{t_{11}} & 0 \\ \overline{t_{12}} & \overline{t_{22}} \end{bmatrix} = \begin{bmatrix} \overline{t_{11}} & 0 \\ \overline{t_{12}} & \overline{t_{22}} \end{bmatrix} \begin{bmatrix} t_{11} & t_{12} \\ 0 & t_{22} \end{bmatrix}.$$

Equate the 11-components to get

$$t_{11}\overline{t_{11}} + t_{12}\overline{t_{12}} = \overline{t_{11}}t_{11}.$$

Thus $t_{12}\overline{t_{12}} = 0$ and $t_{12} = 0$. Assume T is $(n + 1) \times (n + 1)$ and

$$T = \begin{bmatrix} t_{11} & T_{12} \\ 0_{nx1} & T_{22} \end{bmatrix} \text{ where}$$

T_{12} is $1 \times n$ and T_{22} is $n \times n$. Now $TT^* = T^*T$ is

$$\begin{bmatrix} t_{11} & T_{12} \\ 0_{n \times 1} & T_{22} \end{bmatrix} \begin{bmatrix} \overline{t_{11}} & 0_{1 \times n} \\ T_{12}^* & T_{22}^* \end{bmatrix} = \begin{bmatrix} \overline{t_{11}} & 0_{1 \times n} \\ T_{12}^* & T_{22}^* \end{bmatrix} \begin{bmatrix} t_{11} & T_{12} \\ 0_{n \times 1} & T_{22} \end{bmatrix}.$$

The 11-components must be equal and so $T_{12}T_{12}^* = 0$ and $T_{12} = 0_{1 \times n}$. Then the block 22-components must be equal and gives $T_{22}T_{22}^* = T_{22}^*T_{22}$. Since these are upper triangular, the inductive assumption yields T_{22} must be diagonal. This completes the proof. ∎

Example 10.3.5. This matrix is not normal, but the eigenvectors give a diagonal matrix.

$$A = \begin{bmatrix} 1 & 4 \\ 2 & 3 \end{bmatrix} \text{ with}$$

$$A \begin{bmatrix} -2 \\ 1 \end{bmatrix} = (-1) \begin{bmatrix} -2 \\ 1 \end{bmatrix} \text{ and } A \begin{bmatrix} 1 \\ 1 \end{bmatrix} = (5) \begin{bmatrix} 1 \\ 1 \end{bmatrix}.$$

The matrix given by the two eigenvectors has an inverse, but it is not unitary

$$A \begin{bmatrix} -2 & 1 \\ 1 & 1 \end{bmatrix} = \begin{bmatrix} -2 & 1 \\ 1 & 1 \end{bmatrix} \begin{bmatrix} -1 & 0 \\ 0 & 5 \end{bmatrix}.$$

The Schur decomposition is

```
> [U T] = schur(A)
    U =
        -0.8944    -0.4472
         0.4472    -0.8944

    T =
        -1     2
         0     5.
```

10.3.3 Cayley–Hamilton theorem

As motivation consider Example 10.3.2 where

$$A = \begin{bmatrix} 2 & i \\ i & 3 \end{bmatrix} \text{ with } p(\lambda) = \det(A - \lambda I) = \lambda^2 - 5\lambda + 7.$$

Then $p(A) = A^2 - 5A + 7I = 0_{2 \times 2}$. This is true for all $n \times n$ complex matrices. The proof uses the Schur decomposition, and this is illustrated in the above example. Note $A = UTU^*$ and

$$\begin{aligned} p(A) &= p(UTU^*) \\ &= (UTU^*)^2 - 5UTU^* + 7I \\ &= Up(T)U^*. \end{aligned}$$

Then $p(A) = 0_{2\times 2}$ if and only if $p(T) = 0_{2\times 2}$. The upper triangular matrix has the form

$$T = \begin{bmatrix} t_{11} & t_{12} \\ 0 & t_{22} \end{bmatrix} \text{ where } t_{11} = \lambda_1 \text{ and } t_{22} = \lambda_2.$$

Since $p(\lambda) = (\lambda - \lambda_1)(\lambda - \lambda_2)$ with $\lambda_1 = (5 + i\sqrt{3})/2$ and $\lambda_2 = (5 - i\sqrt{3})/2$,

$$\begin{aligned}
p(T) &= (T - \lambda_1 I)(T - \lambda_2 I) \\
&= \left(\begin{bmatrix} t_{11} & t_{12} \\ 0 & t_{22} \end{bmatrix} - \lambda_1 \begin{bmatrix} 1 & 0 \\ 0 & 1 \end{bmatrix} \right) \\
&\quad \left(\begin{bmatrix} t_{11} & t_{12} \\ 0 & t_{22} \end{bmatrix} - \lambda_2 \begin{bmatrix} 1 & 0 \\ 0 & 1 \end{bmatrix} \right) \\
&= \begin{bmatrix} 0 & t_{12} \\ 0 & t_{22} - \lambda_1 \end{bmatrix} \begin{bmatrix} t_{11} - \lambda_2 & t_{12} \\ 0 & 0 \end{bmatrix} \\
&= \begin{bmatrix} 0 & 0 \\ 0 & 0 \end{bmatrix}.
\end{aligned}$$

Theorem 10.3.4. *(Cayley–Hamilton $p(A) = 0_{n\times n}$) Let A be an $n \times n$ complex matrix. If $p(\lambda) = \det(A - \lambda I)$ is the characteristic polynomial, then $p(A) = 0_{n\times n}$.*

Proof. Because $p(A) = Up(T)U^*$, it is sufficient to show $p(T) = 0_{n\times n}$. Since T is upper triangular and the eigenvalues are the diagonal components $\lambda_i = t_{ii}$,

$$p(T) = (T - \lambda_1 I)(T - \lambda_2 I) \cdots (T - \lambda_n I).$$

Each of these factors is upper triangular, and column i in $(T - \lambda_i I)$ has components equal to zero starting at row i. This is illustrated for $n = 3$ where the solid ★ may not be zero

$$\begin{aligned}
p(T) &= \begin{bmatrix} 0 & \star & \star \\ 0 & \star & \star \\ 0 & 0 & \star \end{bmatrix} \begin{bmatrix} \star & \star & \star \\ 0 & 0 & \star \\ 0 & 0 & \star \end{bmatrix} \begin{bmatrix} \star & \star & \star \\ 0 & \star & \star \\ 0 & 0 & 0 \end{bmatrix} \\
&= \begin{bmatrix} 0 & 0 & \star \\ 0 & 0 & \star \\ 0 & 0 & \star \end{bmatrix} \begin{bmatrix} \star & \star & \star \\ 0 & \star & \star \\ 0 & 0 & 0 \end{bmatrix} \\
&= \begin{bmatrix} 0 & 0 & 0 \\ 0 & 0 & 0 \\ 0 & 0 & 0 \end{bmatrix}.
\end{aligned}$$

Multiplying starting from the left, each factor creates a zero column vector.

10.3.4 Exercises

1. Find the Schur decomposition of

$$\begin{bmatrix} 1/2 & 1/2 \\ 1 & 0 \end{bmatrix}.$$

2. Consider the above matrix and find a norm close to the spectral radius.

3. Which of the following matrices are self-adjoint, normal, diagonalizable or symmetric:

$$\begin{bmatrix} 3 & 4 \\ 4 & i \end{bmatrix}, \begin{bmatrix} 3 & 4i \\ 4i & i \end{bmatrix}, \begin{bmatrix} 3 & 4i \\ -4i & i \end{bmatrix} \text{ and } \begin{bmatrix} 3 & 4i \\ -4i & 1 \end{bmatrix}.$$

4. Verify the Cayley–Hamilton theorem for

$$\begin{bmatrix} 3 & 4i \\ -4i & 1 \end{bmatrix}.$$

10.4 Self-Adjoint Differential Operators

This section will extend three properties of matrix operators to differential operators. The matrix properties positive definite $(x^T A x > 0)$, eigenvalue $(Ax = \lambda x)$ and existence $(Ax = d \in N(A^T)^\perp)$ will be extended to differential operators with boundary conditions.

A simple example is $Lu = -u_{xx}$ with boundary conditions $u(0) = 0$ and $u(l) = 0$. Multiply $Lu = -u_{xx}$ by u and integrate by parts to form the inner product

$$< Lu, u >= \int_0^l (u_x)^2 > 0.$$

Note there are an infinite number of eigenvalues $Lu = \lambda u$

$$L \sin(\frac{n\pi}{l}x) = (\frac{n\pi}{l})^2 \sin(\frac{n\pi}{l}x).$$

If the boundary conditions are changed to $u_x(0) = 0$ and $u_x(l) = 0$, then the solution of $Lu = f$ requires

$$< f, 1 >= \int_0^l f(x) 1 = 0.$$

Note 1 is the solution to the homogenous or null boundary value problem $Lv = -v_{xx} = 0$ with $v_x(0) = 0$ and $v_x(l) = 0$. With respect to this inner product f is perpendicular or orthogonal to the null space of the latter boundary value problem for v. There are numerous other examples in Nagle, Saff and Snider, [19, Chapter 11].

10.4.1   Linear operators

Let V and W be vector spaces with norms $\|*\|_v$ and $\|*\|_W$. An $m \times n$ matrix is a mapping or operator from $V = \mathbb{R}^n$ to $W = \mathbb{R}^m$.

Definition. $L : V \to W$ *is called linear if and only if* $L(u+v) = L(u) + L(v)$ *and* $L(cu) = cL(u)$ *for* $c \in \mathbb{C}$ *and* $u, v \in V$.

Example 10.4.1.

1. A complex $m \times n$ matrix $A : \mathbb{C}^n \to \mathbb{C}^m$

2. The differential operator $L : C^2[0\ 1] \to C^0[0\ 1]$ where

$$L(u) = -u_{xx} + u.$$

3. The differential operator $L(u) = -u_{xx} + u^2$ is NOT linear.

4. The differential operator in Example 2 has no boundary condition and is defined on all of $C^2[0\ 1]$. If we impose some boundary condition, say $u(0) = 0$ and $u(1) = 0$, then the domain on L will change to the smaller vector space $V = \{u \in C^2[0\ 1] : u(0) = 0 \text{ and } u(1) = 0\}$.

5. Integration is also linear and gives

$$G(u) = \int_0^1 K(x, y)u(y)dy.$$

If K is continuous on $[0\ 1] \times [0\ 1]$, then $G : C^0[0\ 1] \to C^0[0\ 1]$.

The set of all linear maps from V to W, $L(V, W)$, is also a vector space when addition and scaler multiplication are defined for $F, G \in L(V, W)$

$$(F + G)(x) \equiv F(x) + G(x) \text{ and } (cF)(x) \equiv F(cx).$$

Moreover, $L(V, W)$ has a norm

$$\|F\| \equiv \sup_{x \neq z} \frac{\|F(x)\|_w}{\|x\|_V}.$$

The means $\|F(x)\|_w \leq \|F\| \, \|x\|_V$. This implies bounded linear maps are continuous because

$$\left\|F(x^k) - F(x)\right\|_w = \left\|F(x^k - x)\right\|_w \leq \|F\| \, \left\|x^k - x\right\|.$$

Associated with a linear operator is an adjoint operator. This applies to both matrices and differential operators with various boundary conditions. The transpose of the real $n \times n$ matrix is an example.

Definition. *Let* $L : V \to V$ *be a linear operator on the vector space* V *with inner product* $<,>$. *A linear operator* $L^* : V \to V$ *is called an adjoint to* L *if and only if*

$$< Lu, v >=< u, L^*v > .$$

L is called self-adjoint if $L = L^*$.

Example 10.4.2.

1. A symmetric matrix is self-adjoint on $\mathbb{R}^n$ because $A^T = A$ and

$$< Ax, y >= (Ax)^T y = x^T A^T y = x^T (Ay) =< x, Ay > .$$

2. Let A be an $n \times n$ complex matrix. The conjugate transpose $A^* \equiv [\overline{a_{ji}}] = \overline{A}^T$ is the adjoint of A.

$$< Ax, y >= \left(\overline{Ax}\right)^T y = (\overline{x})^T \overline{A}^T y =< x, A^*y > .$$

If $A = A^*$, the matrix is self-adjoint and is called *hermitian*.

3. Consider the boundary value problem $Lu = -u_{xx} = f$ with boundary conditions $u_x(0) = 0$ and $u_x(1) = 0$. Form the inner product and integrate two times to get

$$\begin{aligned}
< Lu, v >= & u(1)v_x(1) - u(0)v_x(0) \\
& - (v(1)u_x(1) - v(0)u_x(0)) \\
& + < u, -v_{xx} > .
\end{aligned} \tag{10.4.1}$$

Since there are no constraints on $u(1)$ and $u(0)$, one must choose $v_x(1) = 0$ and $v_x(0) = 0$. So the adjoint homogenous boundary value problem is $L^*v = -v_{xx} = 0$ with $v_x(1) = 0$ and $v_x(0) = 0$, and the nonzero solution is $v = 1$. If $-u_{xx} = f$, then

$$< Lu, v >= 0+ < u, L^*v >$$
$$< f, v >=< u, 0 >$$
$$< f, 1 >= 0.$$

4. Consider the boundary value problem $Lu = -u_{xx} - 2u_x - 2u = f$ with boundary conditions $u(0) = 0$ and $u(\pi) = 0$. Form the inner product and integrate two times to get

$$\begin{aligned}
< Lu, v >= & u(\pi)v_x(\pi) - u(0)v_x(0) \\
& - (v(\pi)u_x(\pi) - v(0)u_x(0)) \\
& - 2((u(\pi)v(\pi) - u(0)v(0)) \\
& + < u, -v_{xx} + 2v_x - 2v > .
\end{aligned}$$

Since there are no constraints on $u_x(\pi)$ and $u_x(0)$, one must choose $v(\pi) = 0$ and $v(0) = 0$. So the adjoint homogeneous boundary value problem is

$$L^*v = -v_{xx} + 2v_x - 2v = 0 \text{ with } v(\pi) = 0 \text{ and } v(0) = 0.$$

There is a nonzero solution to the adjoint homogeneous problem $v = e^x \sin(x)$. If $Lu = f$, then f must satisfy

$$< Lu, v >=< u, L^*v >$$
$$< f, v >=< u, 0 >$$
$$< f, e^x \sin(x) >= 0.$$

The condition is known as the Fredholm Alternative and is also sufficient for solution to exist. If there is no nonzero solution to the adjoint problem, then there should be a solution to the inhomogeneous boundary problem.

5. This example and the next illustrate the difference between the two boundary value problems $-u_{xx} + cu = f$ and $-u_{xx} - cu = f$ where $c > 0$ and both with the same boundary conditions $u(0) = 0$ and $u_x(1) = 0$. Consider $L(u) = -u_{xx} + cu = f$. Find the adjoint by using line (10.4.1) and the additional term in $L(u)$

$$< Lu, v >= u(1)v_x(1) - u(0)v_x(0)$$
$$- (v(1)u_x(1) - v(0)u_x(0))$$
$$+ < u, -v_{xx} + cv > . \qquad (10.4.2)$$

Since $u(1)$ and $u_x(0)$ have no constraints, one must choose $v(0) = 0$ and $v_x(1) = 0$ to give $< Lu, v >=< u, Lv >$. So this is self-adjoint and the homogeneous problem $-v_{xx} + cv = 0$ has solution $v = Ae^{\sqrt{c}x} + Be^{-\sqrt{c}x}$. The boundary conditions require $A = 0$ and $B = 0$ and v is the zero function. There is no constraint on f, and the boundary value problem $Lu = f$ has a unique solution.

6. Change the sign on the u term and consider $Lu = -u_{xx} - cu = f$ and the same boundary conditions. The boundary conditions on the adjoint problem are the same and the problem is self-adjoint. However, the homogeneous solution to the adjoint problem may not be the zero function. The homogeneous solution has the form $v = A \sin(\sqrt{c}x)$, and the second boundary condition requires $\cos(\sqrt{c}1) = 0$ or $A = 0$. If $c = (-\pi/2 + n\pi)^2$, then $v = \sin((-\pi/2 + n\pi)x)$ and gives a constraint on f

$$< Lu, v >=< u, Lv >$$
$$< f, v >=< u, 0 >$$
$$< f, \sin((-\pi/2 + n\pi)x) >= 0.$$

In the next subsection we will focus on eigenvalues for differential operator similar to those in the above examples in 3,4 and 6. The last section will focus on self-adjoint and positive definite differential operators similar to Example 5 and more general partial differential equations

$$-(pu_x)_x - (pu_y)_y + cu = f.$$

10.4.2 Sturm-Liouville problem

Many second order linear differential equation can be written, via integrating factors, as $Lu \equiv -(p(x)u_x)_x + q(x)u = f$ or as $Lu = w(x)\lambda u$ for the weighted eigenvalues. The real functions p and w are usually positive and continuous. The boundary conditions can be quite general, but here we assume they are such that the boundary value problem is self-adjoint.

Definition. *The regular Sturm-Liouville problem is*

$$-(p(x)u_x)_x + q(x)u = w(x)\lambda u \text{ on } (a, b)$$
$$a_1 u(a) + a_2 u_x(a) = 0 \text{ with } a_1 a_2 \neq 0 \text{ and}$$
$$b_1 u(b) + b_2 u_x(b) = 0 \text{ with } b_1 b_2 \neq 0.$$

Notation. Usually $q \geq 0$ and the density or weight w is given. Often the above is rewritten as $(p(x)u_x)_x + q(x)u + w(x)\lambda u = 0$. An example is the Legendre polynomials with $p = 1 - x^2$, $q = 0$, $w = 1$ and $\lambda = n(n+1)$.

By using a line similar to (10.4.1) and assuming a_1 or a_2 is not zero, and b_1 or b_2 is not zero, this boundary value problem is self-adjoint with respect to the inner product

$$< Lu, v >= \int_a^b \overline{(Lu)}v = \int_a^b \overline{u}(Lv) =< u, Lv > .$$

The functions $u = u_1 + u_2 i$ and $v = v_1 + v_2 i$ may be complex valued. The boundary conditions hold for both the real and imaginary parts, and the self-adjoint condition $< Lu, v >=< u, Lv >$ is satisfied for real and complex values.

Theorem 10.4.1. *(Orthonormal Eigenfunctions)* *The regular Sturm-Liouville problem has the following properties:*

(i). *The eigenvalues are real.*

(ii). *The eigenvectors with different eigenvalues are orthogonal with respect to the inner product*

$$< f, g >_w = \int_a^b \overline{f}gw.$$

(iii). The eigenvectors can be real and orthonormal

with respect to $< f, g >_w$.

Proof. Let u be an eigenvector $Lu = w(x)\lambda u$. Since p, q and w are real,

$$\overline{Lu} = L\overline{u} = \overline{\lambda} w \overline{u}.$$

Form the inner products:

$$< Lu, u >= \int_a^b \overline{(Lu)} u$$

$$= \int_a^b \overline{\lambda} w \overline{u} u$$

$$= \overline{\lambda} < u, u >_w \quad \text{and}$$

$$< u, Lu >= \int_a^b \overline{u} Lu$$

$$= \int_a^b \overline{u} w \lambda u$$

$$= \lambda < u, u >_w .$$

Since the problem is self-adjoint, these two are equal and we subtract them

$$0 = (\overline{\lambda} - \lambda) < u, u >_w .$$

Because u is an eigenvector, $< u, u >_w \neq 0$ and $\overline{\lambda} - \lambda = 0$.

In order to prove the second item, let u and v be eigenvectors with different real eigenvalues $Lu = \lambda u$ and $Lv = \widehat{\lambda} v$. Form the inner products:

$$< Lu, v >= \int_a^b \overline{(Lu)} v$$

$$= \int_a^b \lambda w \overline{u} v$$

$$= \lambda < u, v >_w \quad \text{and}$$

$$< u, Lv >= \int_a^b \overline{u} Lv$$

$$= \int_a^b \overline{u} w \widehat{\lambda} v$$

$$= \widehat{\lambda} < u, v >_w .$$

The self-adjoint property implies

$$0 = (\lambda - \widehat{\lambda}) < u, v >_w.$$

Since the eigenvalues are different, the other factor must be zero.

The last item requires a real eigenvector. Since the eigenvalues are real, both the eigenvector and its conjugate are eigenvectors with the same eigenvalue

$$Lu = \lambda u \text{ and } \overline{Lu} = L\overline{u} = \lambda \overline{u}.$$

Now $u + \overline{u}$ is real and $L(u + \overline{u}) = Lu + L\overline{u} = \lambda u + \lambda \overline{u} = \lambda(u + \overline{u})$. In order to generate the orthonormal eigenvectors replace the real eigenvectors by

$$u/\sqrt{< u, u >_w}.$$

■

10.4.3 Exercises

1. Confirm the two linear conditions for the examples in Examples 10.4.1.

2. Let A be an $n \times n$ complex matrix and $A = A^*$.

   (a). Show the eigenvalues must be real numbers.

   (b). Show the eigenvectors with different eigenvalues must be orthogonal.

3. Let A be an $n \times n$ complex matrix. Develop a singular value decomposition.

4. Do the two integration by parts to confirm line (10.4.2).

5. Consider Examples 5 and 6 in Examples 10.4.2. Explain why the solutions to $-u_{xx} + cu = 0$ and $-u_{xx} + cu = 0$ are so different.

6. Consider the Sturm-Liouville and verify the self-adjoint condition when $a_2 \neq 0$ and $b_2 \neq 0$.

10.5 Self-Adjoint Positive Definite BVP

This section is the generalization of SPD matrices or *hermitian* matrices (A is an $n \times n$ complex matrix and $A = A^*$). In these cases positive definite means $x^* A x > 0$ and the eigenvalues must be real and positive.

As motivation consider a cooling fin which is attached to a hot mass and other part of the fin is adjacent to a cooler region. The heat diffused from the hot to cooler regions, and this modeled by the Fourier heat law. The model is described in Subsection 1.2.3 and is depicted in Figure 1.2.1. Heat can flow in three possible directions. For certain shapes the flow is dominated by just one direction, and the model simplifies to a steady state boundary value problem.

The *continuous* model is given by the following differential equation and two boundary conditions

$$-(Ku_x)_x + Cu = f, \tag{10.5.1}$$
$$u(0) = u_0 > u_{sur} \text{ and} \tag{10.5.2}$$
$$Ku_x(L) = c(u_{sur} - u(L)). \tag{10.5.3}$$

The boundary condition in (10.5.3) is often called a *derivative or flux or Robin* boundary condition. If $c = 0$, then no heat is allowed to pass through the right boundary. If c approaches infinity and the derivative remains bounded, then (10.5.3) implies $u_{sur} = u(L)$.

The above problem has $C, c \geq 0$ and K and C could be functions of x. If u is replaced by $u - u_0$, then this is becomes a self-adjoint boundary value problem with $p(x) = K$ and $q(x) = C \geq 0$

$$-(pu_x)_x + qu = f \tag{10.5.4}$$
$$u(0) = 0 \text{ and} \tag{10.5.5}$$
$$p(1)u_x(1) + cu(1) = g. \tag{10.5.6}$$

The homogeneous adjoint problem is

$$-(pv_x)_x + qv = 0$$
$$v(0) = 0 \text{ and}$$
$$p(1)v_x(1) + cv(1) = 0.$$

The only solution is the zero function. In order to see this, do one integration by parts in

$$< Lv, v >= -v(1)p(1)v_x(1) + v(0)p(0)v_x(0) +$$
$$+ \int_0^1 p(v_x)^2 + qv^2$$
$$= c(v(1))^2 + \int_0^1 p(v_x)^2 + qv^2 > 0.$$

Since $C, c \geq 0$ and $p > 0$, $v_x = 0$. Because $v(0) = 0$, $v = 0$. This means any nonhomogeneous solution will be a unique solution and any eigenvalue must be positive.

Up to this point the existence of solutions and the nature of the solution function has not been given much attention. Consider the cooling fin with variable materials with different thermal conductivity. If $K(x) = p(x)$ has a jump discontinuity, then the temperature will have a discontinuous derivative $u_x(x_0)$ because of required heat balance $K(x_0-)u_x(x_0-) = K(x_0+)u_x(x_0+)$. Multiply the equation in line (10.5.4) by any "suitable" $v(x)$ and integrate by parts to get

$$-v(1)K(1)u_x(1) + v(0)K(0)u_x(0)+$$

$$+ \int_0^1 Ku_xv_x + Cuv = \int_0^1 fv$$

$$a(u,v) \equiv cu(1)v(1)) + \int_0^1 Ku_xv_x + Cuv = \int_0^1 fv + cgv(1). \qquad (10.5.7)$$

This equation does not require u_{xx} to be continuous or even to exist, but it must hold for all "suitable" functions $v(x)$. The equation in line (10.5.7) is the *weak formulation* of the equations in the boundary value problem (10.5.4-10.5.6). The example at the end of this section will illustrate a weak solution and indicate how solutions to weak equations can be approximated.

The fin problem is analogous to the algebraic system $Ax = d$ where A is symmetric positive definite:

$$x^T Ax > 0 \text{ for } x \neq 0 \text{ and eigenvalues are real and positive.}$$

The solution of $Ax = d$ can be found by using the orthonormal basis of eigenvectors, $Au_i = \lambda_i u_i$

$$d = \sum_{i=1}^{i=n} < u_i, d > u_i$$

$$x = \sum_{i=1}^{i=n} a_i u_i \text{ where } a_i \lambda_i = < u_i, d > .$$

Because $\lambda_i > 0$, one can solve for all the a_i. The summation is finite, but this is not the case for function space example.

Consider the self-adjoint positive definite boundary value problem with real positive eigenvalues and orthonormal eigenvectors (functions). There are an infinite number of eigenvalues. First, find b_i with $1 \leq i \leq K$ so that f is "best" approximated by

$$f \sim \sum_{i=1}^{i=K} b_i u_i .$$

Use the norm associated with $<,>_w$

$$F(b) = < f - \sum_{i=1}^{i=K} b_i u_i, f - \sum_{i=1}^{i=K} b_i u_i >_w .$$

Use the orthonormal property to get

$$F(b) = < f, f >_w - 2\sum_{i=1}^{i=K} b_i < u_i, f >_w + \sum_{i=1}^{i=K} b_i^2.$$

Set the partial derivatives of $F(b)$ equal to zero to find $b_i = < u_i, f >_w$. For this choice of $b_i = < u_i, f >_w$ the *Bessel's inequality* holds

$$< f, f >_w - \sum_{i=1}^{i=K} b_i^2 \geq 0.$$

The solution of $Lu = f$ with the homogeneous boundary conditions is approximated by

$$\sum_{i=1}^{i=K} a_i u_i \text{ where } a_i \lambda_i = <u_i, f>_w$$

The choice of which eigenvalues is best if the largest values are used. The choice of the number K is also of consideration. The nature of convergence depends on the properties of the function f. See volume one in the classic by Courant and Hilbert, [6, Chapter 6].

This analysis can be extended to partial differential equation of similar form

$$Lu = -(pu_x)_x - (pu_y)_y + qu = f \text{ on } \Omega \subset \mathbb{R}^2 \qquad (10.5.8)$$
$$u = 0 \text{ on } \partial\Omega. \qquad (10.5.9)$$

The self-adjoint property follows from the divergence form of Green's theorem acting in place of integration by parts. In Green's theorem let $F_1 = Q$ and $F_2 = -P$ and n be the outward unit normal to the boundary

$$\int\int_\Omega \nabla \cdot F = \int_{\partial\Omega} F \cdot n.$$

Choose $F = up\nabla v$ and $F = vp\nabla u$ and note

$$\nabla \cdot up\nabla v = u\nabla \cdot p\nabla v + \nabla u \cdot p\nabla v \text{ and}$$
$$\nabla \cdot vp\nabla u = v\nabla \cdot p\nabla u + \nabla v \cdot p\nabla u.$$

Put these into Green's theorem to conclude

$$\int\int_\Omega u\nabla \cdot p\nabla v - v\nabla \cdot p\nabla u = \int_{\partial\Omega} (up\nabla v - vp\nabla u) \cdot n = 0.$$

Insert quv to get the self-adjoint property

$$\int\int_\Omega u\nabla \cdot p\nabla v - quv - v\nabla \cdot p\nabla u + qvu = 0$$
$$\int\int_\Omega -u(Lv) + (Lu)v = 0.$$

The positive definite property is obtained by applying the divergence theorem to $F = vp\nabla u$

$$\int\int_\Omega v\nabla \cdot p\nabla u + \nabla v \cdot p\nabla u = \int_{\partial\Omega} vp\nabla u \cdot n = 0. \qquad (10.5.10)$$

Note for $v = u$

$$\int\int_\Omega u\nabla \cdot p\nabla u + \nabla u \cdot p\nabla u = \int_{\partial\Omega} up\nabla u \cdot n = 0. \qquad (10.5.11)$$

Use $Lu = -\nabla \cdot p\nabla u + qu = f$ and $\nabla \cdot p\nabla u = qu - Lu$ in the above

$$\int \int_{\Omega} u(qu - Lu) + \nabla u \cdot p\nabla u = 0.$$

This may be rewritten as

$$\int \int_{\Omega} qu^2 + \nabla u \cdot p\nabla u = \int \int_{\Omega} u(Lu)$$

$$\int \int_{\Omega} qu^2 + pu_x^2 + pu_y^2 = \int \int_{\Omega} uf.$$

If $f = 0$, then the right side must be zero. If $q \geq 0$ and $p > 0$, then the solution of (10.5.8-10.5.9) is the zero function. Also, any solution must be unique. So, define the solution of $Lu = f$ to be the mapping of f to $u \equiv Gf$. This can be represented by an integral operator using Green's functions. Since L is linear, the solution operator must also be linear.

In order to be more precise, further assume $q(x) \geq q_0 > 0$ and $p(x) \geq p_0 > 0$, Then there exists $c_0 > 0$ such that

$$\int \int_{\Omega} uf = \int \int_{\Omega} qu^2 + pu_x^2 + pu_y^2 \geq c_0 \int \int_{\Omega} u^2 + u_x^2 + u_y^2.$$

Let $\|*\|$ and $\|*\|_1$ be the norms associated with the inner products

$$< f, g > \equiv \int \int_{\Omega} fg \text{ and } < f, g >_1 \equiv \int \int_{\Omega} fg + f_x g_x + f_y g_y.$$

Apply the Cauchy inequality to the left side to get

$$\|u\| \, \|f\| \geq c_0 \|u\|_1^2.$$

Since $\|u\| \leq \|u\|_1$,

$$\|u\|_1 \, \|f\| \geq c_0 \|u\|_1^2 \text{ and } \|u\|_1 \leq (1/c_0) \|f\|.$$

This means $\|G(f)\|_1 \leq (1/c_0) \|f\|$. If f is continuous and the solution has two continuous derivatives, then $G : (C^0, \|*\|) \to (C^2, \|*\|_1)$ is a bounded linear operator and must be continuous. So, when $f^k \to f$ with respect to the norm $\|*\|$, then $Gf^k \to Gf$ using the norm $\|*\|_1$

$$\|G(f) - G(f^k)\|_1 = \|G(f - f^k)\|_1 \leq (1/c_0) \|f - f^k\|.$$

The *weak* formulation as in line (10.5.7) can be used in similar problems. Consider the problem in lines (10.5.8-10.5.9). The equation in line (10.5.10) with $\nabla \cdot p\nabla u = qu - Lu$ gives for all "suitable" v

$$\int \int_{\Omega} \nabla v \cdot p\nabla u + quv = \int \int_{\Omega} fv \text{ and}$$

$$a(u, v) \equiv \int \int_{\Omega} v_x pu_x + v_y pu_y + quv = \int \int_{\Omega} fv. \tag{10.5.12}$$

The left side does not require second order derivatives, and the equation (10.5.12) is the *weak* formulation of the original boundary value problem in (10.5.8-10.5.9).

The above boundary value problems have *weak* equations as given in lines (10.5.7) and (10.5.12). Another special case is the SPD matrix where $Au = d$ gives $a(u,v) \equiv v^T Au = v^T d = l(v)$. The bounded property on $a(u,v)$ follows from

$$\left| v^T Au \right| \leq \|v\| \|Au\| \leq \|v\| \|A\| \|u\|.$$

The coercive property follows from

$$\min_{u \neq 0} u^T Au = \min_{\|u\|=1} u^T Au = u_0^T Au_0 > 0 \text{ where}$$

the minimum of the continuous function on the closed bounded set is attained at u_0. The *weak* equation $a(u,v) = l(v)$ has the form $v^T Au = v^T d$ for all $v \in \mathbb{R}^n$. Choose $v = e_i$ the unit direction vectors. Thus, $Au = d$. By Theorem 2.4.4 $Au = d$ is equivalent to $J(u) = \min_{v \in H} J(v)$ with $H = \mathbb{R}^n$ and $J(v) \equiv \frac{1}{2} v^T Av - v^T d$.

These are a special cases of the following *weak equation for all* $v \in V = H$

$$a(u,v) = l(v) \tag{10.5.13}$$

where $a : V \times V \to \mathbb{R}$ is linear in each variable, $l : V \to \mathbb{R}$ is linear and

$$|a(u,v)| \leq c_1 \|u\| \|v\| \quad \text{(bounded)} \tag{10.5.14}$$
$$a(u,v) = a(v,u) \quad \text{(symmetric)} \tag{10.5.15}$$
$$a(u,u) \geq c_0 \|u\|^2 \quad \text{(coercive)} \tag{10.5.16}$$
$$|l(u)| \leq c \|u\| \quad \text{(bounded)}. \tag{10.5.17}$$

In either case the *weak equation* can be shown to be equivalent the minimizing this *"energy" function*

$$J(u) = \min_{v \in H} J(v) \text{ where } J(v) \equiv \frac{1}{2} a(v,v) - l(v). \tag{10.5.18}$$

What is H? $H = V$ is a special inner product vector space, called a *Hilbert space*, with the property every Cauchy sequence will converge:

$$u_k \to u \text{ if and only if } \|u_k - u_m\| < \epsilon \text{ for all } k, m \geq N(\epsilon).$$

Examples include $\mathbb{R}^n, \mathbb{C}^n, L_2$ (the Lebesque measurable functions with $\int u^2 < \infty$) and the Sobolev spaces H^m (with all $k \leq m$ weak derivatives in L_2). This requires further study and should clarify the use of "suitable" function.

The existence of a weak solution can be established by finding a convergent sequence $u_k \to u$ such that

$$J(u) = \min_{v \in H} J(v) \downarrow J(u_k).$$

This is the Lax-Milgram theorem which was established in 1954, see [13]. It was generalized in 1967 by Lions-Stampacchia, see [15] and in 1971 by Babuška, see [1]. Additional constraints on the solution require variational inequalities and finite element approximation; consult the text by Glowinski, [8].

Theorem 10.5.1. *(Lax-Milgram Weak Existence) Consider the weak equation (10.5.13) and the minimum problem (10.5.18). Assume the above (10.5.14-10.5.17) linear properties, the bounded, symmetric and coercive hold.*

(i). The weak equation and the minimum problem are equivalent.

(ii). When a weak solution exists, there is only one solution.

(iii). When $V = H$ is a Hilbert space, there is a solution to (10.5.18).

Proof. The linear conditions and symmetric condition give

$$f(t) \equiv J(u + tv) = J(u) + t(a(u, v) - l(v)) + (1/2)t^2 a(v, v).$$

This is a quadratic function with $f(0) = J(u)$, $f'(0) = a(u, v) - l(v)$ and $f''(0) = a(v, v) > 0$. If (10.5.18) holds, then the minimum of $f(t)$ is attained at $t = 0$ and so $f'(0)$ must be zero. If (10.5.13) holds, then $f(t) = J(u) + (1/2)t^2 a(v, v) \geq J(u) = f(0)$.

The proof of unique solution follows from the coercive condition. Let u and $\widehat{u}$ be two weak solutions

$$a(u, v) = l(v) \text{ and}$$
$$a(\widehat{u}, v) = l(v).$$

Subtract and let $v = u - \widehat{u}$

$$c_0 \|u - \widehat{u}\|^2 \leq a(u - \widehat{u}, u - \widehat{u}) = 0.$$

The first step in the existence is to show $J(v)$ is bounded from below.

$$
\begin{aligned}
J(v) &= \frac{1}{2}a(v, v) - l(v) \\
&\geq \frac{1}{2}c_0 \|v\|^2 - c\|v\| \\
&= \frac{1}{2}c_0(\|v\| - (\frac{c}{c_0}))^2 - \frac{c_0}{2}(\frac{c}{c_0})^2 \\
&\geq -\frac{c_0}{2}(\frac{c}{c_0})^2.
\end{aligned}
$$

The second step is to show the sequence given by $J(u_k) \downarrow d$ is Cauchy and must converge. Use the linear and symmetric properties to derive

$$a(\frac{u_n - u_m}{2}, \frac{u_n - u_m}{2}) = \frac{1}{2}a(u_m, u_m) + \frac{1}{2}a(u_n, u_n)$$
$$-a(\frac{u_n + u_m}{2}, \frac{u_n + u_m}{2})$$
$$= J(u_m) + J(u_n) + l(u_m) + l(u_n)$$
$$-a(\frac{u_n + u_m}{2}, \frac{u_n + u_m}{2})$$
$$= J(u_m) + J(u_n) - 2J(\frac{u_n + u_m}{2}).$$

Use the coercive assumption and the convergence of $J(u_k) \downarrow d$ to get the inequality

$$c_0 \left\| \frac{u_n - u_m}{2} \right\|^2 \le a(\frac{u_n - u_m}{2}, \frac{u_n - u_m}{2}) \le (d + \epsilon) + (d + \epsilon) - 2d.$$

Because we can choose $\epsilon > 0$ as small as we wish, the sequence is Cauchy

$$\|u_n - u_m\|^2 \le 2\epsilon \frac{4}{c_0}.$$

The last step is the show when $u_n \to u$, then $J(u_n) \downarrow J(u) = d$.

$$J(u_n) - J(u) = \frac{1}{2}(a(u_n, u_n) - a(u, u)) - l(u_n) + l(u).$$

Use the linear and bounded properties of $a(u, v)$

$$|a(u_n, u_n) - a(u, u)| = |a(u_n - u, u_n) + a(u, u_n) - a(u, u)|$$
$$\le |a(u_n - u, u_n)| + |a(u, u_n - u)|$$
$$\le c_1 \|u_n - u\| \|u_n\| + c_1 \|u\| \|u_n - u\|.$$

Finally, use the linear and bounded properties of $l(v)$

$$|l(u) - l(u_n)| = |l(u - u_n)| \le c \|u - u_n\|.$$

Since $\|u_n - u\| \to 0$, $J(u_n) - J(u) \to 0$ and $J(u_n) \downarrow d$, we must have $J(u) = d$. ∎

Example 10.5.1. Consider the ordinary differential equation with discontinuous $K(x)$

$$-(K(x)u_x)_x = 1 \text{ and } u(0) = 0 = u(1).$$

Let $K(x) = 1$ for $0 \le x \le 1/2$ and $K(x) = 2$ for $1/2 < x \le 1$. A solution where $u(x)$ and $K(x)u_x(x)$ are continuous at $x = 1/2$ is

$$u(x) = \begin{cases} -\frac{1}{2}x^2 + \frac{5}{12}x & ,0 \le x \le 1/2 \\ -\frac{1}{4}x^2 + \frac{5}{24}x + \frac{1}{24} & ,1/2 < x \le 1. \end{cases}$$

The weak solution requires $a(u,v) = l(v)$ for all "suitable" $v(x)$ with $v(0) = 0 = v(1)$.

$$a(u,v) = \int_0^1 K(x)u_x(x)v_x(x)dx$$

$$= \int_0^{1/2} 1(-x + \frac{5}{12})v_x(x)dx + \int_{1/2}^1 2(-\frac{1}{2}x + \frac{5}{24})v_x(x)dx$$

$$= \int_0^1 (-x + \frac{5}{12})v_x(x)dx$$

$$= (-x + \frac{5}{12})v(x)|_{x=0}^{x=1} - \int_0^1 v(x)(-dx)$$

$$= 0 + \int_0^1 1v(x)dx = l(v).$$

The solution to the weak equation can be approximated by using given "shape" functions, $\phi_j(x)$. An example is a piece-wise linear function

$$\phi_j(x) \equiv \begin{cases} 0 & , x < x_j - h \\ (x - (x_j - h))/h & , x_j - h \le x < x_j \\ -(x - (x_j + h))/h & , x_j \le x < x_j + h \\ 0 & , x_j + h \le x. \end{cases}$$

In order to approximate the weak solution, use a linear combination of $\phi_j(x)$

$$u(x) \approx \sum_{j=1}^{j=n-1} u_j\phi_j(x).$$

Let $v(x)$ be any $\phi_i(x)$ and put these into the weak equation

$$a(\sum_{j=1}^{j=n-1} u_j\phi_j(x), \phi_i(x)) = l(\phi_i(x)).$$

Use the linear property of $a(u,v)$, evaluate $a(u,v)$ and $l(v)$ at the "shape" functions to get

$$\sum_{j=1}^{j=n-1} a(\phi_j(x), \phi_i(x))u_j = l(\phi_i(x)) \qquad (10.5.19)$$

The algebraic system may be solved to give an approximation of the weak solution. This is called the *finite element method*, see [28, Sections 5.1 and 5.2] and [8, Appendix I].

10.5.1 Exercises

1. Show the solution map G must be linear where Gf is the unique solution of the linear boundary value problem $L(u) = f$.

2. Use the Lax-Milgram analysis applied to the BVP

$$Lu = -(pu_x)_x - (pu_y)_y + qu = f \text{ on } \Omega \subset \mathbb{R}^2,$$
$$u = 0 \text{ on } \partial\Omega_0 \text{ and}$$
$$p\nabla u \cdot n + cu = g \text{ on } \partial\Omega \backslash \partial\Omega_0.$$

3. Consider Example 10.5.1 and the finite element method with the piece-wise linear functions. Use $n = 4$ and $h = (1-0)/n$.

(a). Find the algebraic system in line (10.5.19).

(b). Solve it and compare with the weak solution.

11

Iterative Methods

Direct methods such as the many variations of Gauss elimination possibly require large numbers of computations and storage. For example, in partial differential equations for models with three dimensions there may be cubes with 100 partitions in each direction. This gives $n = 10^6$ unknowns and storage for the matrix equal to $n^2 = 10^{12}$. There are three important alternatives to avoid this difficulty. They involve approximating the matrix by splittings, using conjugate directions and minimizing the residual.

11.1   Inverse Matrix Approximations

The first inverse matrix computation is a generalization of the geometric series to matrices. The following result is a continuation of Theorem 10.2.2 and is known as the Neumann lemma. This is illustrated in the MATLAB® code matit_13.m.

Theorem 11.1.1. *(Matrix Geometric Series) Let H be an $n \times n$ real matrix. If there is a matrix norm such that $\|H\| < 1$, then*

(i). $(I - H)^{-1}$ *exists,*

(ii). $I + H + \cdots + H^m$ *converges to* $(I - H)^{-1}$ *with*

$$\|(I - H)^{-1}\| \leq 1/(1 - \|H\|) \ and$$

(iii). H^m *converges to* $0_{n \times n}$.

Proof. If there exists nonzero x such that $(I - H)(x) = 0_{n \times 1}$, then x is an eigenvector with eigenvalue equal to 1. This is a contradiction to $\|x\| = \|Hx\| \leq \|H\| \|x\|$ and $1 \leq \|H\|$ as x is a nonzero vector.

The second item follows from the identity

$$(I - H)(I + H + \cdots + H^m) = I - H^{m+1}.$$

Since $(I - H)$ has an inverse,

$$(I + H + \cdots + H^m) - (I - H)^{-1} = -(I - H)^{-1}H^{m+1}.$$

DOI: 10.1201/9781003304128-11

The norm of the right side is bounded by

$$\left\|(I-H)^{-1}\right\|\left\|H^{m+1}\right\| \leq \left\|(I-H)^{-1}\right\|\|H\|^{m+1}.$$

Since $\|H\| < 1$, the convergence is established. Note

$$\|I + H + \cdots + H^m\| \leq 1 + \|H\| + \cdots + \|H\|^m \leq \frac{1}{1-\|H\|}$$

The last item follow from the above convergence and

$$(I + H + \cdots + H^{m+1}) - (I + H + \cdots + H^m) = H^{m+1}.$$

$\blacksquare$

The second inverse matrix approximation gives a condition on a matrix C so that it will be "near" a nonsingular matrix A and C will also be nonsingular. It is useful for matrices that depend on a parameter $A(x)$, and this will be used in the next chapter. This is often called the perturbation lemma.

Theorem 11.1.2. *(Perturbation of Nonsingular Matrix) Let A and C be $n \times n$ real matrices and assume A has an inverse.*

If $\|A - C\|\,\|A^{-1}\| < 1$, then C has an inverse and

$$\|C^{-1}\| \leq \frac{\|A^{-1}\|}{1 - \|A - C\|\,\|A^{-1}\|}.$$

Proof. Consider $A^{-1}(A - C) = I - A^{-1}C$ and $A^{-1}C = I - (I - A^{-1}C)$. Use the above theorem with H replaced by $I - A^{-1}C$. The assumption $\|A - C\|\,\|A^{-1}\| < 1$ implies $\|H\| < 1$. Therefore, $A^{-1}C = I - H$ has an inverse, $C = A(I - H)$ has an inverse, and

$$C^{-1} = (I - H)^{-1}A^{-1}.$$

Use $\left\|(I - H)^{-1}\right\| \leq 1/(1 - \|H\|)$ and the assumption $\|A - C\|\,\|A^{-1}\| < 1$ to get the desired inequality. $\blacksquare$

The third inverse matrix approximation gives a condition on a matrix B so that it will be "near" A^{-1}. This is useful for matrix splitting methods as described in the next section. In case of the splitting $A = M - N$ use $B = M^{-1}$ so that $I - BA = M^{-1}N$. This is often called the Banach lemma.

Theorem 11.1.3. *(Inverse Matrix Approximations) Let A and B be $n \times n$ real matrices.*

If $\|I - BA\| < 1$, then B has an inverse and

$$\|B^{-1}\| \leq \frac{\|A\|}{1 - \|I - BA\|} \quad and$$

$$\|B^{-1} - A\| \leq \frac{\|A\|}{1 - \|I - BA\|}\|I - BA\|.$$

Proof. Again use Theorem 11.1.1 with $H = I - BA$ and $BA = I - (I - BA)$. Then BA is nonsingular, and consequently both A and B are nonsingular. Moreover,

$$A = B^{-1}(I - H) \text{ and } B^{-1} = A(I - H)^{-1}$$

$$\|B^{-1}\| \le \|A\| \frac{1}{1 - \|H\|} = \frac{\|A\|}{1 - \|I - BA\|}.$$

The second inequality follows from this and $B^{-1} - A = B^{-1}(I - BA)$. ∎

11.1.1 Exercises

1. Consider the matrix the MATLAB® code matit_13.m.

   Execute the code for the two options. Observe convergence and A^k (A^k).

2. Consider the proof of Theorem 11.1.3. Explain why both B and A are nonsingular.

11.2 Regular Splittings for M-Matrices

In this section we assume A and B are nonsingular $n \times n$ real matrices. If the *splitting* $A = B - C$, then $Ax = d$ can be written as

$$Bx = Cx + d \text{ or}$$
$$x = B^{-1}Cx + B^{-1}d \text{ or}$$
$$x = x + B^{-1}(d - Ax) = x + B^{-1}r(x).$$

Splitting iterative methods have the form

$$x^{m+1} = B^{-1}Cx^m + B^{-1}d = x^m + B^{-1}r(x^m).$$

Matrix Splitting Method.

> Let x^0 be an initial guess
> for m = 0, maxm
> > solve $Bx^{m+1} = Cx^m + d$
> > test for convergence
> endloop.

Iterative methods require testing for possible convergence. If one knows the solution u and the iteration, u^m, then simply compute the error $u - u^m$. But, the reason for using an iterative method is to find the solution and so the error cannot be computed. The following are several attempts is judge "convergence":

$$\left\| u^{m+1} - u^m \right\| < \epsilon_1,$$
$$\frac{\left\| u^{m+1} - u^m \right\|}{\left\| u^m \right\|} < \epsilon_2 \text{ (relative error) and}$$
$$\left\| d - Au^m \right\| < \epsilon_3 \text{ (residual error)}.$$

This is all complicated by the choice of norm, ϵ and ill-conditioned problems. Also, the sequence differences may go to zero and still not converge. Recall the harmonic series from the partial sums $S_m = 1 + \frac{1}{2} + \cdots + \frac{1}{m}$ where $S_{m+1} - S_m$ goes to zero and S_m goes to infinity.

Example 11.2.1.

1. Richard. $A = I - (I - A)$ gives $B^{-1}C = I - A$.

2. Jacobi. $A = D - (L + U)$ where D is the diagonal of A, and $-L, -U$ are the strictly lower and upper parts of A. In this case

$$B^{-1}C = D^{-1}(L + U).$$

3. Gauss-Seidel. $A = (D - L) - U$ and $B^{-1}C = (D - L)^{-1}U$.

4. SOR (successive overrelaxation).

$$A = \frac{1}{\omega}(D - \omega L) - \frac{1}{\omega}((1 - \omega)D + \omega U) \text{ with } 1 < \omega < 2.$$

There is a long history of iterative methods starting with Gauss in 1823, Jacobi in 1835 and Seidel in 1846. With the introduction of electronic computation in the middle of 1900s, these methods could be used for larger systems. D. M. Young in 1950, [29], introduced current methods for the analysis of convergence. The books by R. S. Varga, [24], and A. Berman and R. J. Plemmons, [4], give additional analysis of splitting methods.

Iterative methods are usually applied to sparse matrices and only the nonzero components are used. The above matrix descriptions are used in the analysis of convergence. A second description is in terms of the components $A = [a_{ij}]$. The component description of SOR is

$$a_{ii}x_i^{m+1/2} = d_i - \sum_{j<i} a_{ij}x_j^{m+1} - \sum_{j>i} a_{ij}x_j^m$$
$$x_i^{m+1} = (1 - \omega)x_i^m + \omega x_i^{m+1/2}.$$

For sparse matrices most of the terms in the lower and upper sums are zero. A third description comes when a particular computer implementation of the method is done. This involves updating the iteration, stopping criteria and output of the computations. See the MATLAB® code sor3d_13.m.

The convergence is based on Theorem 10.2.2 and Theorem 11.1.1. The following important theorem characterizes convergence properties when $B^{-1}C$ is used in place of A.

Theorem 11.2.1. *(Convergence Chacterizations)* Let A be an $n \times n$ real matrix. The following are equivalent:

   (i). $\rho(A) < 1$.

   (ii). *there exist an norm such that* $\|A\| < 1$,

   (iii). A^m *converges to* $0_{n \times n}$ *and*

   (iv). *for all initial* x^0 *and all* d

$$x^{m+1} = Ax^m + d \text{ converges to } x = Ax + d.$$

Proof. The first and second items have been proven to be equivalent. Theorem 11.1.1 shows the second item implies the third item. Show item four follows from the third item. Use this identity

$$(I + A + \cdots + A^m)(I - A) = I - A^{m+1}$$

to establish the existence of an inverse to $I - A$. This follows from $(I - A)(x) = 0_{n \times 1}$ implies

$$0_{n \times 1} = (I - A^{m+1})(x) = x - A^{m+1}x.$$

Item three gives $x = 0_{n \times 1}$ and $I - A$ has an inverse. Since $I - A$ has an inverse and there is a solution to $x = Ax + d$. Subtract $x^{m+1} = Ax^m + d$ and $x = Ax + d$ to get

$$\begin{aligned}
x^{m+1} - x &= A(x^m - x) \\
&= A(A(x^{m-1} - x)) \\
&\;\;\vdots \\
&= A^{m+1}(x^0 - x).
\end{aligned}$$

Because A^m converges to $0_{n \times n}$, $x^{m+1} - x$ must go to $0_{1 \times n}$. In order to show the first item is a consequence of item four, choose $x^0 = z + x$ where z is the eigenvector with $Az = \lambda z$ and $|\lambda| = \rho(A)$.

$$x^{m+1} - x = A^{m+1}(x^0 - x) = \lambda^{m+1}z.$$

Since z must have at least one nonzero component, $|\lambda| < 1$. ∎

Definition. $A = [a_{ij}]$ *is called strictly diagonally dominant if and only if for all* i

$$|a_{ii}| > \sum_{j \neq i} |a_{ij}|.$$

Theorem 11.2.2. *(Convergent Methods) If A is strictly diagonally dominant matrix, then*

(i). *A is nonsingular,*

(ii). *$\left\|D^{-1}(L + U)\right\|_\infty < 1$ and the Jacobi method converges and*

(iii). *$\rho((D - L)^{-1}U) < 1$ and the Gauss-Seidel method converges.*

Proof. Suppose there is a nonzero vector x with $Ax = 0_{n \times 1}$, and let i be such that $max_j |x_j| = |x_i|$. $Ax = 0_{n \times 1}$ implies $a_{ii}x_i + \sum_{j \neq i} a_{ij}x_j = 0$. This gives a contradiction

$$|a_{ii}x_i| \leq \sum_{j \neq i} |a_{ij}x_j|$$

$$|a_{ii}| \leq \sum_{j \neq i} |a_{ij}| \, |x_j| \, / \, |x_i| \leq \sum_{j \neq i} |a_{ij}| \, 1.$$

Use the component form of the Jacobi splitting

$$x_i^{m+1} = (d_i - \sum_{j \neq i} a_{ij}x_j^m)/a_{ii}$$

The off diagonal components are a_{ij}/a_{ii} and each row sum $\sum_{j \neq i} |a_{ij}| \, / \, |a_{ii}|$ must be less than one. So, $\left\|D^{-1}(L + U)\right\|_\infty < 1$.

Let z be the eigenvector of $(D - L)^{-1}U$ such that $(D - L)^{-1}Uz = \lambda z$, $|\lambda| = \rho((D-L)^{-1}U)$ and choose i with $max_j |z_j| = |z_i|$. Then $Uz = \lambda(D-L)z$ and

$$[Uz]_i = \lambda\left[(D - L)z\right]_i$$

Use $|\ |x| - |y|\ | \leq |x - y|$ and the above

$$|\lambda| \leq \frac{|[Uz]_i|}{|\ |[Dz]_i| - |[Lz]_i|\ |}.$$

Use the strict diagonal dominance to conclude

$$|[Dz]_i| = |a_{ii}| \, |z_i| > \sum_{j < i} |a_{ij}| \, |z_j| + \sum_{j > i} |a_{ij}| \, |z_j|$$

$$|[Dz]_i| - |[Lz]_i| > |[Uz]_i| \text{ and}$$

$$|\lambda| \leq \frac{|[Uz]_i|}{|\ |[Dz]_i| - |[Lz]_i|\ |} < 1.$$

■

Definition. *A splitting $A = B - C$ is called regular if and only if $B^{-1} \geq 0_{n \times n}$ and $C \geq 0_{n \times n}$.*

Recall, an $n \times n$ real M-matrix, A, is defined by requiring off diagonal components to be nonpositive and $A^{-1} \geq 0_{n \times n}$. The goal is to show any regular splitting of an nonsingular M-matrix must be convergent, that is, $\rho(B^{-1}C) < 1$ and, consequently, the iteration $x^{m+1} = B^{-1}Cx^m + B^{-1}d$ will converge to the solution of $Ax = d$. The next theorem is the first step where the matrix H could be $B^{-1}C$.

Theorem 11.2.3. *(Spectral Radius of $H \geq 0$) Let $H \geq 0_{n \times n}$. $(I - H)^{-1} \geq 0_{n \times n}$ if and only if $\rho(H) < 1$.*

Proof. If $\rho(H) < 1$, then for some norm $\|H\| < 1$ and $(I - H)^{-1}$ exists and is limit of $I + H + \cdots + H^m$. Since $H \geq 0_{n \times n}$ each term in the partial sum is nonnegative and $(I - H)^{-1} \geq 0_{n \times n}$.

Let $(I - H)^{-1} \geq 0_{n \times n}$ and $Hz = \lambda z$ where $|\lambda| = \rho(H)$. Since $H \geq 0_{n \times n}$, $H|z| = |\lambda||z|$ and $(I - H)|z| = (1 - |\lambda|)|z|$. Because $(I - H)^{-1} \geq 0_{n \times n}$, $(I - H)^{-1}|z| \geq 0_{n \times n}$ and

$$|z| = (1 - |\lambda|)(I - H)^{-1}|z|.$$

As $|z|$ must have at least one nonzero positive component, this requires $(1 - |\lambda|)$ to be positive. ∎

Theorem 11.2.4. *(M-matrix Chacterization) Let A have nonpositive off diagonal components and let the diagonal of A be D. A is an M-matrix if and only if $\rho(I - D^{-1}A) < 1$ and D has positive diagonal components.*

Proof. Assume A is an M-matrix. If any diagonal component $a_{ii} \leq 0$, then column i has all components ≤ 0. Since $A^{-1} \geq 0_{n \times n}$, A^{-1}(column i)$\leq 0_{n \times 1}$. This contradicts $e_i = A^{-1}$(column i). Let $H = I - D^{-1}A$ and note $H \geq 0_{n \times n}$ and $I - H = D^{-1}A$ with

$$(I - H)^{-1} = A^{-1}D \geq 0_{n \times n}.$$

By the above theorem $\rho(H) < 1$.

Consider the converse and show A is an M-matrix. Because the diagonal components of D are positive,

$$H = I - D^{-1}A \geq 0_{n \times n}.$$

By the above theorem $(I - H)^{-1} \geq 0_{n \times n}$ where $I - H = D^{-1}A$. Then $A = D(I - H)$ has an inverse and

$$A^{-1} = (I - H)^{-1}D^{-1} \geq 0_{n \times n}.$$

∎

If $A = D - L - U$ with $L, U \leq 0_{n \times n}$, then $D^{-1}(L + U) = H$ and $I - H = D^{-1}A$. This means the Jacobi iteration will converge when A is an M-matrix. This is also true for any regular splitting.

Theorem 11.2.5. *(Convergent Regular Splitting)* *If A is an M-matrix and $A = B - C$ is a regular splitting, then $\rho(B^{-1}C) < 1$.*

Proof. Use Theorem 11.2.3 to show $\rho(B^{-1}C) < 1$ where $H = B^{-1}C$. Since this is a regular splitting, $B^{-1} \geq 0_{n \times n}$ and $C \geq 0_{n \times n}$ and then $H \geq 0_{n \times n}$. It remains to show $(I - H)^{-1} \geq 0_{n \times n}$. Note $B^{-1}A = I - H$ and

$$(I + H + \cdots + H^m)(I - H) = I - H^{m+1} \leq I$$
$$(I + H + \cdots + H^m)B^{-1}A \leq I.$$

Since $A^{-1} \geq 0_{n \times n}$,

$$(I + H + \cdots + H^m)B^{-1} \leq A^{-1}.$$

These three matrices have nonnegative components, and represent the matrix product by

$$\sum_j \alpha_{ij}^m \beta_{jl} \leq \gamma_{il}.$$

Each row of B^{-1} must have at least one positive component, say, $\beta_{jL} > 0$. Since the components are nonnegative,

$$\alpha_{ij}^m \beta_{jL} \leq \sum_j \alpha_{ij}^m \beta_{jL} \leq \gamma_{iL}.$$

This means $\alpha_{ij}^m \leq \gamma_{iL}/\beta_{jL}$ is an upper bound and independent of m. Thus, $I + H + \cdots + H^m$ must converge to $(I - H)^{-1}$. Since $H \geq 0_{n \times n}$, $(I - H)^{-1} \geq 0_{n \times n}$. ∎

11.2.1   Exercises

1. Consider the matrix $\begin{bmatrix} 2 & -1 & 0 \\ -1 & 2 & -1 \\ 0 & -1 & 2 \end{bmatrix}$.

   Execute the first few steps of the four splittings Richard, Jacobi, Gauss-Siedel and SOR.

2. SSOR. Do one forward SOR and then one backward SOR sweep. Let $A = D - L - U$ where $L^T = U$ and $0 < \omega < 2$.

$$\frac{1}{\omega}(D - \omega L)x^{m+\frac{1}{2}} = d + \frac{1}{\omega}((1 - \omega)D + \omega U)x^m$$
$$\frac{1}{\omega}(D - \omega U)x^{m+1} = d + \frac{1}{\omega}((1 - \omega)D + \omega L)x^{m+\frac{1}{2}}.$$

   Write this as a single splitting $A = M - N$ with $x^{m+1} = M^{-1}d + M^{-1}Nx^m$. Show

$$M = \frac{1}{\omega}(D - \omega L)(\frac{2 - \omega}{\omega}D)^{-1}\frac{1}{\omega}(D - \omega U).$$

11.3 P-Regular Splittings for SPD Matrices

The main goal in this section is to show the SOR method converges for SPD matrices. This analysis will require the matrix to be SPD and a special splitting.

Definition. *The splitting $A = B - C$ is called P-regular if and only if B^{-1} exists and $B^T + C$ is positive definite.*

Example 11.3.1. Let $A = A^T = D - L - U$ where $L^T = U$. The SOR splitting is

$$A = \frac{1}{\omega}(D - \omega L) - \frac{1}{\omega}((1 - \omega)D + \omega U).$$

Use $L^T = U$ and assume $0 < \omega < 2$.

$$\begin{aligned}
B^T + C &= (\frac{1}{\omega}(D - \omega L))^T + (\frac{1}{\omega}((1 - \omega)D + \omega U)) \\
&= \frac{1}{\omega}D - L^T + \frac{1}{\omega}(1 - \omega)D + U \\
&= \frac{2 - \omega}{\omega}D.
\end{aligned}$$

If A is SPD, then the diagonal components must be positive and $B^T + C$ is positive definite.

Theorem 11.3.1. *(SPD and Convergence) If A is SPD and $A - H^T AH$ is positive definite, then $\rho(H) < 1$.*

Proof. Let x be an eigenvector of H, $Hx = \lambda x$ and $(Hx)^* = \overline{\lambda}x^*$.

$$0 < x^*(A - H^T AH)x = x^* Ax(1 - |\lambda|^2).$$

The eigenvector may have real and imaginary parts $x = u + iv$. Since $A = A^T$,

$$x^* Ax = u^T Au + v^T Av.$$

Since A is SPD, $x^* Ax > 0$. The two positive terms force $1 - |\lambda|^2 > 0$. ∎

Theorem 11.3.2. *(SPD and P-regular) If A is SPD and $A = B - C$ is a P-regular splitting, then $\rho(B^{-1}C) < 1$.*

Proof. Let $H = B^{-1}C$ and use $H = I - B^{-1}A$.

$$\begin{aligned}
A - H^T AH &= A - (I - B^{-1}A)^T A(I - B^{-1}A) \\
&= (B^{-1}A)^T A - (B^{-1}A)^T A(B^{-1}A) + A(B^{-1}A) \\
&= (B^{-1}A)^T [A - A(B^{-1}A) + ((B^{-1}A)^T)^{-1}A(B^{-1}A)].
\end{aligned}$$

Next use $A = A^T$ and $((B^{-1}A)^T)^{-1} = B^T A^{-1}$

$$
\begin{aligned}
A - H^T A H &= (B^{-1}A)^T[A(B^{-1}A)^{-1} - A + B^T](B^{-1}A) \\
&= (B^{-1}A)^T[B - (B - C) + B^T](B^{-1}A) \\
&= (B^{-1}A)^T[B^T + C](B^{-1}A).
\end{aligned}
$$

∎

Example 11.3.2. This is a modified Richard splitting for a SPD matrix

$$
A = \frac{I}{\rho} - (\frac{I}{\rho} - A) = B - C \text{ and the iteration}
$$

$$
x^{m+1} = x^m + \rho(d - Ax^m) = x^m + B^{-1}(d - Ax^m).
$$

In order to show $B^T + C$ is positive definite, consider

$$
x^T(B^T + C)x = \frac{2}{\rho}x^T x - x^T A x = \frac{1}{\rho}(2x^T x - \rho x^T A x).
$$

Since $0 < x^T A x \le \|x\| \, \|Ax\| \le \|x\|^2 \, \|A\|$,

$$
\begin{aligned}
x^T(B^T + C)x &\ge \frac{1}{\rho}(2\|x\|^2 - \rho \|x\|^2 \|A\|) \\
&= \frac{1}{\rho}(2 - \rho \|A\|) \|x\|^2.
\end{aligned}
$$

If $0 < \rho < 2/\|A\|$, then this splitting is P-regular.

Consider the alternative characterization of $Ax = d$ when A is SPD and use a P-regular splitting, see Theorem 2.4.4 and Theorem 10.5.1. The solution of $Ax = d$ is equivalent to the solution of the minimization problem

$$
J(x) = \min_y J(y) \text{ where } J(y) = \frac{1}{2}y^T A y - y^T d.
$$

Write the iteration in residual form

$$
y = x + B^{-1}r(x) \text{ where } y = x^{m+1} \text{ and } x = x^m.
$$

The main observation is that the iteration is decreasing:

$$
\begin{aligned}
J(y) &= J(x) - (B^{-1}r)^T r + \frac{1}{2}(B^{-1}r)^T A(B^{-1}r) \\
&= J(x) - (B^{-1}r)^T B B^{-1}r + \frac{1}{2}(B^{-1}r)^T(B - C)(B^{-1}r) \\
&= J(x) - (B^{-1}r)^T[\frac{B + C}{2}](B^{-1}r).
\end{aligned}
$$

Since $z^T B z = z^T B^T z$ and the splitting is P-regular,

$$
J(x^{m+1}) = J(x^m) - \frac{1}{2}(B^{-1}r(x^m))^T(B^T + C)B^{-1}r(x^m) < J(x^m).
$$

Theorem 11.3.3. *(Convergent Energy) If A is SPD and $A = B - C$ is a P-regular splitting, then*

$$J(x) =\downarrow J(x^{m+1}) \text{ and } x^{m+1} \text{ converges to } x.$$

Proof. Because A is SPD, $J(x)$ is bounded from below. Because $J(x^{m+1}) < J(x^m)$, $J(x^{m+1})$ must converge. We need to show this limit is attained by $J(x)$ where x is the solution of $Ax = d$. Since $A = A^T = B^T - C^T = B - C$, $(B^T + C)^T = B^T + C$. This means $B^T + C$ is SPD and it can be used to define a norm

$$\|x\|_{B^T + C} \equiv (x^T(B^T + C)x)^{1/2}.$$

Write the iteration as $x^{k+1} = x^k + B^{-1}r(x^k)$ and $x^{k+1} - x^k = B^{-1}r(x^k)$ and

$$J(x^{m+1}) = J(x^m) - \frac{1}{2}(B^{-1}r(x^m))^T(B^T + C)B^{-1}r(x^m).$$

Since $J(x^{m+1}) - J(x^m)$ converges to zero, $\|B^{-1}r(x^m)\|_{B^T+C}$ must converge to zero. Then $r(x^m)$ converges to $0_{n \times 1} = d - Ax = r(x)$, that is, $r(x^m) - r(x) = -A(x^m - x)$ converges to $0_{n \times 1}$. Since A has an inverse, x^m converges to x. ∎

Which is the "best" method to use? How does one choose the parameter in the method? These have been extensively studied and are related to the magnitude of the spectral radius of $B^{-1}C$. For special matrices one can show

$$\rho(SOR) < \rho(Gauss - Seidel) < \rho(Jacobi).$$

By carefully choosing ω in the SOR splitting there is significant improvements. For a more detailed discussion see C. D. Meyer, [17, Section 7.10].

The regular splitting can be generalized to weak regular splittings where $B^{-1} \geq 0_{n \times n}$ and $B^{-1}C \geq 0_{n \times n}$. Comparison of these types of splitting is given A. Berman and R. J. Plemmons, [4, Chapter 7].

11.3.1   SOR for diffusion in 3D

The *continuous model for the steady state 3D heat diffusion is*

$$-(Ku_x)_x - (Ku_y)_y - (Ku_z)_z = f \qquad (11.3.1)$$
$$u = g \text{ on the boundary.} \qquad (11.3.2)$$

Let u_{ijl} be the approximation of $u(i\Delta x, j\Delta y, l\Delta z)$ where $\Delta x = L/nx$, $\Delta y = W/ny$ and $\Delta z = H/nz$. Approximate the second order derivatives by the centered finite differences. There are $n \equiv (nx-1)(ny-1)(nz-1)$ equations for n unknowns u_{ijl}. The *discrete finite difference 3D* model for $1 \leq i \leq nx-1$, $1 \leq j \leq ny - 1$, $1 \leq l \leq nz - 1$ is

$$-[K(u_{i+1,j,l} - u_{ijl})/\Delta x - K(u_{ijl} - u_{i-1,j,l})/\Delta x]/\Delta x \qquad (11.3.3)$$
$$-[K(u_{i,j+1,l} - u_{ijl})/\Delta y - K(u_{ijl} - u_{i,j-1,l})/\Delta y]/\Delta y$$
$$-[K(u_{i,j,l+1} - u_{ijl})/\Delta z - K(u_{ijl} - u_{i,j,l-1})/\Delta z]/\Delta z = f(i\Delta x, j\Delta y, l\Delta z).$$

In order to keep the notation as simple as possible, we assume that the number of cells in each direction, nx, ny and nz, are such that $\Delta x = \Delta y = \Delta z = h$ and let $K = 1$. This equation simplifies to

$$6u_{ijl} = f(ih, jh, lh)h^2 + u_{i,j,l-1} + u_{i,j-1,l} + u_{i-1,j,l}$$
$$+u_{i,j,l+1} + u_{i,j+1,l} + u_{i+1,j,l}. \tag{11.3.4}$$

Equation (11.3.4) suggests the use of the SOR algorithm where there are three nested loops within the SOR loop. The u_{ijl} are now stored in a 3D array, and either $f(ih, jh, lh)$ can be computed every SOR sweep, or $f(ih, jh, lh)$ can be computed once and stored in a 3D array. The classical order of ijl is to start with $l = 1$ (the bottom grid plane) and then use the classical order for ij starting with $j = 1$ (the bottom grid row in the grid plane l). This means the l-loop is the outermost, j-loop is in the middle, and the i-loop is the innermost loop.

Classical Order SOR Method (11.3.4).

    choose nx, ny, nz such that h = L/nx = W/ny = H/nz
    for m = 1,maxm
        for l = 1,nz
            for j = 1,ny
                for i = 1,nx
                    $utemp = (f(ih, jh, lh) * h * h$
                        $+u(i - 1, j, l) + u(i + 1, j, l)$
                        $+u(i, j - 1, l) + u(i, j + 1, l)$
                        $+u(i, j, l - 1) + u(i, j, l + 1))/6$
                  $u(i, j, l) = (1 - w) * u(i, j, l) + w * utemp$
                endloop
            endloop
        endloop
        test for convergence
    endloop.

11.3.2 MATLAB® implementation of SOR

The MATLAB® code sor3d_13.m illustrates the 3D steady state cooling fin problem with the finite difference discrete model given in (11.3.3) where $f(x, y, z) = 0.0$. The following parameters were used $nx = 10$, $ny = 20$, and $nz = 30$ for 4959 unknowns. In sor3d_13.m the initialization and boundary conditions are defined in lines 8–17. The SOR loop is in lines 21–42, where the lji-nested loop for all the interior nodes is executed in lines 23–38. The test for SOR convergence is in lines 39–41. Line 46 lists the SOR iterations needed for convergence, and line 51 has the MATLAB® command $slice(u, [5\ 10\ 15\ 20], 10, 10)$, which generates a color-coded 3D plot of the temperatures within the cooling fin.

MATLAB® Code sor3d_13.m

```
1    % This is the SOR solution of a 3D problem.
2    % Assume steady state heat diffusion.
3    % Given temperature on the boundary.
4    clear; clf(figure(1));
5    %
6    %   Input Data
7    %
8    w     = 1.8;
9    eps   = 0.001;
10   maxit = 200;
11   L = 1.0; nx = 10; dx = L/nx; x = 0:dx:L;
12   W = 2.0; ny = 20; dy = W/ny; y = 0:dy:W;
13   H = 3.0; nz = 30; dz = H/nz; z = 0:dz:H;
14   raijl = 1./( 2/(dx*dx) + 2/(dy*dy) + 2/(dz*dz));
15   nunk = (nx-1)*(ny-1)*(nz-1);
16   u = 70.*ones(nx+1,ny+1,nz+1);   % initial guess
17   u(1,:,:) = 200.;                % hot boundary at x = 0
18   %
19   %   Execution of SOR
20   %
21   for iter = 1:maxit;             % begin SOR
22      numi = 0;
23      for l = 2:nz
24         for j = 2:ny
25            for i = 2:nx
26               temp = (u(i-1,j,l)/(dx*dx) + u(i,j-1,l)/(dy*dy)...
27                      + u(i,j,l-1)/(dz*dz) + u(i+1,j,l)/(dx*dx)...
28                      + u(i,j+1,l)/(dy*dy) + u(i,j,l+1)/(dz*dz))...
29                      *raijl;
30               temp = (1. - w)*u(i,j,l) + w*temp;
31               error = abs(temp - u(i,j,l));
32               u(i,j,l) = temp;
33               if error<eps
34                  numi = numi + 1;
35               end
36            end         %  end i loop
37         end            %  end j loop
38      end               %  end l loop
39      if numi==nunk
40         break
41      end
42   end                  %  end sor loop
43   %
44   %   Output via slice
45   %
46   iter                         % iterations for convergence
47   [yy,xx,zz] = meshgrid(y, x, z);
```

```
48     xvec = [5*dx 10*dx ];
49     yvec = [5*dy 10*dy 15*dy];
50     zvec = [10*dz];
51     slice(yy,xx,zz, u, yvec,xvec,zvec)    % creates color 3D plot
52     axis([0 W 0 L 0 H])
53     colorbar
54     pause(5)
55     axis image
```

The SOR parameters $w = 1.5, 1.6, 1.7$ and 1.8 were used with the convergence criteria $eps = 0.001$, and this resulted in convergence after $67, 49, 49$ and 69 iterations, respectively. Figure 11.3.1 indicates the steady state temperature is a mass whose boundary temperature is 200 at $x = 0.0$ and temperature is 70 at the other five boundaries.

11.3.3   Exercises

Consider the MATLAB® code sor3d_13.m.

1. Execute the code for variable ω. Observe the number of iterations required for "convergence."

2. Execute the code for variable error *eps*. Observe the number of iterations required for "convergence."

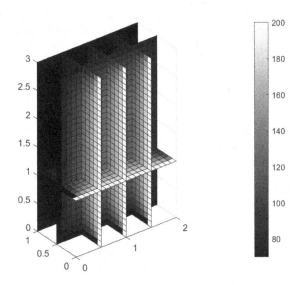

FIGURE 11.3.1
Steady state temperature.

3. Execute the code for variable unknowns nx, ny and nz. Observe the number of iterations required for "convergence."

4. Modify the code to include a nonzero function $f(x, y)$.

5. Consider the splitting

$$
A = \begin{bmatrix} A_{11} & A_{12} & A_{13} \\ A_{21} & A_{22} & A_{23} \\ A_{31} & A_{32} & A_{33} \end{bmatrix} = \begin{bmatrix} B_1 & 0 & A_{13} \\ 0 & B_2 & A_{23} \\ A_{31} & A_{32} & A_{33} \end{bmatrix} - \begin{bmatrix} C_1 & -A_{12} & 0 \\ -A_{21} & C_2 & 0 \\ 0 & 0 & 0 \end{bmatrix}
$$

When will it be a convergent P-regular splitting?

11.4 Conjugate Gradient for SPD Matrices

The *conjugate gradient method* has three basic components: steepest descent in multiple directions, conjugate directions and preconditioning. The multiple direction version of steepest descent ensures the largest possible decrease in the "energy." The conjugate direction ensures that solution of the reduced algebraic system is done with a minimum amount of computations. The preconditioning modifies the initial problem so that the convergence is more rapid. There are a number of variations on this, and the initial analysis was done in 1952 by M. Hestenes and E. Stiefel, [10].

The steepest descent method hinges on the fact that for SPD matrices the algebraic system $Ax = d$ is equivalent to minimizing a real-valued function $J(x) = \frac{1}{2}x^T A x - x^T d$. For more details, see the Subsections 2.4.1 and 2.4.2 with Exercises 6, 7 and 8. Make an initial guess, x, for the solution and move in some direction p so that the new x, $x^+ \equiv x + cp$, will make $J(x^+)$ a minimum. The direction, p, of steepest descent, where the directional derivative of J is largest, is given by $p = -r$ where $r = r(x) \equiv d - Ax$. Next choose the c so that $f(c) \equiv J(x + cr)$ is a minimum, and this is $c = r^T r / r^T A r$. In the steepest descent method only the current residual is used. If a linear combination of all the previous residuals were to be used, then $J(\widehat{x}^+)$ would be smaller than $J(x^+)$ for the steepest descent method.

For multiple directions the new x should be the old x plus a linear combination of the all the previous residuals

$$
x^{m+1} = x^m + c_0 r^0 + c_1 r^1 + \cdots + c_m r^m.
$$

This can be written in matrix form where R is $n \times (m + 1)$, $m \ll n$ and is formed by the residual column vectors

$$
R \equiv \begin{bmatrix} r^0 & r^1 & \cdots & r^m \end{bmatrix}.
$$

Then c is an $(m+1) \times 1$ column vector of the coefficients in the linear combination

$$x^{m+1} = x^m + Rc. \tag{11.4.1}$$

Choose c so that $J(x^m + Rc)$ is the smallest possible.

$$J(x^m + Rc) = \frac{1}{2}(x^m + Rc)^T A(x^m + Rc) - (x^m + Rc)^T d. \tag{11.4.2}$$

Because A is symmetric, we may use the algebraic identity

$$J(y) = J(x) - (y-x)^T r(x) + \frac{1}{2}(y-x)^T A(y-x) \tag{11.4.3}$$

with $x = x^m$ and $y = x^m + Rc$

$$
\begin{aligned}
J(x^m + Rc) &= J(x^m) - c^T R^T(d - Ax^m) + \frac{1}{2}c^T(R^T AR)c \\
&= J(x^m) - c^T R^T r^m + \frac{1}{2}c^T(R^T AR)c.
\end{aligned} \tag{11.4.4}
$$

Now choose c so that $-c^T R^T r^m + \frac{1}{2}c^T(R^T AR)c$ is a minimum.

If $R^T AR$ is symmetric positive definite, then use the discrete equivalence theorem. In this case x is replaced by c and the matrix A is replaced by the $(m+1) \times (m+1)$ matrix $R^T AR$. Since A is assumed to be symmetric positive definite, $R^T AR$ will be symmetric and positive definite if the columns of R are linearly independent ($Rc = 0$ implies $c = 0$). In this case c is $(m+1) \times 1$ and will be the solution of the reduced algebraic system

$$(R^T AR)c = R^T r^m. \tag{11.4.5}$$

The purpose of using the conjugate directions is to ensure the matrix $R^T AR$ is easy to invert. The ij-component of $R^T AR$ is $(r^i)^T Ar^j$, and the i-component of $R^T r^m$ is $(r^i)^T r^m$. $R^T AR$ would be easy to invert if it were a diagonal matrix, and in this case for i not equal to j $(r^i)^T Ar^j = 0$. This means the column vectors would be "perpendicular" (orthogonal) with respect to the inner product given by $x^T Ay$ where A is symmetric positive definite.

Here we may apply the Gram-Schmidt process. For two directions r^0 and r^1 this has the form

$$p^0 = r^0 \text{ and } p^1 = r^1 + bp^0. \tag{11.4.6}$$

Now, b is chosen so that $(p^0)^T Ap^1 = 0$

$$
\begin{aligned}
(p^0)^T A(r^1 + bp^0) &= 0 \\
(p^0)^T Ar^1 + b(p^0)^T Ap^0 &=
\end{aligned}
$$

and solve for

$$b = -(p^0)^T Ar^1 / (p^0)^T Ap^0. \tag{11.4.7}$$

By the steepest descent step in the first direction

$$x^1 = x^0 + cr^0 \text{ where}$$
$$c = (r^0)^T r^0 / (r^0)^T A r^0 \text{ and} \tag{11.4.8}$$
$$r^1 = r^0 - cAr^0.$$

The definitions of b in (11.4.7) and c in (11.4.8) yield the following additional equations

$$(p^0)^T r^1 = 0 \text{ and } (p^1)^T r^1 = (r^1)^T r^1. \tag{11.4.9}$$

Moreover, use $r^1 = r^0 - cAr^0$ in $b = -(p^0)^T A r^1 / (p^0)^T A p^0$ and in $(r^1)^T r^1$ to show

$$b = (r^1)^T r^1 / (p^0)^T p^0. \tag{11.4.10}$$

These equations allow for a simplification of (11.4.5) where R is now formed by the column vectors p^0 and p^1

$$\begin{bmatrix} (p^0)^T A p^0 & 0 \\ 0 & (p^1)^T A p^1 \end{bmatrix} \begin{bmatrix} c_0 \\ c_1 \end{bmatrix} = \begin{bmatrix} 0 \\ (r^1)^T r^1 \end{bmatrix}.$$

Thus, $c_0 = 0$ and $c_1 = (r^1)^T r^1 / (p^1)^T A p^1$. From (11.4.1) with $m = 1$ and r^0 and r^1 replaced by p^0 and p^1

$$x^2 = x^1 + c_0 p^0 + c_1 p^1 = x^1 + 0 p^0 + c_1 p^1. \tag{11.4.11}$$

For the three direction case we let $p^2 = r^2 + b p^1$ and choose this new b to be such that $(p^2)^T A p^1 = 0$ so that $b = -(p^1)^T A r^2 / (p^1)^T A p^1$. Use this new b and the previous arguments to show $(p^0)^T r^2 = 0$, $(p^1)^T r^2 = 0$, $(p^2)^T r^2 = (r^2)^T r^2$ and $(p^0)^T A p^2 = 0$. Moreover, one can show $b = (r^2)^T r^2 / (p^1)^T p^1$. The equations give a 3×3 simplification of the new (11.4.5)

$$\begin{bmatrix} (p^0)^T A p^0 & 0 & 0 \\ 0 & (p^1)^T A p^1 & 0 \\ 0 & 0 & (p^2)^T A p^2 \end{bmatrix} \begin{bmatrix} c_0 \\ c_1 \\ c_2 \end{bmatrix} = \begin{bmatrix} 0 \\ 0 \\ (r^2)^T r^2 \end{bmatrix}.$$

Thus, $c_0 = c_1 = 0$ and $c_2 = (r^2)^T r^2 / (p^2)^T A p^2$. From (11.4.1) with $m = 2$ and r^0, r^1 and r^2 replaced by p^0, p^1 and p^2

$$x^3 = x^2 + c_0 p^0 + c_1 p^1 + c_2 p^2 = x^2 + 0 p^0 + 0 p^1 + c_2 p^3. \tag{11.4.12}$$

Fortunately, this process continues, and one can show by mathematical induction that the reduced matrix in the new (11.4.5) will always be a diagonal matrix and the right side will have only one nonzero component, namely, the last component, see [20, Section 8.6]. Thus, the use of conjugate directions substantially reduces the amount of computations, and the previous search direction vectors do not need to be stored.

In the following description the conjugate gradient method corresponds to the case where the preconditioner is $M = I$. One common preconditioner is

SSOR where the SOR scheme is executed in a forward and then a backward sweep. If $A = D - L - L^T$ where D is the diagonal part of A and $-L$ is the strictly lower triangular part of A, then M is

$$M = (D - wL)(1/((2 - w)w))D^{-1}(D - wL^T).$$

The solve step is relatively easy because there is a lower triangular solve, a diagonal product and an upper triangular solve. If the matrix is sparse, then these solves will also be sparse solves. Other preconditioners include *incomplete LU* and *incomplete domain decomposition* where select components of the matrix are set equal to zero.

Preconditioned Conjugate Gradient Method.

> Let x^0 be an initial guess
> $r^0 = d - Ax^0$
> solve $M\hat{r}^0 = r^0$ and set $p^0 = \hat{r}^0$
> for m = 0, maxm
> $c = (\hat{r}^m)^T r^m / (p^m)^T A p^m$
> $x^{m+1} = x^m + cp^m$
> $r^{m+1} = r^m - cAp^m$
>   test for convergence
>   solve $M\hat{r}^{m+1} = r^{m+1}$
> $b = (\hat{r}^{m+1})^T r^{m+1} / (\hat{r}^m)^T r^m$
> $p^{m+1} = \hat{r}^{m+1} + bp^m$
> endloop.

11.4.1 MATLAB® implementations of CG

MATLAB® has a number of conjugate gradient implementations:

> \>> help pcg
> pcg Preconditioned Conjugate Gradients Method.
> equations A*X=B for X = pcg(A,B) attempts to solve the system
> of linear for X. The N-by-N coefficient matrix A must be symmetric
> and positive definite and the right hand side column vector B must
> have length N.

The following code cgfull_13.m is for demonstration purposes and uses a full matrix. In this code the convergence test was $\left\| r(x^k) \right\| \le 10^{-4}$ and this was attained in 23 iterations. Another unlisted code, precg_13.m with ssor.m, uses a sparse matrix and the SSOR preconditioner. With only steepest descent convergence occurs in more than 200 iterations, with no preconditioning CG convergence takes 26 iterations, and with SSOR preconditioner convergence takes 11 iterations.

The cgfull_13.m implementation of the conjugate gradient method uses a full matrix and no preconditioner. The matrix comes from the finite difference model of $-u_{xx} - u_{yy} = f$. The coefficient matrix is symmetric positive definite, and so, one can use this particular scheme. Here the partial differential equation has right side equal to $200(1 + sin(\pi x)sin(\pi y))$ and the solution is required to be zero on the boundary of $(0, 1) \times (0, 1)$. Lines 9–32 define the coefficient matrix AA. The conjugate gradient method is executed in lines 51–67. All the conjugate directions are stored in the array PP. It is not necessary to do this, but it enables one to see that the conjugate directions are orthogonal as is indicated in line 72. If one changes the partial differential equation to $-u_{xx} - u_{yy} + au_x = f$, then the matrix will no longer be symmetric and the direction vectors will no longer be orthogonal. To see this experiment with the indicated changes in lines 12 and 14. Also, one can turn off the conjugate directions by allowing line 60 to use the residual and not the conjugate direction. In this case this is the steepest descent method and orthogonality and convergence will fail. Graphical output is given lines 73–86.

MATLAB® Codes cgfull_13.m

```
1    %  Solves a block tridiagonal SPD from the finite difference
2    %  method.It is applied to the partial differential equation
3    %    - u_xx  - u_yy = f(x,y) with zero boundary conditions.
4    %  Uses the conjugate gradient method with the full matrix.
5    clear;
6    %
7    %  Input the coefficient matrix
8    %
9    n = 10;
10   A = zeros(n);
11   for i = 1:n
12       A(i,i) = 4;     % 4 + 4/n;
13       if (i>1)
14           A(i,i-1) = -1;    % -1 - 4/n;
15       end
16       if (i<n)
17           A(i,i+1) = -1;
18       end
19   end
20       I  = eye(n);
21       AA = zeros(n*n);
22       for i = 1:n
23           newi = (i-1)*n +1;
24           lasti = i*n;
25           AA(newi:lasti,newi:lasti) = A;
26           if (i>1)
27               AA(newi:lasti,newi-n:lasti-n) = -I;
28           end
29           if (i<n)
30               AA(newi:lasti,newi+n:lasti+n) = -I;
```

```
31          end
32        end
33    N = n*n;
34    h = 1./n;
35    u = zeros(N,1);    % initial guess
36    r = zeros(N,1);
37    p = zeros(N,1);
38    q = zeros(N,1);
39    for j= 1:n
40       for i = 1:n
41          I = (j-1)*n + i;
42          r(I)= h*h*200*(1+sin(pi*(i-1)*h)*sin(pi*(j-1)*h));
43       end
44    end
45    err = 1.0;
46    m   = 0;
47    rho = 0.0;
48    %
49    %  Begin the conjugate gradient method
50    %
51    while ((err>.0001)*(m<200))
52       m = m+1;
53       oldrho = rho;
54       rho = r'*r;
55       if (m == 1) % use p for the conjugate gradient method
56           p = r; PP = p;
57       else
58           p = r + (rho/oldrho)*p; PP = [PP p];
59       end
60       % p = r;   % use this p = r for steepest descent method
61       q = AA*p;
62       alpha = rho/(p'*q);
63       u     = u + alpha*p;
64       r     = r - alpha*q;
65       err   = norm(r);
66       reserr(m) = err;
67    end
68    %
69    % Output
70    %
71    m                          % iterations needed for convergence
72    PP(:,1:9)'*AA*PP(:,1:9)    % check orthogonality
73    figure(1)
74    semilogy(reserr)
75    xlabel('iterations')
76    ylabel('norm(r)')
77    for j = 1:n
78       for i = 1:n
79          I = (j-1)*n + i;
```

```
80          u2d(i,j) = u(I);
81        end
82      end
83      figure(2)
84      uu2d = zeros(n+2);
85      uu2d(2:n+1,2:n+1) = u2d;
86      mesh(uu2d)
```

The connection with the *Krylov vectors* $A^i r(x^0)$ evolves from expanding the conjugate gradient loop. The first iteration is

$$x^1 = x^0 + \alpha_0 p^0 = x^0 + \alpha_0 r^0$$
$$r^1 = r^0 - \alpha_0 A p^0 = r^0 - \alpha_0 A r^0$$
$$p^1 = r^1 + \beta_0 p^0 = r^1 + \beta_0 r^0.$$

The next iterations are

$$
\begin{aligned}
x^2 &= x^1 + \alpha_1 p^1 \\
&= x^1 + \alpha_1 \left(r^1 + \beta_0 r^0 \right) \\
&= x^0 + \alpha_0 r^0 + \alpha_1 \left(r^0 - \alpha_0 A r^0 + \beta_0 r^0 \right) \\
&= x^0 + c_0 r^0 + c_1 A r^0 \\
&\vdots \\
x^m &= x^0 + c_0 r^0 + c_1 A r^0 + \cdots + c_{m-1} A^{m-1} r^0.
\end{aligned}
$$

An alternative definition of the conjugate gradient method is to choose the coefficients of the Krylov vectors so as to minimize $J(x)$

$$J(x^{m+1}) = \min_c J(x^0 + c_0 r^0 + c_1 A r^0 + \cdots + c_m A^m r^0).$$

Definition. *The alternate conjugate gradient method is*

$$x^{m+1} = x^0 + \sum_{i=0}^{m} c_i A^i r^0$$

where $r^0 \equiv d - A x^0$ and $c_i \in \mathbb{R}$ are chosen so that

$$J(x^{m+1}) = \min_y J(y)$$
$$y \in x^0 + K_{m+1} \text{ and}$$
$$K_{m+1} \equiv \{ z \mid z = \sum_{i=0}^{m} c_i A^i r^0, c_i \in \mathbb{R} \}.$$

The utility of the Krylov approach to both the conjugate gradient and the generalized residual methods is a very nice analysis of convergence properties. These are based on the following algebraic identities. Let x be the solution of $Ax = d$ so that $r(x) = d - Ax = 0$, and use the identity in line (11.4.3) for symmetric matrices

$$J(x^{m+1}) - J(x) = \frac{1}{2}(x^{m+1} - x)^T A(x^{m+1} - x)$$
$$= \frac{1}{2}\left\|x^{m+1} - x\right\|_A^2.$$

Now write the next iterate in terms of the Krylov vectors

$$x - x^{m+1} = x - (x^0 + c_0 r^0 + c_1 A r^0 + \cdots + c_m A^m r^0)$$
$$= x - x^0 - (c_0 r^0 + c_1 A r^0 + \cdots + c_m A^m r^0)$$
$$= x - x^0 - (c_0 I + c_1 A + \cdots + c_m A^m) r^0.$$

Note $r^0 = d - Ax^0 = Ax - Ax^0 = A(x - x^0)$ so that

$$x - x^{m+1} = x - x^0 - (c_0 I + c_1 A + \cdots + c_m A^m) A(x - x^0)$$
$$= (I - (c_0 I + c_1 A + \cdots + c_m A^m) A)(x - x^0)$$
$$= (I - (c_0 A + c_1 A + \cdots + c_m A^{m+1}))(x - x^0).$$

Thus

$$2(J(x^{m+1}) - J(x)) = \left\|x^{m+1} - x\right\|_A^2 \le \left\|q_{m+1}(A)(x - x^0)\right\|_A^2 \quad (11.4.13)$$

where $q_{m+1}(z) = 1 - (c_0 z + c_1 z^2 + \cdots + c_m z^{m+1})$. One can make appropriate choices of the polynomial $q_{m+1}(z)$ to establish a number of important properties, see [11, Chapter 2]. Although this is an iterative method and usually the size of the matrix is very large, for nonsingular matrices the solution will be attained within $m = n$ iterations. The proof follows from the Cayley-Hamilton theorem where $p(\lambda) = \det(A - \lambda I)$ is the polynomial of order n and $p(0) = \det(A) \neq 0$. In this case $p(A) = 0_{n \times n}$ and choose $q_n(A) = p(A)/\det(A)$.

11.4.2   Exercises

Consider the MATLAB® code cgfull_13.m.

1. Execute the code for variable n. Observe the number of iterations required for "convergence."

2. Observe the orthogonal matrix in line 72 and the iterations for convergence in figure(1).

   (a). Turn off the conjugate directions by changing line 60.

   (b). Use conjugate directions but consider the non-symmetric case by adjusting lines 10 and 12.

11.5　Generalized Minimum Residual

If A is not a SPD matrix, then the conjugate gradient method cannot be directly used. One alternative is to replace $Ax = d$ by the normal equations $A^T A x = A^T d$, which may be ill-conditioned and subject to significant roundoff errors. Another approach is to try to minimize the real-valued residual function $R(x) \equiv r(x)^T r(x)$ in place of quadratic function $J(x) \equiv \frac{1}{2} x^T A x - x^T d$ for the SPD case. As in the conjugate gradient method, this will be done on the Krylov space. This approach is attributed to Y. Saad and M. Shultz in 1986, [22].

Definition. *The generalized minimum residual method (GMRES) is*

$$x^{m+1} = x^0 + \sum_{i=0}^{m} \alpha_i A^i r^0$$

where $r^0 \equiv d - Ax^0$ and $\alpha_i \in \mathbb{R}$ are chosen so that

$$R(x^{m+1}) = \min_y R(y)$$

$$y \in x^0 + K_{m+1} \text{ and}$$

$$K_{m+1} \equiv \{ z \mid z = \sum_{i=0}^{m} c_i A^i r^0, c_i \in \mathbb{R} \}.$$

Like the conjugate gradient method the Krylov vectors are very useful for the analysis of convergence. Consider the residual after $m + 1$ iterations

$$
\begin{aligned}
d - Ax^{m+1} &= d - A(x^0 + \alpha_0 r^0 + \alpha_1 A r^0 + \cdots + \alpha_m A^m r^0) \\
&= r^0 - A(\alpha_0 r^0 + \alpha_1 A r^0 + \cdots + \alpha_m A^m r^0) \\
&= (I - A(\alpha_0 I + \alpha_1 A + \cdots + \alpha_m A^m)) r^0.
\end{aligned}
$$

Thus

$$\left\| r^{m+1} \right\|_2^2 \leq \left\| q_{m+1}(A) r^0 \right\|_2^2 \tag{11.5.1}$$

where $q_{m+1}(z) \equiv 1 - (\alpha_0 z + \alpha_1 z^2 + \cdots + \alpha_m z^{m+1})$.

Similar to the conjugate gradient method one can make appropriate choices of the polynomial $q_{m+1}(z)$ to establish a number of important properties, see [11, Chapter 2]. Although this is an iterative method and usually the size of the matrix is very large, for nonsingular matrices the solution will be attained within $m = n$ iterations. The proof also follows from the Cayley-Hamilton theorem where $p(\lambda) = \det(A - \lambda I)$ is the polynomial of order n and $p(0) = \det(A) \neq 0$. In this case $p(A) = 0_{n \times n}$ and choose $q_n(A) = p(A)/\det(A)$. However, unlike the conjugate gradient method, the generalized minimum residual method does require the storage of the residuals. One alternative is to restart the method after a fixed number of residuals and use the last step as a new initial starting point.

The Krylov space of vectors has the nice property that $AK_m \subset K_{m+1}$. This allows one to reformulate the problem of finding the α_i

$$A(x^0 + \sum_{i=0}^{m-1} \alpha_i A^i r^0) = d$$

$$Ax^0 + \sum_{i=0}^{m-1} \alpha_i A^{i+1} r^0 = d$$

$$\sum_{i=0}^{m-1} \alpha_i A^{i+1} r^0 = r^0. \tag{11.5.2}$$

Let bold $\mathbf{K}_m$ be the $n \times m$ matrix of Krylov vectors

$$\mathbf{K}_m = \begin{bmatrix} r^0 & Ar^0 & \cdots & A^{m-1}r^0 \end{bmatrix}.$$

The equation in (11.5.2) has the form

$$A\mathbf{K}_m \alpha = r^0 \text{ where} \tag{11.5.3}$$

$$A\mathbf{K}_m = A \begin{bmatrix} r^0 & Ar^0 & \cdots & A^{m-1}r^0 \end{bmatrix} \text{ and}$$

$$\alpha = \begin{bmatrix} \alpha_0 & \alpha_1 & \cdots & \alpha_{m-1} \end{bmatrix}^T.$$

The equation in (11.5.3) is a least squares problem for $\alpha \in \mathbb{R}^m$ where $A\mathbf{K}_m$ is an $n \times m$ matrix.

In order to efficiently solve this sequence of least squares problems, we construct an orthonormal basis of K_m one column vector per iteration. Let $V_m = \{v_1, v_2, ..., v_m\}$ be this basis, and let bold $\mathbf{V}_m$ be the $n \times m$ matrix whose columns are the basis vectors

$$\mathbf{V}_m = \begin{bmatrix} v_1 & v_2 & \cdots & v_m \end{bmatrix}.$$

Since $AK_m \subset K_{m+1}$, each column in $A\mathbf{V}_m$ should be a linear combination of columns in $\mathbf{V}_{m+1}$. This allows one to construct $\mathbf{V}_m$ one column per iteration by using the modified Gram-Schmidt process.

Let the first column of $\mathbf{V}_m$ be the normalized initial residual

$$r^0 = bv_1$$

where $b = ((r^0)^T r^0)^{\frac{1}{2}}$ is chosen so that $v_1^T v_1 = 1$. Since $AK_1 \subset K_2$, A times the first column should be a linear combination of v_1 and v_2

$$Av_1 = v_1 h_{11} + v_2 h_{21}.$$

Find h_{11} and h_{21} by requiring $v_1^T v_1 = v_2^T v_2 = 1$ and $v_1^T v_2 = 0$ and assuming $Av_1 - v_1 h_{11}$ is not the zero vector

$$h_{11} = v_1^T Av_1,$$
$$z = Av_1 - v_1 h_{11},$$
$$h_{21} = (z^T z)^{\frac{1}{2}} \text{ and}$$
$$v_2 = z/h_{21}.$$

For the next column

$$Av_2 = v_1 h_{12} + v_2 h_{22} + v_3 h_{32}.$$

Again require the three vectors to be orthonormal and $Av_2 - v_1 h_{12} - v_2 h_{22}$ is not the zero vector to get

$$h_{12} = v_1^T Av_2 \text{ and } h_{22} = v_2^T Av_2,$$
$$z = Av_2 - v_1 h_{12} - v_2 h_{22},$$
$$h_{32} = (z^T z)^{\frac{1}{2}} \text{ and}$$
$$v_3 = z/h_{32}.$$

Continue this and represent the results in matrix form

$$AV_m = V_{m+1} H \quad \text{where} \tag{11.5.4}$$
$$AV_m = \begin{bmatrix} Av_1 & Av_2 & \cdots & Av_m \end{bmatrix},$$
$$V_{m+1} = \begin{bmatrix} v_1 & v_2 & \cdots & v_{m+1} \end{bmatrix},$$

$$H = \begin{bmatrix} h_{11} & h_{12} & \cdots & h_{1m} \\ h_{21} & h_{22} & \cdots & h_{2m} \\ 0 & h_{32} & \cdots & h_{3m} \\ & & \ddots & \vdots \\ 0 & 0 & & \\ 0 & 0 & 0 & h_{m+1,m} \end{bmatrix},$$

$$h_{i,m} = v_i^T Av_m \text{ for } i \leq m, \tag{11.5.5}$$
$$z = Av_m - v_1 h_{1,m} \cdots - v_m h_{m,m} \neq 0,$$
$$h_{m+1,m} = (z^T z)^{\frac{1}{2}} \text{ and} \tag{11.5.6}$$
$$v_{m+1} = z/h_{m+1,m}. \tag{11.5.7}$$

Here A is $n \times n$, V_m is $n \times m$, and H is an $(m+1) \times m$ *upper Hessenberg matrix* ($h_{ij} = 0$ when $i > j+1$). This allows for the easy solution of the least squares problem (11.5.3).

Theorem 11.5.1. *(GMRES Reduction) The solution of the least squares problem (11.5.3) is given by the solution of the least squares problem*

$$H\beta = e_1 b \tag{11.5.8}$$

where e_1 is the first unit vector, $b \equiv ((r^0)^T r^0)^{\frac{1}{2}}$ and $AV_m = V_{m+1} H$.

Proof. Since $r^0 = bv_1$, $r^0 = \mathbf{V}_{m+1}e_1 b$. The least squares problem in (11.5.3) can be written in terms of the orthonormal basis

$$A\mathbf{V}_m \beta = \mathbf{V}_{m+1}e_1 b.$$

Use the orthonormal property in the expression for

$$\widehat{r}(\beta) = \mathbf{V}_{m+1}e_1 b - A\mathbf{V}_m \beta = \mathbf{V}_{m+1}e_1 b - \mathbf{V}_{m+1}H\beta$$

$$\begin{aligned}
(\widehat{r}(\beta))^T \widehat{r}(\beta) &= (\mathbf{V}_{m+1}e_1 b - \mathbf{V}_{m+1}H\beta)^T (\mathbf{V}_{m+1}e_1 b - \mathbf{V}_{m+1}H\beta) \\
&= (e_1 b - H\beta)^T \mathbf{V}_{m+1}^T \mathbf{V}_{m+1}(e_1 b - H\beta) \\
&= (e_1 b - H\beta)^T (e_1 b - H\beta).
\end{aligned}$$

Thus the least squares solution of (11.5.8) will give the least squares solution of (11.5.3) where $\mathbf{K}_m \alpha = \mathbf{V}_m \beta$. ∎

If $z = Av_m - v_1 h_{1,m} \cdots - v_m h_{m,m} = 0$, then the next column vector v_{m+1} cannot be found and

$$A\mathbf{V}_m = \mathbf{V}_m H(1:m, 1:m).$$

Now $H = H(1:m, 1:m)$ must have an inverse and $H\beta = e_1 b$ has a solution. This means

$$\begin{aligned}
0 &= r^0 - A\mathbf{V}_m \beta \\
&= d - Ax^0 - A\mathbf{V}_m \beta \\
&= d - A(x^0 + \mathbf{V}_m \beta).
\end{aligned}$$

So, $x^0 + \mathbf{V}_m \beta$ is a solution to $Ax = d$.

If $z = Av_m - v_1 h_{1,m} \cdots - v_m h_{m,m} \neq 0$, then $h_{m+1,m} = (z^T z)^{\frac{1}{2}} \neq 0$ and $A\mathbf{V}_m = \mathbf{V}_{m+1}H$. Now H is an upper Hessenberg matrix with nonzero components on the subdiagonal. This means H has full column rank so that the least squares problem in (11.5.8) can be solved by the QR factorization of $H = QR$. The normal equation for (11.5.8) gives

$$\begin{aligned}
H^T H\beta &= H^T e_1 b \text{ and} \\
R\beta &= Q^T e_1 b.
\end{aligned} \tag{11.5.9}$$

The QR factorization of the Hessenberg matrix can easily be done by Givens rotations. An implementation of the GMRES method can be summarized by the following algorithm.

GMRES Method

> let x^0 be an initial guess for the solution
> $r^0 = d - Ax^0$ and $V(:,1) = r^0/((r^0)^T r^0)^{\frac{1}{2}}$

for m = 1, maxm
$$V(:, m+1) = AV(:, m)$$
compute columns $m + 1$ of V_{m+1} and H in (11.5.4)–(11.5.7)
(use modified Gram-Schmidt)
compute the QR factorization of H
(use Givens rotations)
test for convergence
solve (11.5.8), (11.5.9) for β
$$x^{m+1} = x^0 + V_{m+1}\beta$$
endloop.

11.5.1   MATLAB® implementations of GMRES

>> help gmres
gmres Generalized Minimum Residual Method. X = gmres(A,B) attempts to solve the system of linear equations A*X A*X=B for X. The N-by-N coefficient matrix A must be square and the right hand side column vector B must have length N. This uses the unrestarted method with MIN(N,10) total iterations.

The MATLAB® codes gmresfull_13.m and pcgmres.m are for a two variable partial differential equation with both first order derivatives and second order derivatives. The discrete problem is obtained by using centered differences for the second order derivatives and upwind differences for the first order derivatives. The code gmresfull_13.m is listed here and uses the full matrix. The sparse matrix implementation, pcgmres.m, is not listed and uses the SSOR preconditioner. The calculations given in Figure 11.5.1 were done by gmresfull_13.m with no preconditioning required 85 iterations. The SSOR preconditioner used in pcgmres.m gave convergence in 17 iterations.

The code gmresfull_13.m is initialized in lines 11–37, the GMRES loop is done in lines 39–92, and the output is generated in lines 103–117. The modified Gram-Schmidt orthogonalization in lines 60–67, and Givens rotations are done in lines 69–89. Upon exiting the GMRES loop the upper triangular solve is done in line 95, and the approximate solution $x^0 + \mathbf{V}_{k+1}\beta$ is generated in the loop 97–99.

MATLAB® Codes gmresfull_13.m

```
1    clear;
2    % Solves a block tridiagonal non SPD from the
3    % finite difference method applied to
4    % - u_xx - u_yy + a1 u_x + a2 u_y + c u = f(x,y)
5    % with zero boundary conditions.
6    % Uses the generalized minimum residual method.
7    % Define the NxN coefficient matrix AA where N = n^2.
```

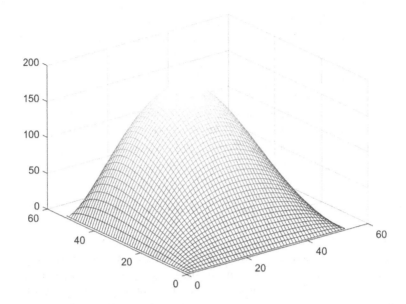

FIGURE 11.5.1
GMRES output.

```
8    %
9    % Input Data
10   %
11   n = 50; N = n*n;
12   errtol = 0.001;
13   kmax = 200;
14   A = zeros(n); a1 = 10; a2 = 1; c = 1; h = 1./(n+1);
15   for i = 1:n
16     A(i,i) = 4 + c*h*h + a1*h + a2*h;
17     if (i>1)
18       A(i,i-1) = -1 - a1*h ;
19     end
20     if (i<n)
21       A(i,i+1) = -1;
22     end
23   end
24   I = eye(n);
25   II = I + I*a2*h;
26   AA = zeros(N);
27   for i = 1:n
28     newi = (i-1)*n + 1;
29     lasti = i*n;
```

```
30      AA(newi:lasti,newi:lasti) = A;
31      if (i>1)
32        AA(newi:lasti,newi-n:lasti-n) = -II;
33      end
34      if (i<n)
35        AA(newi:lasti,newi+n:lasti+n) = -I;
36      end
37    end
38    %
39    % Execution of gmres
40    %
41    u = zeros(N,1);
42    r = zeros(N,1);
43    for j = 1:n
44      for i = 1:n
45        I = (j-1)*n + i;
46        r(I)= h*h*2000*(1+sin(pi*(i-1)*h)*sin(pi*(j-1)*h));
47      end
48    end
49    rho = (r'*r)^.5;
50    errtol = errtol*rho;
51    g = rho*eye(kmax+1,1);
52    v(:,1) = r/rho;
53    k = 0;
54    % Begin gmres loop.
55    while((rho > errtol) & (k < kmax))
56      k = k+1;
57      % Matrix vector product.
58      v(:,k+1) = AA*v(:,k);
59      % Begin modified GS. May need to reorthogonalize.
60      for j = 1:k
61        h(j,k) = v(:,j)'*v(:,k+1);
62        v(:,k+1) = v(:,k+1) - h(j,k)*v(:,j);
63      end
64      h(k+1,k) = (v(:,k+1)'*v(:,k+1))^.5;
65      if (h(k+1,k) ~= 0)
66        v(:,k+1) = v(:,k+1)/h(k+1,k);
67      end
68      % Apply old Givens rotations to h(1:k,k).
69      if k>1
70        for i=1:k-1
71          hik = c(i)*h(i,k)-s(i)*h(i+1,k);
72          hipk = s(i)*h(i,k)+c(i)*h(i+1,k);
73          h(i,k) = hik;
74          h(i+1,k) = hipk;
```

```
75          end
76        end
77        normh = norm(h(k:k+1,k));
78        % May need better Givens implementation.
79        % Define and apply new Givens rotations to h(k:k+1,k).
80        if normh ~= 0
81          c(k) = h(k,k)/normh;
82          s(k) = -h(k+1,k)/normh;
83          h(k,k) = c(k)*h(k,k) - s(k)*h(k+1,k);
84          h(k+1,k) = 0;
85          gk = c(k)*g(k) - s(k)*g(k+1);
86          gkp = s(k)*g(k) + c(k)*g(k+1);
87          g(k) = gk;
88          g(k+1) = gkp;
89        end
90        rho = abs(g(k+1));
91        mag(k) = rho;
92      end
93      % End of gmres loop.
94      % h(1:k,1:k) is upper triangular matrix in QR.
95      y = h(1:k,1:k)\g(1:k);
96      % Form linear combination.
97      for i = 1:k
98        u(:) = u(:) + v(:,i)*y(i);
99      end
100     %
101     % Output via semilogy and mesh
102     %
103     k
104     figure(1)
105     semilogy(mag)
106     xlabel('iterations')
107     ylabel('rho')
108     for j = 1:n
109       for i = 1:n
110         I = (j-1)*n + i;
111         u2d(i,j) = u(I);
112       end
113     end
114     figure(2)
115     uu2d = zeros(n+2);
116     uu2d(2:n+1,2:n+1) = u2d;
117     mesh(uu2d)
```

11.5.2 Exercises

Consider the MATLAB® code gmresfull_13.m.

1. Execute the code for variable n and *errtol*. Observe the number of iterations required for "convergence."

2. Execute the code for variable $a1, a2$ and c. Observe the graphic output.

3. Modify gmresfull_13.m to restart after some fixed number of iterations, say 20. Use the last estimate for the solution to be the new initial starting point in the restarted gmres method. Experiment with the number of restarts and compare with the original gmresfull_13.m.

12

Nonlinear Problems and Least Squares

The first two sections will focus on approximating solution to nonlinear systems. The methods of successive approximations or Picard's method and Newton's method will be studied for systems of equations. In the case of one variable these methods are introduced in calculus as an application of the mean value theorems.

The Levenberg-Marquardt method will be used to approximate the solutions to nonlinear least squares problems. Application to parameter identification of differential equations will be illustrated. In particular, the elementary models of epidemics will be discussed. This includes SIR and SIRD (death is allowed). Two limitations of this model are nonhomogeneous population and time-dependent parameters. Here the matrices may be ill-conditioned, and the effects of uncertain data are noted.

12.1 Picard Approximation

Nonlinear problems can be formulated as a *fixed point* of a function $x = g(x)$, or equivalently, as a *root* of $f(x) \equiv x - g(x) = 0$. This is a common problem that arises in computations, and a more general problem is to find n unknowns when n equations are given. In one unknown these methods are

$$x^{k+1} \equiv g(x^k) \qquad \text{(Picard) and}$$
$$x^{k+1} \equiv x^k - f(x^k)/f'(x^k) \quad \text{(Newton)}.$$

Nonlinear problems can have more than one solution. For example, consider finding $\pm\sqrt{2}$ via either the fixed point with $g(x) = x + 2 - x^2$ or the root with $f(x) = 2 - x^2$. The Picard fails, and Newton works for good initial x^0! This can be explained by the contractive property

$$|g(x) - g(y)| \le r\,|x - y| \text{ where } r < 1.$$

For $g(x) = x + 2 - x^2$ the parameter r is estimated by $|g'(x)| = |1 - 2x|$. For x^0 near either solution this is larger than one. Newton can be viewed as a fixed point problem with $g(x) = x - f(x)/f'(x)$. Consider $f(x) = 2 - x^2 = 0$. The derivative of $f(x)$ is $-2x$, note $g'(x) = -1/x^2 + 1/2$ so that $g(x)$ is contractive near either root.

DOI: 10.1201/9781003304128-12

The Picard method was introduced in 1890. The Newton method was first considered in 1669, and in 1697 by Raphson and for systems in 1740 by Simpson.

Let $D \subset \mathbb{R}^n$ be *closed* (all limit points must be in D), *bounded* (all elements $\|x\| \leq B$) and *convex* (all line segments, $x + t(y - x)$ with $0 \leq t \leq 1$, must be in D). The objective is to find a fixed point $x \in D$ of $G : D \to \mathbb{R}^n$.

Picard Method for $G(x) = x$.
$$x^0 \in D$$
$$for \ k = 0, maxk$$
$$x^{k+1} = G(x^k)$$
$$\text{test for convergence}$$
$$end$$

Example 12.1.1. Consider a splitting of $Ax = d$ and the iterative method

$$x^{k+1} = B^{-1}Cx^k + B^{-1}d \equiv G(x^k).$$

The goal is to find the fixed point $x = G(x)$. This requires

$$\|G(x) - G(y)\| = \|B^{-1}C(x - y)\|$$
$$\leq \|B^{-1}C\| \, \|x - y\|$$
$$< 1 \, \|x - y\| \, .$$

Example 12.1.2. Consider the system of ordinary differential equations

$$x' = f(t, u) \in \mathbb{R}^n.$$

The Euler-trapezoid method requires the solution at each time step of

$$\frac{x - \overline{x}}{\Delta t} = \frac{f(t, x) + f(\overline{t}, \overline{x})}{2} \in \mathbb{R}^n.$$

$$x = \frac{\Delta t}{2} \left(f(t, x) + f(\overline{t}, \overline{x}) \right) + \overline{x} \equiv G(x).$$

Example 12.1.3. Consider a steady state heat diffusion with temperature dependent conductivity and heat source or sink. Here the matrix and the right side will depend on the unknown

$$A(u)u = f(u).$$

If the matrix has an inverse, then this may be written as a fixed point

$$u = A(u)^{-1}f(u) \equiv G(u).$$

Definition. *Let* $D \subset \mathbb{R}^n$ *and* $G : D \to \mathbb{R}^n$. *G is called contractive if and only if for all* $x, y \in D$

$$\|G(x) - G(y)\| \leq r \, \|x - y\| \ \text{and} \ r < 1.$$

The above examples can lead to contractive maps. Example 12.1.1 just requires $\left\| B^{-1}C \right\| < 1$. In Example 12.1.2 may require the time step to be small in the case where $f(u) = [f_i(u_i)]$ has bounded derivatives. A special case of Example 12.1.3 is

$$Au = f(u) = [f_i(u_i)]$$
$$u = A^{-1}[f_i(u_i)] \equiv G(u).$$

The contractive property can be estimated by

$$\begin{aligned}
\|G(u) - G(v)\|_\infty &= \left\| A^{-1}[f_i(u_i)] - A^{-1}[f_i(v_i)] \right\|_\infty \\
&\leq \left\| A^{-1} \right\|_\infty \left\| [f_i(u_i) - f_i(v_i)] \right\|_\infty \\
&\leq \left\| A^{-1} \right\|_\infty \max \left| \frac{df_i}{du_i} \right| \|u - v\|_\infty .
\end{aligned}$$

A particular illustration is given in the code piccool.m at the end of this section.

In order to determine the contractive property on closed bounded convex set D, define

$$f(t) \equiv g_i(x + t(y - x)) \text{ where } G(x) = [g_i(x)] .$$

By the chain rule

$$f'(t) = \sum_j g_{ix_j}(x + t(y - x))(y_j - x_j).$$

Next integrate from $t = 0$ to $t = 1$

$$\begin{aligned}
g_i(y) - g_i(x) &= f(1) - f(0) \\
&= \int_0^1 \sum_j g_{ix_j}(x + t(y - x))(y_j - x_j).
\end{aligned}$$

Let $z(t) = x + t(y - x)$ and define the Jacobian matrix $G' \equiv [g_{ix_j}]$. The above is

$$g_i(y) - g_i(x) = \int_0^1 (G'(z(t))(y - x))_i dt. \tag{12.1.1}$$

Use the infinity norm and assume g_{ix_j} are bounded on D

$$\begin{aligned}
\|G(y) - G(x)\|_\infty &\leq \int_0^1 \|G'(z(t))\|_\infty dt \, \|y - x\|_\infty \\
&\leq \max_D \|G'\|_\infty \|y - x\|_\infty .
\end{aligned}$$

This is summarized in the Lemma.

Lemma 12.1. *Let $D \subset \mathbb{R}^n$ be closed, bounded and convex. If the partial derivative g_{ix_j} are continuous on D, then*

$$\|G(y) - G(x)\|_\infty \leq \max_D \|G'\|_\infty \|y - x\|_\infty .$$

Theorem 12.1.1. *(Contraction and Fixed Point) Let $D \subset \mathbb{R}^n$ be closed, bounded and convex. If $G : D \to D$ $(G(x) \in D$ when $x \in D)$ and is contractive, then*

(i). *for all $x^0 \in D$ the iteration $x^{k+1} = G(x^k) \in D$ converges to $x \in D$,*

(ii). *$G(x) = x$ and*

(iii). *x is the only fixed point.*

Proof. The convergence will be established by showing the iteration is Cauchy via a comparison with the geometric series. First note the contractive property gives

$$\begin{aligned}
\left\|x^{k+1} - x^k\right\|_\infty = \left\|G(x^k) - G(x^{k-1})\right\|_\infty &\leq r \left\|x^k - x^{k-1}\right\|_\infty \\
&\leq r^2 \left\|x^{k-1} - x^{k-2}\right\|_\infty \\
&\vdots \\
&\leq r^k \left\|x^1 - x^0\right\|_\infty .
\end{aligned}$$

Let $m = k + l$, use the triangle inequality and the above

$$\begin{aligned}
\left\|x^m - x^k\right\|_\infty &\leq \left\|x^m - x^{m-1}\right\|_\infty + \cdots + \left\|x^{k+1} - x^k\right\|_\infty \\
&\leq \left(r^{m-1} + \cdots + r^k\right) \left\|x^1 - x^0\right\|_\infty .
\end{aligned}$$

Since $r < 1$, the geometric series must a converge and is Cauchy. This forces the iteration $x^{k+1} = G(x^k)$ to be Cauchy in $\mathbb{R}^n$ and therefore it must converge. Since D is closed, the limit must also be in D.

The proof that the limit is a fixed point follows from

$$\begin{aligned}
x - G(x) &= x - x^{k+1} + x^{k+1} - G(x) \\
&= x - x^{k+1} + G(x^k) - G(x).
\end{aligned}$$

Use the contractive property to get

$$\begin{aligned}
\|x - G(x)\|_\infty &\leq \left\|x - x^{k+1}\right\|_\infty + \left\|G(x^k) - G(x)\right\|_\infty \\
&\leq \left\|x - x^{k+1}\right\|_\infty + r \left\|x^k - x\right\|_\infty .
\end{aligned}$$

The convergence of implies $x = G(x)$.

The unique fixed point also follows from the contractive property. If x and y are both fixed points, then $x = G(x)$ and $y = G(y)$ and

$$\|x - y\|_\infty = \|G(x) - G(y)\|_\infty \leq r \|x - y\|_\infty < \|x - y\|_\infty .$$

This is a contradiction if $x \neq y$.   ∎

Example 12.1.4. Solve the $n = 2$ nonlinear system

$$x_1 = 1 + \Delta t(-x_1^2 + ax_2) \text{ and}$$
$$x_2 = 2 + \Delta t(-x_1^2 + ax_2).$$

We are seeking a fixed point where

$$G(x) = \begin{bmatrix} 1 + \Delta t(-x_1^2 + ax_2) \\ 2 + \Delta t(-x_1^2 + ax_2) \end{bmatrix} = \begin{bmatrix} x_1 \\ x_2 \end{bmatrix} \text{ and}$$

$$G'(x) = \Delta t \begin{bmatrix} -2x_1 & a \\ b & -2x_2 \end{bmatrix}.$$

For $a = 2$ and $b = 3$, the Picard iteration will converge provided $\Delta t \leq 0.10$. See the code picard2d.m.

12.1.1 MATLAB® code piccool.m

Consider glowing hot wire which is gaining heat from electrical current and is loosing heat from radiation. A nonlinear model involves Fourier heat law and Stefan-Boltzmann law.

$$-(Ku_x)_x = A\varepsilon\sigma_{sb}(u_{sur}^4 - u^4) + S \text{ where}$$
$$K = K(u) \text{ is thermal conduction}$$
$$A = \text{surface area}$$
$$\varepsilon = \text{emissivity}$$
$$\sigma_{sb} = Stefan\text{-}Boltzmann \text{ constant}$$
$$S = \text{heat source.}$$

In order to simplify the calculation and modeling, we have kept $K = 1$, $A = 0.01$, $\varepsilon = 1$, $\sigma_{sb} = 5.67e - 8$ and $S = 12000$. The boundary conditions are fixed at $u(0) = 300 = u(1)$.

This code uses two functions as indicated in lines 7–12. The matrix is defined in lines 13–32. Lines 35–59 are the Picard iteration loop. The right side for $f(u)$ is computed in lines 35–46, and the linear solve step is done in line 47. The stopping test is given by lines 48 and 56–58. In this case 46 iterations were required to meet this test and the maximum temperature was 1535. The results of the first four iterations is given in Figure 12.1.1 and indicate a maximum temperature equal to 1452 for four iterations.

```
1    clear; clf
2    % This code is for a nonlinear ODE.
3    % Stefan radiative heat loss is modeled.
4    % Picard's method is used.
5    % The linear steps are solved by A\d.
6    %
```

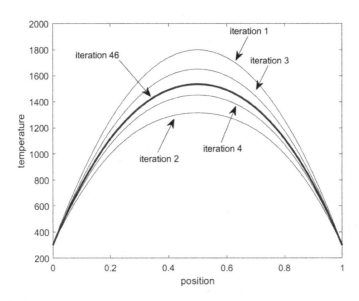

FIGURE 12.1.1
Picard iterations.

```
7    %    function fcool = fcool(u)
8    %    fcool = .01*(5.67e-8)*(300^4 - u^4)+ 12000;
9    %
10   %    function fcoolp = fcoolp(u)
11   %    fcoolp = .01*(5.67e-8) *(-4)*u^3;
12   %
13   uo = 300.;
14   n = 39;
15   h = 1./(n+1);
16   A = zeros(n);
17   for i = 1:n        % compute A
18     if i==1
19        A(i,i) = 2;
20        A(i,i+1) = -1;
21     else
22        if i<n
23           A(i,i) = 2;
24  ~        A(i,i-1) = -1;
25           A(i,i+1) = -1;
26        else
27           A(i,i) = 2;
28           A(i,i-1) = -1;
29        end;
```

```
30        end
31     end
32     F = zeros(n,1);
33     u = ones(n,1)*uo;
34     % Begin Picard iteration
35     for m = 1:20
36       for i = 1:n        % compute f(u)
37         if i==1
38           F(i) = fcool(u(i))*h*h + uo;
39         else
40           if i<n
41             F(i) = fcool(u(i))*h*h ;
42           else
43             F(i) = fcool(u(i))*h*h + uo;
44           end;
45         end
46       end
47       newu = A\F;        % solve linear system
48       error = norm(u - newu);
49       u = newu;
50       [m error]
51       uu = [300 u' 300];
52       x = 0:h:1;
53       plot(x,uu)
54       hold on
55       pause
56       if error < 0.00001
57         break;
58       end
59     end         % End Picard iteration
60     m
61     error
```

12.2 Newton Method

In this section we assume the solution $x^* \in \mathbb{R}^n$ of $F(x^*) = 0_{n \times 1}$ exists and $F(x) = [F_i(x)]$ has continuous partial derivative F_{ix_j} in a ball centered at x^*, D. Newton's method is based on the approximation

$$F(x^*) - F(x^k) \simeq F'(x^k)(x^* - x^k).$$

Use $F(x^*) = 0_{n \times 1}$, let x^* on the right side be replaced by the next iterate, and replace the approximation with equality to get

$$0_{n \times 1} - F(x^k) = F'(x^k)(x^{k+1} - x^k).$$

If $F'(x^k)$ has an inverse, then one can solve for the next iterate. Some difficulties include the initial approximation x^0, the computation of the Jacobian matrix x^0 and the solution step. However, once the these are resolved the iterates often have quadratic convergence.

Newton Method for $F(x) = 0_{n \times 1}$.
$\quad$ choose x^0 "close" to root
$\quad$ *for* $k = 0, maxk$
$\quad\quad$ compute $F'(x^k)$ and $F(x^k)$
$\quad\quad$ solve $F'(x^k)\Delta x = F(x^k)$
$\quad\quad$ $x^{k+1} = x^k - \Delta x$
$\quad\quad$ test for convergence
$\quad$ *end*

Example 12.2.1. Consider $n = 2$ for the possible intersection(s) of a circle and hyperbola

$$F_1 = x_1^2 + x_2^2 - 1 = 0 \text{ and}$$
$$F_1 = x_1^2 - x_2^2 + d = 0.$$

There can be no solutions $(d > 1)$, two solutions $(d = 1)$ or four solutions $(0 < d < 1)$. Which solution you get depends of the initial approximation. The x^k cannot be on the axes because $det(F'(x^k)) = -8x_1 x_2 = 0$ where

$$F'(x) = \begin{bmatrix} 2x_1 & 2x_2 \\ 2x_1 & -2x_2 \end{bmatrix}.$$

Example 12.2.2. Consider solving the ordinary differential equation $-u_{xx} = f(u)$. Use the finite difference method to solve the approximation of $f(u) + u_{xx} = 0$

$$F_i(u_{i-1}, u_i, u_{i+1}) \equiv (\Delta x)^2 f(u_i) + u_{i-1} - 2u_i + u_{i+1} = 0.$$

In this case the Jacobian matrix $F'(u)$ will be tridiagonal, and this is illustrated in the code newtcool.m and the end of this section.

The local and quadratic convergence will be established by Lemma 12.1 and Lemma 12.2. These play the same function as the mean value theorems for single variable Newton method analysis.

Lemma 12.2. *Let* $D \subset \mathbb{R}^n$ *be closed and convex. If* $F(x)$ *has continuous partial derivatives,* $F \in C^1(D)$, *and*

$$\|F'(y) - F'(x)\|_\infty \leq \gamma \|y - x\|_\infty$$

then
$$\left\| F(y) - F(x) - F'(x)(y-x) \right\|_{\infty} \le \frac{1}{2}\gamma \left\| y - x \right\|_{\infty}^{2}$$

Proof. Use line (12.1.1) with $G = F$, $z(t) = x + t(y-x)$ and $F'(x)(y-x) = \int_0^1 F'(x)(y-x)dt$ to get

$$F(y) - F(x) - F'(x)(y-x) = \int_0^1 (F'(z(t)) - F'(x))(y-x)dt$$

$$\left\| F(y) - F(x) - F'(x)(y-x) \right\|_{\infty} \le \int_0^1 \left\| F'(z(t)) - F'(x) \right\|_{\infty} dt \, \left\| y - x \right\|_{\infty}.$$

Use the assumption to bound the integral

$$\left\| F(y) - F(x) - F'(x)(y-x) \right\|_{\infty} \le \int_0^1 t \left\| y - x \right\|_{\infty} \gamma dt \, \left\| y - x \right\|_{\infty}$$

$$= \frac{1}{2}\gamma \left\| y - x \right\|_{\infty}^2.$$

∎

Theorem 12.2.1. *(Local Convergence to Root)* Let $D(x^*, \delta) = \{y : \|y - x^*\|_{\infty} \le \delta\}$. Consider the Newton iterates. If $F(x^*) = 0_{n \times 1}, F \in C^1(D(x^*, \delta)), F'(x^*)$ is nonsingular and

$$\|F'(x) - F'(x^*)\|_{\infty} \le \gamma \|x - x^*\|_{\infty}, \qquad (12.2.1)$$

then there exists $\varepsilon > 0$ with $\varepsilon \le \delta$ such that
(i). *(local convergence) x^k converges to x^* when $x^0 \in D(x^*, \epsilon)$, and*
(ii). *(quadratic convergence) there exists C so that*

$$\left\| x^{k+1} - x^* \right\|_{\infty} \le C \left\| x^k - x^* \right\|_{\infty}^2$$

Proof. In order for Newton iterations to be defined, the inverse of $F'(x)$ must exist for x near the root x^*. This follows from the perturbation result in Theorem 11.1.2 with $A = F'(x^*)$ and $C = F'(x)$. Assumption $\|A - C\| \, \|A^{-1}\| < 1$ is satisfied because of line (12.2.1) and restricting the δ in the ball $D(x^*, \delta)$. Choose δ so that
$$\left\| F'(x)^{-1} \right\| \le 2 \left\| F'(x^*)^{-1} \right\|.$$

In order to show the quadratic convergence, use $F(x^*) = 0_{n \times 1}$, Lemma 12.2 and the assumption in line (12.2.1).

$$\begin{aligned} G(x) - x^* &= G(x) - G(x^*) \\ &= (x - F'(x)^{-1}F(x)) - (x^* - F'(x^*)^{-1}F(x^*)) \\ &= (x - x^*) - F'(x)^{-1}F(x) \\ &= F'(x)^{-1}[F'(x)(x - x^*) - (F(x) - F(x^*))]. \end{aligned}$$

Use the infinity norm

$$\|G(x) - x^*\|_\infty \le \left\|F'(x)^{-1}\right\|_\infty \left\|F'(x)(x - x^*) - (F(x) - F(x^*))\right\|_\infty$$
$$\le \left\|F'(x)^{-1}\right\|_\infty \frac{\gamma}{2} \|x - x^*\|_\infty^2$$
$$\le 2 \left\|F'(x^*)^{-1}\right\|_\infty \frac{\gamma}{2} \|x - x^*\|_\infty^2$$
$$\le C_1 \|x - x^*\|_\infty^2 \text{ where } C_1 = \gamma \max_{x \in D(x^*, \delta)} \left\|F'(x^*)^{-1}\right\|_\infty$$

Further restrict $D(x^*, \delta)$ so that $C_1 \|x - x^*\|_\infty \le C_1\epsilon < 1$. In this case

$$\|G(x) - x^*\|_\infty \le C_1\epsilon \|x - x^*\|_\infty \text{ and } r \equiv C_1\epsilon < 1.$$

This means G maps back into $D(x^*, \epsilon)$.

Provided the starting point $x^0 \in D(x^*, \epsilon)$ each $G(x^k) = x^{k+1} \in D(x^*, \epsilon)$ and

$$\left\|x^{k+1} - x^*\right\|_\infty \le C_1 \left\|x^k - x^*\right\|_\infty^2$$
$$\le r \left\|x^k - x^*\right\|_\infty$$
$$\vdots$$
$$\le r^k \left\|x^0 - x^*\right\|_\infty.$$

Since $r < 1$, the Newton iterates converge and converge quadratically.   ∎

12.2.1 MATLAB® code newtcool.m

This solves the same problem as in piccool.m in the previous section. Here we use Newton's method with $F(u) = [F_i(u_{i-1}, u_i, u_{i+1})]$ as indicated in Example 12.2.2. Let $h = \Delta x = 1.0/(n+1)$ and $1 \le i \le n$ for the unknowns u_i

$$F_i(u_{i-1}, u_i, u_{i+1}) \equiv h^2 f(u_i) + u_{i-1} - 2u_i + u_{i+1} = 0 \text{ with}$$
$$u_0 = 300 \text{ and } u_{n+1} = 300.$$

Because each component of $F(u)$ depends on three unknowns, the Jacobian matrix $F'(u)$ will be tridiagonal

$$F'(u) = \begin{bmatrix} h^2 f'(u_1) - 2 & 1 & 0 & 0 & 0 \\ 1 & h^2 f'(u_2) - 2 & 1 & 0 & 0 \\ 0 & 1 & \ddots & & \vdots \\ 0 & 0 & & \ddots & 1 \\ 0 & 0 & \cdots & 1 & h^2 f'(u_n) - 2 \end{bmatrix}.$$

The Newton algorithm is executed in lines 20–52. Lines 22–39 compute $F(u)$ and $F'(u)$. The solve step is done on line 40, and line 41 has the Newton update. There are two possible "tests" for convergence as indicated in lines 42 and 49–51. The Newton method converges in 6 iterations is must faster than the Picard method which took 46 iterations.

```
1    clear; clf
2    % This code is for a nonlinear ODE.
3    % Stefan radiative heat loss is modeled.
4    % Newton's method is used.
5    % The linear steps are solved by A\d.
6    %
7    %    function fcool = fcool(u)
8    %    fcool = .01*(5.67e-8)*(300^4 - u^4)+ 12000;
9    %
10   %    function fcoolp = fcoolp(u)
11   %    fcoolp = .01*(5.67e-8) *(-4)*u^3;
12   %
13   uo = 300.;
14   n = 39;
15   h = 1./(n+1);
16   FP = zeros(n);
17   F = zeros(n,1);
18   u = ones(n,1)*uo;
19   % Begin Newton iteration
20   for m = 1:20
21   % Compute Jacobian matrix
22     for i = 1:n
23       if i==1
24         F(i) = fcool(u(i))*h*h + u(i+1) - 2*u(i) + uo;
25         FP(i,i) = fcoolp(u(i))*h*h - 2;
26         FP(i,i+1) = 1;
27       else
28         if i<n
29           F(i) = fcool(u(i))*h*h + u(i+1) - 2*u(i) + u(i-1);
30           FP(i,i) = fcoolp(u(i))*h*h - 2;
31           FP(i,i-1) = 1;
32           FP(i,i+1) = 1;
33         else
34           F(i) = fcool(u(i))*h*h - 2*u(i) + u(i-1) + uo;
35           FP(i,i) = fcoolp(u(i))*h*h - 2;
36           FP(i,i-1) = 1;
37         end
38       end
39     end
40     du = FP\F; % Solve linear system
41     u = u - du;
```

```
42        error = norm(F);   %  error = norm(du,Inf)
43        [m error]
44        uu = [300 u' 300];
45        x = 0:h:1;
46        plot(x,uu)
47        pause
48        hold on
49        if error < 0.00001
50           break;
51        end
52     end % End Newton iteration
53     m
54     error
```

Remark. If the $F'(x)$ is not sparse, then the computation or approximation of the partial derivatives can be costly. Another concern is making the initial iterate, x^0, suitably close to the root, x^*.

12.3   Levenberg-Marquardt Method

The objective is to identify n parameters from m measured data at time or space. The model is usually an ODE or a PDE. Consider the following terms:

$y \in \mathbb{R}^m$ measured data vector with $m > n$,

$p \in \mathbb{R}^n$ is the parameter vector,

$f(p,t)$ real solution from a DE with given parameters and

error or *residual* $\equiv y_i - f(p,t_i)$.

Nonlinear Least Squares Problem.

Find p so that the residual in a minimum

$$\min \sum_{i=1}^{m} (y_i - f(p,t_i))^2.$$

Use a Linear Approximation.

$f(p,t_i) \simeq f(p^0,t_i) + f'(p^0,t_i)(p - p^0)$ and

$f'(p^0,t_i) \equiv [f_{p_j}(p^0,t_i)]$ is an $m \times n$ matrix of partial derivatives.

Find p so that the residual in a minimum

$$\min \sum_{i=1}^{m} ((y_i - f(p^0,t_i)) - f'(p^0,t_i)(p - p^0))^2.$$

This is a least squares problem $A\Delta p = d$ where

$$A \equiv [f_{p_j}(p^0, t_i)] \text{ is } m \times n,$$
$$\Delta p = p - p^0 \text{ and } d = [y_i - f(p^0, t_i)].$$

Solve this and let p be the next iterate $p^1 = p$. Continue this to get the Gauss-Newton method.

Gauss-Newton Method. Let $A \equiv [f_{p_j}(p^k, t_i)]$.

$$p^{k+1} = p^k + \Delta p \text{ where } A^T A\Delta p = A^T(y - f(p^k, t_i)).$$

Notation. Let $F(p) \equiv y - f(p)$. $F'(p) = 0 - f'(p)$ and so the least squares problem is

$$F'(p)^T F'(p)\Delta p = -F'(p)^T(F(p)).$$

If $F'(p)$ has full column rank, then

$$p^{k+1} = p^k - (F'(p^k)^T F'(p^k))^{-1} F'(p^k)^T(F(p^k)). \tag{12.3.1}$$

This method does not always converge to the solution of nonlinear least squares problem. The following example has $m = 2$ and $n = 1$ with $y = 0_{2 \times 1}$.

Example 12.3.1. Let $f : \mathbb{R}^1 \to \mathbb{R}^2$

$$f(p) = \begin{bmatrix} p + 1 \\ ap^2 + p - 1 \end{bmatrix}.$$
$$r = y - f(p)$$
$$\min r^T r = \min(\frac{1}{2}(p+1)^2 + \frac{1}{2}(ap^2 + (p-1))^2).$$

Define $G(p) = r^T r$ and note $G'(0) = 0$ and $G''(0) = 2(1 - a) > 0$ for $a < 1$. Thus $p = 0$ is a solution. However, the Gauss-Newton method fails for $a < -1$ as is illustrated by the following calculation which oscillates between $+1$ and -1.

```
clear
a = -1.1; p = 0.1;
for k = 1:100
    F = -[p+1; a*p^2+p-1];
    Fp = -[1; 2*a*p+1];
    newp = p - (Fp'*Fp)/(Fp'*F);
    p = newp;
    pp(k) = p;
end
plot(pp)
```

The Levenberg-Marquardt method has two enhancements: the α and λ real parameters. These can be adjusted within the iteration so to accelerate or to minimize the residual. The parameters were introduced by K. Levenberg in 1944, [14], and by D. Marquardt in 1963, [16]; see [20, Section 8.5]. Let $y \in \mathbb{R}^m$ be the data and let $f : \mathbb{R}^n \longrightarrow \mathbb{R}^m$ be a function of the unknown parameters p. Approximate the solution of

$$\min_p (y - f(p))^T (y - f(p)).$$

The α and λ parameters are inserted in line (12.3.1) of the Gauss-Newton method to obtain the Levenberg-Marquardt method.

Levenberg-Marquardt Method.
>   Let $F(p) = y - f(p)$.
> $p^0 \in \mathbb{R}^n$ initial estimate for the parameters
>   *for* $k = 0, maxk$
>>   choose α_k and λ_k
>>   compute $F(p^k)$ and the $m \times n$ matrix $F'(p^k)$
>>   solve the least squares problem
>
>$$(F'(p^k)^T F'(p^k) + \lambda_k I_n)\Delta p = F'(p^k)^T F(p^k)$$
>
>> $p^{k+1} = p^k - \alpha_k \Delta p$
>>   test for convergence
>   *end*

Remark. In the following examples the parameters α_k and λ_k were constants determined by experimentation. MATLAB® optimization toolbox with fminunc and others are much more robust solvers.

Example 12.3.2. Return price model in Section 3.2. The price prediction based on six previous observations was a discrete model, which could be viewed as an approximation of a first order differential equation

$$u' = c(p_{\min} - u) \text{ and } u(0) = 2080.$$

Here the parameters are c and $p_{\min}$. The solution of the differential equation is

$$\begin{aligned}
u(t) &= (2080) - pmin)\ exp(ct)\ +\ pmin \\
&= (2080) - p_2)\ exp(p_1 t)\ +\ p_2.
\end{aligned}$$

There are six measurements for $u(t_i)$ to approximate the two parameters. The matrix in the Levenberg-Marquardt algorithm is 6×2. The computer code levmarqprice.m is in the next section. One can use the results in price_expdata.m as a starting point for the code in levmarqprice.m.

Example 12.3.3. Consider over damped mass spring system where damping, spring constant, initial position and initial velocity are four unknown parameters. The model has the form

$$mx'' + cx' + kx = 0$$

If the position is measured at m time steps, find these parameters.

$$f(p, t) = p_1 \exp(p_2 t) + p_3 \exp(p_4 t).$$

The interested reader should download levmarqheat.m. Here there are eight measurements for $x(t_i)$ to approximate the four parameters. The matrix in the Levenberg-Marquardt algorithm is 8×4.

12.3.1 MATLAB® code levmarqprice.m

Line 18 contains the initial estimate for the parameters. Here one could have used the results of the finite difference model in price_expdata.m. Line 20 contains constant values for α and λ. In more sophisticated implementations, these are adjusted within the algorithm's loop in lines 21–32. The algorithm converged in only 11 iterations, and this is in contrast to levmarqheat.m. The outputs for the residual and price at time equal to 8 are 40.2620 and 1812.6 from price_expdata.m, and 31.5225 and 1813.3 from levmarqprice.m. The graphical output is given in Figure 12.3.1.

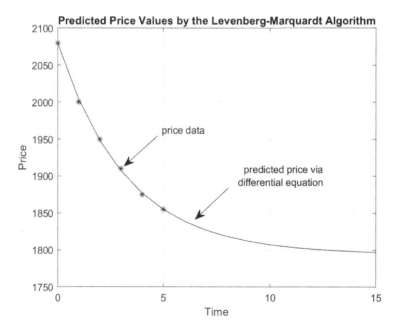

FIGURE 12.3.1
Levenberg-Marquardt prediction of price.

```
1    % This code illustrates the Levenberg-Marquardt algorithm
2    % for curve fitting to price data (or Newton cooling).
3    % Consider the first order ODE: u' = c(pmin - u) and
            u(0) = 2080.
4    % Suppose c and pmin are unknown, and measured values for
            u(t_i)
5    % are known where i = 1:m. Approximate the two unknown
6    % parameters from the measured data.
7    %
8    % f_i(p1,p2,t) = (2080) - pmin) exp(ct) + pmin
9    %                =(2080 - p(2)) exp(p(1)t) + P(2)
10   % F(p1,p2,t_i) = udata(t_i) - f_i(p1,p2,t)
11   % FP is an mx2 matrix with components f_i,_p(1) and
            f_i,_p(2).
12   %
13   clear; clf;
14   % Define "data"
15   tdata = 0:1:5; tdata = tdata';
16   udata = [ 2080 2000 1950 1910 1875 1855]';
17   % Initial guess for parameters
18   p = [ -1 1500]';
19   % Begin Levenberg-Marquardt
20   lam = 0.01; alpha = 1.0;
21   for lm = 1:20
22     F = udata -((2080 - p(2))*exp(p(1)*tdata) + p(2));
23     FP = -[(2080 - p(2))*exp(p(1)*tdata).*tdata...
              -exp(p(1)*tdata)+1];
24
25     newp = p - alpha*(FP'*FP + lam*eye(2))\(FP'*F);
26     error = norm(newp - p);
27     p = newp;
28     if error < 0.00001
29         break
30     end
31   % pause
32   end
33   lm
34   error
35   % Compare computed and initial parameters
36   plot(tdata,udata,'*')
37   hold on
38   time = [0:1:15];
39   newu = (2080 - p(2))*exp(p(1)*time) + p(2);
40   plot(time,newu, 'r')
41   title('Predicted Price Values by the ...
            Levenberg-Marquardt Algorithm')
```

```
42    xlabel('Time')
43    ylabel('Price')
44    pmin = p(2)
45    c = p(1)
46    display('Predicted price at 8 = ')
47    newu(9)
48    r = udata - newu(1:6)'
49    rTr = r'*r
50    cond(FP)
51    S = svd(FP)
```

```
>> levmarqprice
lm =
     11
error =
   3.5374e-06
pmin =
   1.7942e+03
c =
  -0.3095
Predicted price at 8 =
ans =
   1.8183e+03
r =
         0
   -3.9384
    1.8776
    2.8369
   -2.1060
   -0.0493
rTr =
   31.5225
ans =
   2.3261e+03
S =
  675.1095
    0.2902
```

12.4   SIRD Epidemic Models

The SIRD model has four differential equations for four unknown functions of time. The functions are $S(t)$ = susceptible, $I(t)$ = infected, $R(t)$ = recovered and $D(t)$ = death populations. If there are no deaths, this is called the SIR model. This was introduced by W. Kernmack and A. McKendrick (1927).

There are four primary components that govern the growth of an infected population: probability of transmission, the degree of susceptibility, number of contacts with an infected and duration of contacts. These components determine the size of the three parameters, a, b_r and b_d, in the following system of differential equations.

SIRD Model.

$$\frac{dS}{dt} = -aSI,$$

$$\frac{dI}{dt} = aSI - b_r I - b_d I,$$

$$\frac{dR}{dt} = b_r I \text{ and}$$

$$\frac{dD}{dt} = b_d I.$$

The parameter a is often called the "effective contact" because it has the primary components of probability, susceptibility and contacts. The parameters b_r and b_d are the removal rates from the infected to either the recovered or death populations. Their reciprocals determine the durations of the infected. Here we assume the four primary components are fixed, and the three parameters are constants.

The populations move from S to I and then to either R or D. S is a decreasing function of time, and R and D are increasing functions of time. The total population must be constant because the sum of the derivatives is zero. The equation of the infected may be written as

$$\frac{dI}{dt} = I(aS - b_r - b_d) \text{ and } I \text{ is positive.}$$

So, if the second factor is positive (negative), then I will increase (decrease). This may also be written as

$$\frac{aS(t)}{b_r + b_d} > 1 \text{ (or } < 1) \text{ for } I \text{ to increase (or decrease).}$$

Initially, the infection will increase only if $I(0) > 0$ and $aS(0) - b_r - b_d > 0$. The infected will be a maximum at time $t_{\max}$ where $aS(t_{\max}) - b_r - b_d = 0$, which follows from

$$\frac{dI}{dt}(t_{\max}) = 0 \text{ and } \frac{d^2 I}{dt^2}(t_{\max}) < 0.$$

An alternate analysis is to determine if a single infected will infect more than one susceptible. This can be broken into product of three factors:
(new infections per contact)
(contacts per time) and
(duration time per infected).
In the SIR model where $b_d = 0$, this is $a(1/b) > 1$.

Because this model is nonlinear, one must approximate the solution using numerical methods. In the next subsection this is done using Runge-Kutta variable step size method. The first calculation illustrates the effect of decreasing the effective contact parameter, see Figure 12.4.1. Note, the maximum of the infection decreases and moves to the right. The other calculations in the next section determine the three parameters resulting from additional observations.

12.4.1 MATLAB® code sird_parid.m

The primary objective of this code is to illustrate how the Levenberg-Marquardt algorithm can be used to approximate the three parameters in the SIRD model. This is done in lines 117–156. For each of the four functions there are 12 data points. So, the objective function is a 48×1 column vector and the derivative matrix is 48×3. The columns in the derivative matrix are partial derivatives, and they are approximated by finite differences (can be tricky).

Lines 37–68 solves the SIRD model for given parameters, and this was used to generate Figure 12.4.1. Lines 70–92 define test data (would be measured values of the four functions). Lines 98–116 give a first estimate of three

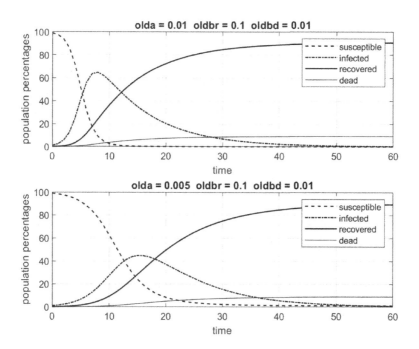

FIGURE 12.4.1
Variable contact parameter.

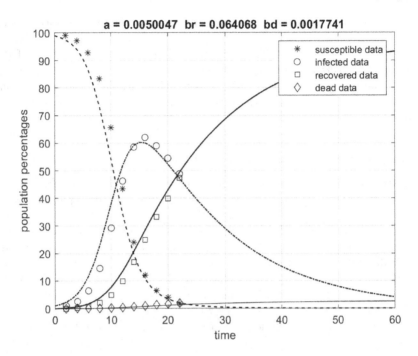

FIGURE 12.4.2
Parameter identification.

parameters using the first order finite difference approximation of the SIRD system. This is gives a least squares problem, which is solved in line 111. These are used as an initial estimate for the more accurate Levenberg-Marquardt algorithm.

The implementation requires many evaluations of the objective function and the derivative matrix. These are done using the higher order ode45 implementation of the variable step size Runge-Kutta method. See lines 117, 131,136, 141 and 149.

The output is given in Figure 12.4.2 where the data points are indicated by the isolated points. The three computed parameters are listed at the top of the figure. The four curves were computed using these parameters. The reader will find it interesting to experiment with different numbers of data points.

```
1   % This code uses least squares to identify three parameters
2   % in the SIRD model:
3   %    S_t = -a SI,
4   %    I_t = a SI - (br + bd) I,
5   %    R_t = br I and
6   %    D_t = bd I where
7   %         a = "contagious" (effective contact) parameter,
8   %         br = "recovery" parameter and
```

```
9    %            bd = "death" parameter.
10   % The data is given in the vectors Sd, Id Rd and Dd,
11   % and they are adjusted by a random variable.
12   % The data is used in the finite difference approximation:
13   %    (S_i+1 - S_i-1)/(2 dt) = -a S_i I_i,
14   %    (I_i+1 - I_i-1)/(2 dt) = a S_i I_i - (br + bd) I_i,
15   %    (R_i+1 - R_i-1)/(2 dt) = br I_i an
16   %    (D_i+1 - D_i-1)/(2 dt) = bd I_i.
17   % Least squares is used to compute the coefficients.
18   % The variable data points can be used.
19   %
20   % function [t y] = sirdid
21   %    global olda oldbr oldbd so io ro do t0 y0 tend
22   %    y0 = [so io ro do];
23   %    to = 0;
24   %    tf = 60;
25   %    opts = odeset('RelTol',1e-8,'AbsTol',1e-8);
26   %    [t y] = ode45('ypsirdid',[to:1:tf],y0,opts);
27   %
28   % function ypsirdid = ypsirdid(t,y)
29   %    global olda oldbr oldbd so io ro do t0 y0 tend
30   %    ypsirdid(1) = -olda*y(1)*y(2);
31   %    ypsirdid(2) = olda*y(1)*y(2) - (oldbr + oldbd)*y(2);
32   %    ypsirdid(3) = oldbr*y(2);
33   %    ypsirdid(4) = oldbr*y(2);
34   %    ypsirdid = [ypsirdid(1) ypsirdid(2)...
35   %                ypsirdid(3) ypsirdid(4)]';
36   %
37   clear; clf(figure(1)); clf(figure(2));
38   global olda oldbr oldbd so io ro do t0 y0 tend
39   figure(1)
40   %
41   olda = 0.01; oldbr = 0.100; oldbd = 0.010;
42   io = 1;
43   ro = 0;
44   do = 0;
45   so = 100 - io - ro - do;
46   t0 = 0; tend = 60;
47   [t y] = sirdid;
48   subplot(2,1,1)
49   plot(t,y(:,1),'b', t,y(:,2),'m', t,y(:,3),'r', t,y(:,4),'k')
50   title(['olda = ' ,num2str(olda),' oldbr = ' ,num2str(oldbr),...
51       ' oldbd = ' ,num2str(oldbd)]);
52   ylabel('population percentages')
53   xlabel('time')
54   legend('susceptible', 'infected', 'recovered', 'dead');
55   grid on
56   %
57   % Decrease "effective contact"
```

```
58    %
59    olda = 0.005; oldbr = 0.100; oldbd = 0.010;
60    [t y] = sirdid;
61    subplot(2,1,2)
62    plot(t,y(:,1),'b', t,y(:,2),'m', t,y(:,3),'r', t,y(:,4),'k')
63    title(['olda = ' ,num2str(olda),' oldbr = ' ,num2str(oldbr),...
64          ' oldbd = ' ,num2str(oldbd)]);
65    ylabel('population percentages')
66    xlabel('time')
67    legend('susceptible', 'infected', 'recovered', 'dead');
68    grid on
69    %
70    % Test Data
71    %
72    nd = 12;
73    td = 2:2:(2*nd);
74    %td = 1:nd;
75    Id = [1.00 2.56 6.37 14.68 29.20 46.35 58.19 61.64 ...
76          59.35 54.50 48.88 43.32 38.15];
77    Rd = [0.00 0.22 0.77 2.09 4.93 9.93 16.92 24.91 32.94 ...
78          40.07 47.29 53.37 58.75];
79    Dd = [0.000 0.008 0.031 0.083 0.197 0.396 0.675 0.994 1.315 ...
80          1.615 1.888 2.131 2.345];
81    Sd = 100 - Id - Rd - Dd;
82    rvec = rand(1,nd);
83    Id(1:nd) = Id(1:nd) + Id(1:nd).*(2*rvec - 1)/100;
84    rvec = rand(1,nd);
85    Rd(1:nd) = Rd(1:nd) + Rd(1:nd).*(2*rvec - 1)/100;
86    rvec = rand(1,nd);
87    Dd(1:nd) = Dd(1:nd) + Dd(1:nd).*(1*rvec )/50;
88    Sd = 100 - Id - Rd - Dd;
89    io = Id(1);
90    ro = Rd(1);
91    do = Dd(1);
92    so = 100 - io - ro - do;
93    %
94    % Parameter ID From First Order Finite Difference
95    %
96    for i = 2:1:nd-1
97      ii = (i-1)*4;
98      d(ii) = (Sd(i+1) - Sd(i-1))/(td(i+1) - td(i-1));
99      d(ii+1) = (Id(i+1) - Id(i-1))/(td(i+1) - td(i-1));
100     d(ii+2) = (Rd(i+1) - Rd(i-1))/(td(i+1) - td(i-1));
101     d(ii+3) = (Dd(i+1) - Dd(i-1))/(td(i+1) - td(i-1));
102     A(ii,1) = -Sd(i)*Id(i); A(ii,2) = 0; A(ii,3) = 0;
103     A(ii+1,1)= Sd(i)*Id(i); A(ii+1,2) = -Id(i); ...
104     A(ii+1,3) = -Id(i);
105     A(ii+2,1)= 0.0; A(ii+2,2) = Id(i); A(ii+2,3) = 0;
106     A(ii+3,1)= 0.0; A(ii+3,2) = 0; A(ii+3,3) = Id(i);
```

```
107    end
108    %
109    meas = nd - 2;
110    m = 4*meas + 1;
111    x = A(2:m,:)\d(2:m)'; % solves the least squares
112    olda = x(1);
113    oldbr = x(2);
114    oldbd = x(3);
115    display('Parameters from finite difference')
116    [olda oldbr oldbd]
117    [t y] = sirdid; % SIRD with new parameters
118    dd = [ Sd(1:nd)'; Id(1:nd)'; Rd(1:nd)'; Dd(1:nd)'];
119    sol = [y(td,1); y(td,2); y(td,3); y(td,4)]; % computed
120    FF = dd - sol; % residual
121    display('Norm of residual')
122    norm(FF)
123    %
124    % Parameter ID from Levenberg-Marquardt Algorithm Using ODE45
125    %
126    lam = 0.1; alpha = 4.0; p = [x(1) x(2) x(3)]';
127    %pause
128    for lm = 1:100
129      da = 0.001;
130      olda = p(1) + da;,oldbr = p(2); oldbd = p(3);
131      [t yp] = sirdid;
132      solp = [yp(td,1); yp(td,2); yp(td,3); yp(td,4)];
133      FFa = (solp - sol)/da;
134      dbr = 0.01;
135      oldbr = p(2) + dbr; olda = p(1); oldbd = p(3);
136      [t yp] = sirdid;
137      solp = [yp(td,1); yp(td,2); yp(td,3); yp(td,4)];
138      FFbr = (solp - sol)/dbr;
139      dbd = 0.001;
140      oldbd = p(3) + dbd; olda = p(1); oldbr = p(2);
141      [t yp] = sirdid;
142      solp = [yp(td,1); yp(td,2); yp(td,3); yp(td,4)];
143      FFbd = (solp - sol)/dbd;
144      FFP = -[FFa FFbr FFbd];
145      newp = p - alpha*(FFP'*FFP + lam*eye(3))\FFP'*FF;
146      error = norm(newp - p)/norm(p);
147      p = newp;
148      olda = p(1); oldbr = p(2); oldbd = p(3);
149      [t y] = sirdid;
150      sol = [y(td,1); y(td,2); y(td,3); y(td,4)];
151      FF = dd - sol;
152      norm(FF);
153      if error < 0.00001
154        break
155      end
```

```
156   end
157   display('Parameters from Levenberg-Marquardt')
158   olda = p(1);
159   oldbr = p(2);
160   oldbd = p(3);
161   [olda oldbr oldbd]
162   display('Norm of residual from Levenberg-Marquardt')
163   norm(FF)
164   error;
165   lm
166   %
167   figure(2)
168   plot(td(1:1:meas+1),Sd(1:1:meas+1),'*',td(1:1:meas+1),...
169   Id(1:1:meas+1),'o',...
170   td(1:1:meas+1),Rd(1:1:meas+1),'s', td(1:1:meas+1),...
171   Dd(1:1:meas+1),'d')
172   hold on
173   % plot(td,Sd,'x', td,Id,'x', td,Rd,'x', td,Dd,'x')
174   [t y] = sirdid;
175   plot(t,y(:,1),'b' ,t,y(:,2),'m' ,t,y(:,3),'r' ,t,y(:,4),'k')
176   title(['a = ' ,num2str(olda),' br = ' ,num2str(oldbr),...
177              ' bd = ' ,num2str(oldbd)]);
178   ylabel('population percentages')
179   xlabel('time')
180   legend('susceptible data', 'infected data',...
181              'recovered data', 'dead data');
182   grid on
183   svd(A(2:m,:));
184   cond(A(2:m,:));
185   svd(FFP);
186   cond(FFP);
```

12.5   The Cumulated Infection Version of SIRD

The motivation for introducing the cumulated infection model is to use "smoother" data than is given by daily data for the four SIRD functions of time. Figure 12.5.1 displays the daily infected and death data for the US COVID-19. Note the jagged appearance. By taking the integral or cumulated sum of the data one obtains a continuous and strictly increasing function of time. This section shows the four functions of time can be expressed as

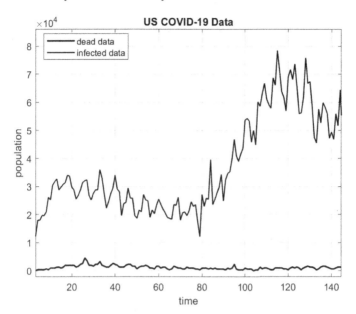

FIGURE 12.5.1
US COVID-19 data.

function of the cumulated infection, which evolves from a nonlinear differential equation, CI-equation, with the three parameters of the SIRD model.

Consider the first equation in the SIRD model.

$$\frac{1}{S}\frac{dS}{dt} = -aI$$

$$\int_{t_0}^{t} \frac{1}{S}\frac{dS}{dt} = -a\int_{t_0}^{t} I = -aC(t)$$

$$\ln(S(t)/(S(t_0)) = -aC(t).$$

Now solve for $S(t), R(t)$ and $D(t)$ in terms of the cumulated infection $C(t)$:

$$S(t) = S(t_0)\exp(-aC(t)),$$
$$R(t) = R(t_0) + b_r C(t) \text{ and}$$
$$D(t) = D(t_0) + b_r C(t).$$

In order to find $C(t)$ without explicitly knowing I, use the fact that $S + I + R + D$ is a constant, say *popmax*. Solve for I and use the above

$$
\begin{aligned}
I &= popmax - S - R - D \\
&= popmax - S(t_0)\exp(-aC(t)) \\
&\quad - (R(t_0) + b_r C(t)) - (D(t_0) + b_d C(t)).
\end{aligned}
$$

$C(t)$ is the integral of I, and therefore the derivative of $C(t)$ is I. This gives the nonlinear differential equation for the cumulative infection.

Cumulative Infection Equation, CI-equation.

$$
\begin{aligned}
\frac{dC}{dt} &= popmax - S(t_0)\exp(-aC(t)) \\
&\quad - (R(t_0) + b_r C(t)) - (D(t_0) + b_d C(t)) \\
&= I(t_0) + S(t_0)(1 - \exp(-aC(t))) \\
&\quad - b_r C(t) - b_d C(t) \text{ and} \\
C(t_0) &\text{ is given.}
\end{aligned}
$$

If the initial values $I(t_0), R(t_0), D(t_0)$ and the cumulative function at $C(t_0)$ as well as the three parameters are known, then one can solve the CI-equation for $C(t)$ with $t \geq t_0$. Furthermore, by using cumulative infection and cumulative death data near to t_0, the three parameters can be identified.

12.5.1   US COVID-19: An aggregated model

The SIRD model assumes the parameters are constant with respect to the particular population set and the time interval. This is not the case with respect to the entire US population. The COVID-19 virus varies with population location and time. This model is an attempt to "lump" or "aggregate" these local and time intervals into a more global (with respect to US) model. One approach would be to approximate the three SIRD parameters for all locations and time intervals, and then to form some sort of weighted average. The approach here is to use the cumulated infection and cumulated death data for the US and nonlinear parameter identification to approximate the three SIRD parameters.

In addition to the three parameters, two implicit parameters will include "death delay" and "effective population size." The death delay is related to the time gap between infection and time of death. The effective population is initially small and then increases as the population moves and spreads the virus.

Figure 12.5.2 has COVID-19 infected data given by the jagged curve. The first 21 days of data were used with the CI-equation and parameter identification using the Levenberg-Marquardt algorithm via sird_paridcuscovid2.m. This code uses the data recorded in UScovid19.m, and both these are a work-in-progress! The predicted parameters are at the top of the figure, and the solid

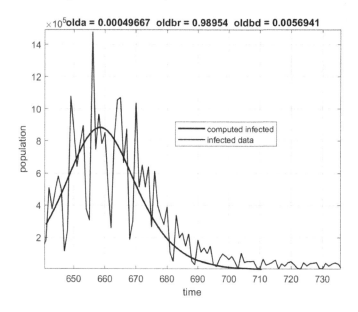

FIGURE 12.5.2
Covid projection with 21-day data.

bell-shaped curve are the predicted infections. Over the course of COVID-19 many similar calculations were done. Generally, 21 days of data would give somewhat accurate predictions for about 21 days. The introduction of a new variate or new protocols did cause uncertain/inaccurate predictions.

Bibliography

[1] I. Babuška, "Error bound for the finite element method," *Numer. Math.*, 16, pp. 322–333, 1971.

[2] James Balama, hypatia.gz, www.math.uri.edu/~jbaglama/#Software.

[3] Jörn Behrens and Randy LeVeque, "Modeling and simulating tsunamis with an eye to hazard mitigation," *SIAM Review*, vol. 44, no. 4, May, 2011.

[4] A. Berman and R. J. Plemmons, *Nonnegative Matrices in the Mathematical Sciences*, SIAM, Philadelphia, 1994.

[5] M. W. Berry and M. Browne, *Understanding Search Engines: Mathematical Modeling and Text Retrieval, Second Edition*, SIAM, Philadelphia, 2005.

[6] R. Courant and D. Hilbert, *Methods of Mathematical Physics, Vol. 1*, Interscience Publishers Inc., NY, 1953.

[7] Nicolas Gillis, "Learning with Nonnegative Matrix Factorizations," *SIAM News*, vol. 52, no. 5, June, 2019.

[8] Roland Glowinski, *Numerical Methods for Nonlinear Variational Problems*, Springer-Verlag, NY, 1982.

[9] Gene H. Golub and Charles E. Van Loan, *Matrix Computations*, 2nd ed., John Hopkins University Press, London, 1989.

[10] M. Hestenes and E. Stiefel. "Methods of conjugate gradients for solving linear systems," *J. Res. Nat. Bur. Standards*, 49, pp. 409–436, 1952.

[11] C. T. Kelley, *Iterative Methods for Linear and Nonlinear Equations*, SIAM, Philadelphia, 1995.

[12] Serge Lang, *Linear Algebra*, Addison-Wesley, London, 1966.

[13] P. D. Lax and A. N. Milgram, "Parabolic equations," *Ann. Math. Studies*, pp. 167–190, 1954.

[14] K. Levenberg, "A method for the solution of certain nonlinear problems in least squares," *Quart. Appl. Math*, 2, pp. 164–168, 1944.

[15] J. L. Lions and G. Stampacchia, "Variational inequalities," *Comm. Pure. Appl. Math*, 20, pp. 493–519, 1967.

[16] D. Marquardt, "An algorithm for least squares estimation of nonlinear parameters," *SIAM J. Appl. Math*, 11, pp. 431–441, 1963.

[17] Carl D. Meyer, *Matrix Analysis and Applied Linear Algebra*, SIAM, Philadelphia, 2000.

[18] Cleve Moler, *Experiments with MATLAB®*, MathWorks, 2011.

[19] Nagle, Saff and Snider, *Fundamental of Differential Equations and Boundary Value Problems, Fourth Edition*, Pearson/Addison Wesley, New York, 2004.

[20] J. M. Ortega and W. C Rheinboldt, *Iterative Solutions of Nonlinear Equations in Several Variables*, Academic Press, New York, 1970.

[21] Lothar Reichel, www.math.kent.edu/~reichel/course/researdh.html.

[22] Y. Saad and M. Schultz, "GMRES a generalized minimal residual algorithm for solving nonsymmetric linear sytems," *SIAM J. Sci. Statist. Comput.*, 7, pp. 856–869, 1986.

[23] Gilbert Strang, *Introduction to Linear Algebra, Fourth Edition*, SIAM, Wellesley, MA, 2009.

[24] R. S. Varga, *Matrix Iterative Analysis*, Prentice-Hall, Engelwood Cliiffs, NJ, 1962.

[25] C. R. Vogel, *Computation Methods for Inverse Problems*, SIAM, Philadelphia, 2002.

[26] Robert E. White, *Elements of Matrix Modeling and Computations with MATLAB®*, Chapman Hall/CRC, Boca Raton, 2007.

[27] Robert E. White, "Nonlinear least squares algorithm for identification of hazards," *Congent Mathematics*, vol. 2, no. 1, 15 December 2015. https://www.cogentoa.com/article/10.1080/23311835.2015.1118219

[28] Robert E. White, *Computational Mathemaics: Models, Methods and Analysis with MATLAB®, Second Edition*, Chapman Hall/CRC, Boca Raton, 2016.

[29] D. M. Young, *Iterative Methods for Solving Partial Differential Equations of Elliptic Type*, Ph.D. Thesis, Harvard University, Cambridge, MA, 1950.

Index

adjoint operator, 230
Applications
    circuit
        three-loop, 71, 123
        two-loop, 9
    epidemic, 294
    fluid flow, 126
    hazard identification, 201
    heat conduction, 18
        2D cooling fin, 73
        cooling fin, 18, 235
        in a wire, 29
    image compression, 175
    noise filter, 185
    search engine, 179
    truss
        five-bar, 55
        six-bar, 37, 124
        two-bar, 8
    visualization, 7
array operations, 6

basis, 96, 213
block Gauss elimination, 119
boundary conditions
    derivative, 20, 236
    Dirichlet, 20
    flux, 20, 236
    Neumann, 20
    Robin, 20, 236

Cauchy inequality, 130, 216
    Cauchy-Bunyakovsky-Schwarz, 4
Cayley-Hamilton, 228
CBS inequality, 5
Cholesky factorization, 84
classical Gram–Schmidt, 141

condition number, 196
conjugate gradient method, 259
continuous models
    cooling fin in 1D, 19, 236
    heat in 3D, 255
contractive, 278
Cramer's rule, 58, 63

determinant, 60
diagonalizable matrix, 225
direct sum, 130
dot product, 4

eigenvalue, 147
    Gerschgorin circles, 151
    power iteration, 151
    QR iteration, 152
eigenvector, 147
energy function, 240
equilibrium equations, 118

finite differences
    cooling fin in 1D, 20
    heat in 3D, 255
finite element method, 243
fixed point, 277
fixed variable, 106, 108
Fourier heat law, 18, 19
free variable, 106, 108
full column rank, 83

Gauss elimination, 35, 76
Gauss transform, 51
Gauss–Jordan method, 45
Gauss-Newton method, 289
Givens transform, 142, 270
GMRES, 267, 270
Gram-Schmidt, 217

Gram-Schmidt modified, 268

heat transfer coefficient, 19
homogeneous solution, 111
Householder transform, 143

ill-conditioned problems, 24
inconsistent algebraic system, 110
inconsistent system, 106
inner product, 215
inverse matrix, 42, 64

Jacobian matrix, 279

Krylov vectors, 265

Lax-Milgram, 241
least squares, 89
least squares problem, 270
Levenberg-Marquardt method, 290
linear operator, 230
linearly independent, 212
LU factorization, 76

M-matrix, 78
MATLAB® codes
    bridge.m, 55
    cgfull_13.m, 263
    circuit3.m, 72
    fin1d_13.m, 21
    gauss_el.m, 39
    gauss_elim_13.m, 39
    gmresfull_13.m, 271
    GScomplex.m, 218
    hazidsvd1.m, 201
    Image1dsvd.m
        Setup1dsvd.m, 185
    imagusa.m
        letteru.m, letters.m, lettera.m,
            175
    inv_mat.m, 53
    levmarqheat.m, 291
    levmarqprice.m, 291
    matit_13.m, 245
    newtcool.m
        fcool.m, fcoolp.m, 286

pcgmres.m, 271
picard2d.m
    gpic2d.m, 281
piccool.m
    fcool.m, fcoolp.m, 281
power_qrone.m, 152
price_expdata.m, 93
qr_col.m, 136
qr_row.m, 136
sengine.m, senginesparse.m, 179
sird_parid.m
    sirdid.m, ypsirdid.m, 295
sird_paridcuscovid2.m
    UScovid19.m, sirdidc.m,
        ypsirdidc.m, 302
sor3d_13.m, 257
support.m, 40
svd_ex.m, 166
svdimage.m
    microchip.jpg, 175
svdviaqr.m, 157
trid_13.m, 21, 28, 53
tridblock_13.m, 74
matrix
    augmented, 6, 33
    block elementary, 69
    diagonal, 24
    diagonalizable, 153
    elementary, 33
    identity, 41
    inconsistent, 34
    inverse, 42
    lower triangular, 25
    M-matrix, 78
    row echelon form, 108
    strictly diagonally dominant, 82
    symmetric positive definite
        (SPD), 82
    tridiagonal, 28
    upper Hessenberg, 269
    upper triangular, 25
matrix-matrix products, 16
matrix-vector products, 13
modified Gram–Schmidt, 136
modified Gram-Schmidt, 135

multiple solutions, 111

Newton method, 284
norm, 4, 219, 224
normal equations, 89
normal matrix, 225
null space, 96

orthonormal, 217
orthonormal basis, 138

particular solution, 111
Picard method, 278
pivot component, 108
preconditioned conjugate gradient, 262
preconditioners
  incomplete domain decomposition, 262
  incomplete LU, 262
  SSOR, 252
projection to subspace, 100
pseudoinverse, 189

QR factorization, 270

range space, 96
rank of matrix, 108
root, 277
row echelon form matrix, 108
row operation, 33

Schur complement, 70, 83
Schur decomposition, 221
Sherman–Morrison–Woodbury, 55
SOR method
  3D space, 256
SPD, 82, 83, 85
spectral radius of matrix, 149
spectrum of matrix, 149
splitting iterative method, 247
Splittings
  Gauss-Seidel, 248
  Jacobi, 248
  P-regular, 253
  regular, 251

Richard, 248
SOR, 248
SSOR, 252
SSOR, 262
strictly diagonally dominant, 52, 250
Sturm-Liouville, 233
SVD
  full, 161, 164
  small, 161
  truncated, 169

Theorems in Chapter 1
  1.1.1 vector space properties, 3
  1.1.2 dot product properties, 4
  1.1.3 norm properties, 4
  1.2.1 matrix-vector properties, 14
  1.2.2 matrix-matrix properties, 17
  1.3.1 triangular solves, 27
  1.5.1 properties of inverse matrix, 49
  1.5.2 diagonal dominant, 52
  1.6.1 Cramer's rule, 63
  1.6.2 inverse via determinant, 64
  1.6.3 inverse matrix equivalance, 65
Theorems in Chapter 10
  10.1.1 vector subspace, 210
  10.2.1 inner product properties, 216
  10.2.2 norm properties, 220
  10.3.1 Schur decomposition, 223
  10.3.2 special matrix norm, 224
  10.3.3 normal matrix characterization, 226
  10.3.4 $P(A) = 0$, 228
  10.4.1 orthonormal eigenfunctions, 233
  10.5.1 weak solution existence, 241
Theorems in Chapter 11
  11.1.1 matrix geometric series, 245

11.1.2 perturbation of
nonsingular, 246
11.1.3 inverse matrix
approximation, 246
11.2.1 convergence
chacterizations, 249
11.2.2 convergent methods, 250
11.2.3 spectral radius of H>=0,
251
11.2.4 M-matrix chacterization,
251
11.2.5 convergent regular
splitting, 252
11.3.1 SPD and convergence, 253
11.3.2 SPD and P-regular, 253
11.3.3 convergent energy, 255
11.5.1 GMRES reduction, 269
Theorems in Chapter 12
12.1.1 contraction and fixed
point, 280
12.2.1 local convergence to root,
285
Theorems in Chapter 2
2.1.1 inverse matrix properties,
70
2.1.2 Schur complement
existence, 70
2.2.1 LU existence, 77
2.3.1 M-matrix properties, 79
2.3.2 M-matrix and LU, 80
2.3.3 Schur complement and
M-matrix, 81
2.4.1 properties of SPD matrix,
83
2.4.2 Schur complement and
SPD, 83
2.4.3 Cholesky factors and SPD,
85
2.4.4 energy equivalence, 86
Theorems in Chapter 3
3.1.1 least squares via normal
equation, 90
3.1.2 full rank and normal
equations, 91
3.3.1 basis properties, 97

3.4.1 projection to subspace, 100
3.4.2 projection and least
squares, 101
Theorems in Chapter 4
4.2.1 inconsistent row echelon
form, 110
4.2.2 general solution of Ax=d,
111
4.3.1 bases for N(A) and R(A),
117
4.3.2 equal ranks, 117
Theorems in Chapter 5
5.1.1 orthogonal subspace
properties, 130
5.2.1 fundamental
decomposition, 131
5.2.2 projections and smallest
LS, 132
5.3.1 full rank and A=QR, 134
5.4.1 orthonormal basis, 139
5.5.1 Householder transform,
144
Theorems in Chapter 6
6.1.1 real eigenvalues, 148
6.1.2 positive eigenvalues, 149
6.3.1 diagonalizable matrix, 154
Theorems in Chapter 7
7.1.1 singular values exist, 162
7.2.1 SVD exists, 164
7.3.1 norm of A, 170
Theorems in Chapter 9
9.1.1 singular values identities,
190
9.1.2 pseudoinverse identities,
190
9.2.1 minimal least squares, 193
9.3.1 norm of pseudoinverse, 197
9.3.2 relative error for LS, 198
Tikhonov-Phillips regularzation, 185
triangle inequality, 5

vector
addition, 2
augmentation, 2
column, 1

dot product, 4
norm, 4
orthogonal, 4
row, 2
scalar product, 2
transpose of a column, 2

vector space, 3, 209
vector subspace, 4, 96,
    210

weak equation, 240
weak solution, 239

Printed in the United States
by Baker & Taylor Publisher Services